STUDY GUIDE

for the Fourth Edition of

Keeton/Gould's

Biological Science

STUDY GUIDE

for the Fourth Edition of

Keeton/Gould's

Biological Science

CAROL HARDY MC FADDEN

W · W · NORTON & COMPANY · NEW YORK · LONDON

ISBN 0-393-95388-2

W. W. Norton & Company, Inc., 500 Fifth Avenue, New York, NY 10110
W. W. Norton & Company Ltd., 37 Great Russell Street, London WC1B 3NU

4 5 6 7 8 9 0

CONTENTS

TO THE STUDENT

Research has shown that almost 90 percent of the material a student hears or reads is forgotten within a few months—a most discouraging fact! Fortunately, retention can be increased by working with the material and by explaining it to others. This *Study Guide* gives you the opportunity to do just that; it requires you to use the material you are studying. To answer many of the questions you must synthesize, analyze, and interpret information. The *Study Guide* is designed to help you focus on the important concepts you need in order to learn the material presented in *Biological Science*, Fourth Edition.

Each chapter of the *Study Guide* has eight parts: A General Guide to the Reading, Key Concepts, Objectives, Key Terms, Summary, Key Diagrams, Questions, and Answers.

A General Guide to the Reading   You should begin your study of each chapter in the text by reading the general guide to the reading in the corresponding chapter of the *Study Guide*. This is a helpful guide to the text chapter, pointing out areas of difficulty, figures that are particularly useful, sections that are important for understanding subsequent chapters and other related material, and providing hints for learning. You will want to keep this part of the *Study Guide* handy as you do your reading in the text.

Key Concepts   Each chapter contains a list of the important concepts, which provide a framework for facts that the chapter presents.

Objectives   Each chapter provides clear objectives. These objectives describe what you should be able to do when you have mastered the material. They are not all-inclusive; they are directed toward the chapter's major concepts. They suggest a plan for study, define the goals toward which you should aim, and help you organize the content of the chapter.

Once you have read the text chapter, using the general guide in the reading, take a look at the key concepts and objectives. Then go back over the text, trying to keep the key concepts in mind. At the end of each section, stop and try to relate what you have read to the corresponding concept; then review the objectives, and see if you could meet them. Next, reread any sections that are still unclear. The page references given for each concept will make it easy to find the relevant sections in the text.

Key Terms The *Study Guide* provides a list of key terms for each chapter. You will want to study this list and make sure you are familiar with the terms, since they are important. Again, page references make it easy to locate in the text any terms that are unfamiliar.

Summary The *Study Guide* offers a summary of each chapter in the text. Since many of the chapters are long, the summaries can help you master the material.

Key diagrams A series of coloring exercises has been included to increase your retention of anatomical detail. Although coloring may at first seem to be a "kindergarten" activity, research has shown that the act of coloring actually enhances learning. Coloring the name and associated structures with the same color forces you to make a visual connection and frees you from rote memorization. When you do the exercises, first color in the outlined name (title), then the corresponding structure. The reference diagrams in the text are cited should you need guidance.

Questions In most chapters the questions are divided into three sections: "Testing recall," "Testing knowledge and understanding," and "For further thought." The "recall" questions require you to remember the basic facts and vocabulary presented in the text. The questions are varied: true or false, matching, fill-in-the-blank, and multiple-choice. When you have finished reading the chapter in the text, do the questions in this section and check your answers; the results will help you identify areas that need further study. Page references to the text are given for each question, so you can look up the material you have missed. Once you answer the "recall" questions correctly, you can go on to the next section.

In general, the questions in "Testing knowledge and understanding" are more demanding. You will need to recall information and also to analyze and integrate it. Most of the questions in this section are multiple choice. You should choose the *best*, most specific answer from among the alternatives given. When you choose the answer you think is correct, be sure you understand why each one of the other possibilities is wrong. Once again, the page references next to the questions will help you find the relevant material in the text. All of the questions have been tested and evaluated; they are drawn from fourteen years of examinations in the introductory biology course at Cornell University.

Answers Answers for the "recall" and "knowledge" questions are provided at the end of each chapter.

Acknowledgments

I am indebted to Kraig Adler, Carolyn Eberhard, Neil Campbell, Antonie Blackler, Paul R. Ecklund, Carl Hopkins, and Robert Turgeon for many of the questions used in the *Study Guide*; and to Thomas Stewart and John Yuen, who spent many hours checking over the questions and making suggestions for improving the *Study Guide*. I also thank Joseph Calvo, from whom I learned much about writing objectives. Particular appreciation is due the students in introductory biology at Cornell University for their comments and suggestions, which have often led to improvements in the *Study Guide*.

I am especially grateful to the late William T. Keeton, who gave me the opportunity to develop the *Study Guide* for *Biological Science*, and from whom I learned the pleasure of teaching introductory biology.

Finally, I thank all the people at W. W. Norton who have been so helpful to me in the preparation of this book, especially James Jordan and Esther Jacobson.

I hope the *Study Guide* will help you in your study of biology. Your comments and suggestions would be most welcome.

Ithaca, New York
June 1986   Carol Hardy McFadden

INTRODUCTION

A GENERAL GUIDE TO THE READING

Chapter 1 is designed to give you a preliminary idea of the way science works and of where the major organizing principles of biology have come from. The pages that follow in the *Study Guide* will outline these ideas and give you ample opportunity to test your recall and understanding of various details. As you read Chapter 1 in your text, you will find the following major topics presented.

1 The flow of information. One of the themes of the text is the flow of chemical information in the cell. This topic is introduced for the first time in Figure 1.2 (p. 2); you may want to spend some time studying this diagram.

2 The scientific method. Pay particular attention to the discussion of this method (pp. 2–6). You will see the method at work in the many experiments described throughout the book.

3 The development of modern science. A brief summary of the development of scientific thought from the ancient Greeks up through Isaac Newton is presented. The beginnings of modern astronomy and physics are discussed.

4 Darwin's theory of evolution. Understanding this theory (pp. 12–20) is of crucial importance, since the concept is a unifying theme of the text. The theory underlies the presentation of many topics throughout the text and will be discussed at greater length in Chapters 33 and 34. Pay special attention to the summary on page 20.

KEY CONCEPTS

1 Every living organism has a set of instructions resident in its genes that directs its metabolism, organization, and reproduction, and is the raw material upon which evolution acts. (p. 2)

2 All science is concerned with the material universe, seeking to discover facts about it and to fit these facts into theories or laws that will clarify the relationships among them. (pp. 3–6)

3 Organisms change with time; the course of this evolutionary change is determined by natural selection. (pp. 12–20)

OBJECTIVES

After studying this chapter and reflecting on it, you should be able to carry out the following objectives.

1 Discuss the scientific method and its applications and limitations. (pp. 2–6)

2 Briefly describe the contribution of each of the following to the development of scientific thought: Aristotle, Albertus Magnus, Thomas Aquinas, Roger Bacon, Nicolaus Copernicus, Galileo Galilei, Johannes Kepler, and Isaac Newton. (pp. 8–11)

3 Contrast Jean Baptiste de Lamarck's explanation for the mechanism of evolutionary change with that of Charles Darwin. (pp. 13–15)

4 Discuss the two parts of Darwin's theory of evolution and give the five basic assumptions upon which it rests. Indicate the types of evidence that Darwin used in formulating his theory. (pp. 17–21)

KEY TERMS

The following terms are important in this chapter; you should become familiar with them.

scientific method (p. 2)
hypothesis (p. 3)
theory (p. 4)
controlled experiment (p. 4)
natural selection (p. 16)

SUMMARY

Biology is the science of living things. All living organisms are chemically complex, are highly organized, utilize energy, undergo development, and reproduce. In addition, every living organism has a set of instructions resident in its genes that directs its metabolism, organization, and reproduction, and is the raw material upon which evolution acts.

All science is concerned with the material universe, seeking to discover facts about it and to fit these facts into theories or laws that will clarify their relationships.

Scientific method The scientific method involves formulating a question and making repeated, careful observations in an attempt to answer it, and then using these observations to generate a *hypothesis*. Hypotheses are then tested, using controlled experiments. If the test results do not confirm a hypothesis, the hypothesis must be altered to conform with the evidence, or rejected. If all evidence continues to support the new idea, it may eventually become widely accepted and called a *theory*. The insistence on testability in science imposes limitations on what it can do. Science cannot make value and moral judgments.

The rise of modern biological science The physical sciences underwent explosive growth during the sixteenth and seventeenth centuries, but biological science did not enter its modern era until 1859 when Charles Darwin proposed his theory of evolution by natural selection. To this day the theory remains one of the most important unifying principles in all biology.

Darwin's theory Darwin postulated that all organisms living today have descended by slow, gradual changes from ancient ancestors quite unlike themselves. The course of this evolutionary change is determined by natural selection. Darwin's theory depends upon five basic assumptions:

1 Many more individuals are born in each generation than will survive and reproduce.
2 There is variation among individuals.
3 Individuals with certain characteristics have a better chance of surviving and reproducing than individuals with other characteristics.
4 At least some of the characteristics resulting in differential reproduction are heritable.
5 Long spans of time are available for slow, gradual change.

QUESTIONS

Choose the one best answer.

1 A general statement that is a tentative causal explanation for a group of observations is referred to as a(n)

   a theory.   d prediction.
   b law.   e control.
   c hypothesis.   (p. 3)

2 The first scientist to promote the use of the scientific method and the controlled experiment was

   a Thomas Aquinas.
   b Roger Bacon.
   c Nicolaus Copernicus.
   d Galileo Galilei.
   e Johannes Kepler.   (p. 8)

3 Which one of the following biologists is *mismatched* with his work?

   a Jean Baptiste de Lamarck—organisms have a natural tendency towards perfection
   b Andreas Vesalius—circulation of blood
   c Antony van Leeuwenhoek—discovered microorganisms
   d Louis Pasteur—disproved spontaneous generation
   e Joseph Lister—effectiveness of antiseptics   (p. 11)

4 For consistency with the theory of natural selection as the agent of evolution, it is necessary to postulate that

   a in each generation, all individuals well adapted for their environment live longer than those not so well adapted.
   b the deaths of individuals occur completely at random with respect to the environment.
   c some of the reproductive success of individual organisms is heritable.
   d most deaths of individual organisms occur soon after birth.
   e more than half of the poorly adapted individuals must be eliminated by natural selection in each generation.
                                    (p. 20)

5 In formulating the theory of evolution, Charles Darwin collected and synthesized data on a variety of subjects. Which one of the following is *not* a line of evidence used by Darwin in formulating his theory?

   a the existence of fossils of extinct animals
   b the similarities of living organisms
   c creation of the earth in 4004 B.C.
   d gradual morphological changes in fossils over time
   e structural changes in domestic animals and plants   (pp. 18–20)

ANSWERS

   1 c   4 c
   2 b   5 c
   3 b

SOME SIMPLE CHEMISTRY

A GENERAL GUIDE TO THE READING

Chemistry is basic to all biology. You will return again and again to the concepts presented in this chapter and the next, so take the time to learn them now. The following topics are of central importance in Chapter 2.

1   Atomic structure. You will want to be familiar with the basic structure of an atom (pp. 27–30) since understanding atomic structure is prerequisite to understanding the material on isotopes, energy levels in atoms, radioactivity, formation of chemical bonds (pp. 33–38), and the material in organic chemistry (Chapter 3).

2   Chemical bonds. You will want to learn about ionic and covalent bonds, particularly the latter. The concept of electronegativity and its relationship to polar and nonpolar bonds is important for you to know (pp. 33–38).

3   Biologically important weak bonds. Pay particular attention to this material (pp. 38–41); understanding weak bonds will help you understand many of the other biological concepts that follow, including the special properties of water, protein structure, the structure of DNA, the activity of enzymes, and membrane structure.

4   Properties of water. The sections "Water as a solvent" and "Special physical properties of water" (pp. 41–45) show why water is fundamental to life as we know it. Grasp of the concepts presented in these sections is essential for understanding the relative solubilities of various organic molecules (pp. 50, 55, 58, 59, 60) and the structure of the cell membrane (p. 97). The phenomena of capillarity and the cohesion of water molecules will be referred to again in Chapter 12.

KEY CONCEPTS

1 Living organisms are integral parts of the physical universe and must obey the fundamental laws of chemistry and physics. (pp. 25–46)

2 Covalent bonds are strong bonds; they can be broken apart only with relatively large amounts of energy; hence covalently bonded molecules are stable. (pp 36–38)

3 In a polar covalent bond the charge is distributed unequally among the atoms involved; in a nonpolar covalent bond the charge is shared equally. Bonds to atoms such as nitrogen and oxygen (which, you will remember, have valences of 3 and 2, respectively) are often polar, because these atoms tend to attract electrons more strongly; bonds between carbon atoms and between carbon and hydrogen are essentially nonpolar. (pp. 37–38)

4 Weak chemical bonds play a crucial role in stabilizing the shape of many of the large molecules found in living matter and in holding them together. (pp. 40–41)

5 Water molecules are polar and have a strong tendency to form hydrogen bonds with one another; the consequent ordering of the molecules gives water many of its special properties, making it the medium of life. (pp. 40, 41–45)

OBJECTIVES

After studying this chapter and reflecting on it, you should be able to carry out the following objectives.

1 Describe the structure of an atom, using the terms proton, neutron, electron, mass number (atomic mass), and orbital. Indicate what is meant by electrons in an "excited state" and those in a "ground state." (pp. 27–29)

2 Explain what an isotope is and give two important physical properties of isotopes that make them useful for biological research. Describe what is meant by the "half-life" of a radioactive isotope. (pp. 27, 31–32)

3 Using diagrams such as Figures 2.8 (p. 33) and 2.9 (p. 34), explain what an ion is, and how it forms. Describe an ionic bond. Indicate whether, and why or why not, sodium chloride exists as a molecule when in solution. (pp. 33–35)

4 Explain what is meant by pH. Specify the pH of the material within most living cells and indicate whether this is acidic or basic. Describe what happens if the pH changes appreciably from this value. Give the name used for compounds that resist changes in pH. (pp. 35–36)

5 Describe a covalent bond and explain how it differs from an ionic bond. Relate the structure of an atom to its chemical properties and to the type of chemical bond it forms. (pp. 36–38)

6 Using diagrams such as Figures 2.14 (p. 38) and 2.16 (p. 39), explain the difference between nonpolar covalent bonds and polar covalent bonds. In a molecule like the following, indicate which of the bonds marked by arrows are polar covalent bonds and which are nonpolar covalent bonds. (pp. 37–38)

$$
\begin{array}{c}
\quad\ \ \text{H} \\
\quad\ \ |\ \ \downarrow\quad\ \downarrow \\
\text{H} - \text{C} - \text{O} - \text{H} \\
\quad\ \ |\ \ \leftarrow \\
\text{H} - \text{C} - \text{H} \\
\quad\ \ | \\
\quad\ \ \text{H}
\end{array}
$$

Indicate whether, and why or why not, a sharp distinction exists between ionic, polar covalent, and nonpolar covalent bonds. (pp. 37–38)

7 Explain the crucial role of weak chemical bonds (such as hydrogen bonds, van der Waals interactions, and hydrophobic interactions) in the organization of living materials. (pp. 39–41)

8 Describe the special physical properties of water. In doing so, draw two water molecules in a way that illustrates a hydrogen bond; specify the number of hydrogen bonds a single water molecule can form with other water molecules in ice; explain why water is a good solvent for ionic and polar compounds, but not for hydrophobic substances; and show the basis for the high surface tension of water and for capillarity. (pp. 41–45)

9 Indicate why the inorganic molecules O_2 and CO_2 are basic to life, and name the principal source of each of these molecules. (p. 46)

KEY TERMS

The following terms are important in this chapter; you should become familiar with them.

atom (p. 26)
atomic nucleus (p. 27)
proton (p. 27)
neutron (p. 27)
atomic number (p. 27)
mass number (p. 27)
isotope (p. 27)
electron (p. 27)
orbital (p. 29)
alpha particles (p. 32)
beta particles (p. 32)
gamma irradiation (p. 32)
half-life (p. 32)
ion (p. 34)
ionic bond (p. 34)
electrostatic attraction (p. 34)
valence (p. 34)
ionization (p. 34)
molecule (p. 34)
acid (p. 35)
base (p. 35)
pH (p. 35)
buffer (p. 36)
nonpolar covalent bond (p. 37)
polar covalent bond (p. 37)
electronegativity (p. 38)
hydrogen bond (p. 40)
hydrophobic (p. 40)
hydrophobic interactions (p. 40)
van der Waals interactions (p. 40)
hydration (p. 42)
hydrophilic (p. 42)
bound water (p. 43)
surface tension (p. 43)
capillarity (p. 44)
heat capacity (p. 45)
heat of vaporization (p. 45)

SUMMARY

The matter of the universe is composed of a limited number of substances called *elements*. Atoms, the basic chemical units of matter, are made up of a nucleus containing positively charged *protons* and uncharged *neutrons*, surrounded by a cloud of negatively charged electrons. In an atom the number of protons is equal to the number of electrons. Isotopes are atoms of the same element that contain different numbers of neutrons. Some isotopes are radioactive; they give off various particles to reach a more stable form. Electrons are in constant motion outside the nucleus. The region in which an electron will probably be found depends on its energy level and is called the orbital of the electron. An atom can contain a maximum of two electrons at the lowest energy level and eight at the next. The chemical properties of elements are determined largely by the number of electrons in their outer shell. Most atoms are particularly stable when their outer shell contains eight electrons. Atoms tend to form complete outer shells by reacting with other atoms.

The attractive force that holds atoms together is a *chemical bond*. Chemical bonds may be *ionic* or *covalent*. An ionic bond results from an electrostatic attraction between a positively charged ion and a negatively charged ion. Such bonds are strong in the solid state, but relatively weak in aqueous solutions. Two important classes of ionic compounds are *acids* and *bases*. Acidity and alkalinity (basicity) are measured in terms of *pH*, a value reflecting the concentration of hydrogen ions. Living matter is extremely sensitive to changes in pH.

Covalent bonds, in which electrons are shared between atoms, are strong; hence covalently bonded molecules are stable.

When the electrons in a covalent bond are shared equally, the bond is *nonpolar;* when the electrons are pulled closer to one element than to the other, the charge is distributed unequally, and the bond is *polar*.

Two types of weak bonds, *hydrogen bonds* and *hydrophobic interactions*, are important in stabilizing the shape of many of the large, complex molecules found in living matter.

The water molecule is polar as a result of the polarity of bonds within the molecule and the V-shaped arrangement of its atoms. Each water molecule can form hydrogen bonds with four other water molecules, permitting an orderly arrangement of molecules. Because of the polarity of the molecule, water is an excellent solvent for many chemicals. Ionic substances and nonionic polar substances dissolve easily in water; electrically neutral and nonpolar substances are not soluble in water.

Molecules may be classified into two types: organic and inorganic. *Organic* compounds always contain the element carbon and are the principal materials that make up living systems. All other compounds are said to be *inorganic*; these molecules are usually smaller and less complex. Water, oxygen, and carbon dioxide are three inorganic molecules basic to life.

KEY DIAGRAMS

1 Below is a diagram of an atom that has four electrons in its outer energy level. When struck by a photon of light, an electron absorbs energy and moves to a higher energy level. The "excited" electron quickly reemits the absorbed energy and returns to its original energy level. In the diagram below, color each name and the appropriate structure(s) with the same color. (Reference: Figure 2.2, p. 28)

K ELECTRONS

L ELECTRONS

M ELECTRON
(EXCITED
ELECTRON)

PROTONS

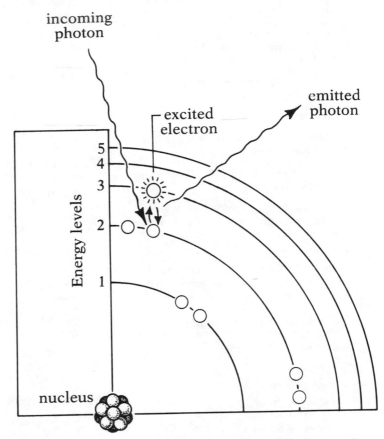

2 Below are diagrams of the four atoms that make up the bulk of living matter. The symbol, atomic number, and mass number (atomic weight) are given. For each, determine the number of electrons in the K level and L level and draw in the electrons in their appropriate energy levels using different colors for each energy level. (Reference: Figure 2.7, p. 31)

K ELECTRONS
L ELECTRONS

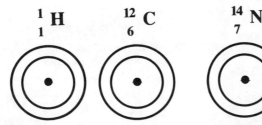
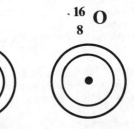

8 • CHAPTER 2

QUESTIONS

Testing recall

Decide whether the following statements are true or false, and correct the false statements.

1 The isotope ^{14}C is heavier than ^{12}C because its atomic nucleus contains two additional protons. (p. 27)

2 The number of electrons at the outermost energy level determines the chemical properties of an atom. (p. 30)

3 Ions are formed when an atom loses or gains electrons. (p. 34)

4 All isotopes are radioactive. (p. 32)

5 In solution, ionic bonds are stronger than covalent bonds. (pp. 34, 37)

6 When a solution is acidic, its pH is greater than 7. (p. 35)

7 The orderly three-dimensional arrangement of molecules in pure water is due primarily to the formation of ionic bonds between water molecules. (p. 40)

8 The marked polarity of water molecules makes water an excellent solvent for many important classes of chemicals. (pp. 41–42)

Testing knowledge and understanding

Choose the one best answer.

9 The chemical properties of an atom are primarily determined by the

a number of protons.
b number of electrons.
c number of neutrons.
d atomic weight.
e number of isotopes.   (p. 30)

10 A nonpolar covalent bond occurs

a when one atom has a greater affinity for electrons than another.
b when the constituent atoms attract the electrons equally.
c when an electron from one atom is completely transferred to another atom.
d between atoms whose outer energy levels are complete.
e when the molecule becomes ionized.   (p. 37)

11 Which one of the following kinds of biologically important bonds requires the most energy to break it in aqueous solution?

a van der Waals interaction
b covalent bond
c hydrophobic interaction
d hydrogen bond
e ionic bond   (p. 37)

12 When matter is broken down into units indivisible by ordinary chemical means, the resulting structure is the

a proton.   d molecule.
b atom.   e nucleus.   (p. 26)
c neutron.

Questions 13–18 concern the following atoms.

a $^{19}_{9}F$ d $^{39}_{18}Ar$
b $^{12}_{6}C$ e $^{28}_{14}Si$
c $^{40}_{20}Ca$

13 How many protons does fluorine (a) have?

a 9   c 19
b 10   d 26   (p. 27)

14 Which one of the above would be relatively inert? (p. 30)

15 Which one of the above would have two electrons in its outer orbital? (p. 29)

16 Which one of the above would be a strong electron acceptor? (p. 39)

17 Which two of the above have similar chemical properties?

a a and c   d b and e
b b and c   e c and e
c a and d   (p. 30)

18 How many electrons would a Ca^{++} ion have?

a 2   d 22
b 18   e 40
c 20   (p. 34)

19 In an electronically neutral atom, how many electrons would be at the *outermost* energy level or shell if the atom had an approximate atomic weight of 12 and an atomic number of 6?

a 0
b 2
c 4
d 6
e some other number   (p. 27)

20 Which one of the following would be the strongest electron donor?

a $^{12}_{6}C$ d $^{32}_{16}S$
b $^{19}_{9}F$ e $^{39}_{19}K$
c $^{14}_{7}N$ (p. 34)

21 Atoms X and Y readily combine together. Atom X has an atomic number of 13 and atom Y has an atomic number of 17. Which one of the following molecules would they form? (Subscripts refer to the number of atoms; no subscript indicates 1 atom.)

a XY
b X_3Y
c X_2Y_3
d XY_2
e XY_3 (p. 33)

22 For a hypothetical atom in which the number of protons is X, the number of neutrons is Y, and the number of electrons is Z, the atomic *number* would be

a X.
b Y.
c Z.
d the sum of X and Y.
e the sum of Y and Z. (p. 27)

23 In an ionic bond

a electrons are shared.
b there is a mutual attraction between two electrically neutral atoms. No electron transfer is involved.
c there is a mutual attraction between two charged atoms. Electron transfer is involved.
d there is an unequal sharing of electrons. (p. 39)

24

The bond indicated by the arrow is

a a hydrogen bond.
b a covalent bond.
c a polar bond.
d both a and c.
e both b and c. (p. 37)

25 Solution A has a pH of 4 and solution B has a pH of 8. The hydrogen ion concentration of A is _____ times that of B.

a 10,000
b 1,000
c 4
d 0.001
e 0.0001 (p. 35)

26 Solution A has a pH of 8 and solution B has a pH of 2. Which one of the following statements is *correct*?

a Solution A is basic and solution B is acidic.
b Solution A is acidic and solution B is basic.
c Solution A has more hydrogen ions than solution B.

d Solution A is basic and solution B is neutral.
e two of the above are correct. (p. 35)

27 Here is a molecule composed of carbon, nitrogen, oxygen, and hydrogen. The hydrogen atoms have not been drawn in. Which one of the following is the correct chemical formula?

a C_2NOH_3
b C_2NOH_4
c C_2NOH_5
d C_2NOH_6
e C_2NOH_7 (p. 31)

28 The cohesiveness among water molecules is due largely to

a hydrogen bonds.
b polar covalent bonds.
c nonpolar covalent bonds.
d hydrophobic interactions.
e van der Waals interactions. (p. 40)

Questions 29–30 refer to the following diagrams of water molecules.

29 The bond (labeled x) from O to H is a(n)

a polar covalent bond.
b nonpolar covalent bond.
c hydrogen bond.
d hydrophobic interaction.
e ionic bond. (p. 40)

30 When ice melts, breaks occur in the bond(s) marked

a x.
b y.
c both x and y. (p. 41)

ANSWERS

Testing recall

1 false—two additional neutrons
2 true
3 true
4 false—some
5 false—ionic bonds are weaker
6 false—less than 7
7 false—hydrogen bonds
8 true

Testing knowledge and understanding

9	b	15	c	21	e	26	a
10	b	16	a	22	a	27	d
11	b	17	d	23	c	28	a
12	b	18	b	24	b	29	a
13	a	19	c	25	a	30	b
14	d	20	e				

Chapter 3

THE CHEMISTRY OF LIFE

A GENERAL GUIDE TO THE READING

This chapter, like the preceding one, presents several ideas that are central to an understanding of modern biology. Organic chemistry is difficult for many students, primarily because the molecules seem so complex. They really are not; organic molecules are composed of relatively few elements, and you will soon learn to recognize the key features of the various kinds of molecules. As you do your reading, remember that you do not have to memorize the structures shown in the figures; most are provided only as examples of particular classes of molecules. If you focus instead on what makes each class unique, you will find this chapter much easier and more meaningful. As you read Chapter 3 in your text, you will want to concentrate on the following topics.

1 Carbohydrates, lipids. Read carefully the sections on these compounds (pp. 49–57); the material they present will be referred to repeatedly throughout the book.

2 Proteins. The discussion of proteins (pp. 57–68) is crucial; you need to finish this chapter with an understanding of the three-dimensional shape of proteins and the relationship between their shape and their function. You will find that changes in protein shape are involved in many cellular activities that you will be studying.

3 Nucleic acids (pp. 68–70) will be discussed in more detail in later chapters; at this point you need only a general understanding of what they are.

4 Free energy. The concept of free energy (pp. 71–73) must be understood because it lays the groundwork for material presented in Chapters 4, 7, and 8. The terms ΔG, exergonic, and endergonic should be learned.

5 The equilibrium constant (K_{eq}). Read carefully the material on pages 73–74 so you understand what the equilibrium constant represents. Note also the convention regard-

ing the length of arrows in the forward and backward reactions and their relationship to K_{eq}. This material, though somewhat difficult, is necessary for future use. K_{eq} and ΔG will be used to describe many of the chemical reactions in the text, particularly in Chapters 7 and 8. Understanding the concepts here will greatly facilitate learning the later material, and provide an intuitive understanding of how chemical reactions are controlled in the cell.

6 Enzymes. Careful study of this section (pp. 76–82) is essential since enzymes and enzymatic activity will be referred to repeatedly throughout the text. Figures 3.40 and 3.41 are particularly helpful in understanding enzymatic activity.

KEY CONCEPTS

1 Both living and nonliving matter are made up of the same fundamental particles. The only difference between living and nonliving things seems to be in the way these basic materials are organized. (see Chapter 2, p. 25)

2 All complex organic molecules are composed of many simpler building-block molecules, bonded together by condensation reactions; they can be broken down to their building-block molecules by hydrolysis. (pp. 47–70)

3 Proteins may act as important enzymes in chemical reactions or as structural components of cells. (p. 57)

4 The amino acid content and sequence of a protein determine its three-dimensional shape; the R groups of amino acids largely determine the structure and properties of proteins. Of particular importance is the fact that some R groups are hydrophilic, while others are hydrophobic. Weak bonds between these groups play a crucial role in determining the three-dimensional structure of proteins. Alteration in the shape of a protein can lead to a change in its biological function. (pp. 57–60)

5 Nucleic acids make up the genes that determine the structural and functional characteristics of living things. (pp. 68–70)

6 The course of a chemical reaction depends on whether the free energy in the covalent bonds in the reactants is greater or less than

the free energy in the covalent bonds in the products. If the reaction results in products with less free energy in the covalent bonds than the reactants possessed, the reaction is exergonic and will proceed spontaneously. If, however, the covalent bonds of the products have more free energy than the reactants, the reaction will require a net input of energy and is said to be endergonic. (p. 72)

7 In a chemical reaction, when the forward reaction is just counterbalanced by the back reaction, the two processes are in equilibrium. The equilibrium ratio of the products to the reactants is the equilibrium constant (K_{eq}). Exergonic reactions have a negative ΔG and a K_{eq} greater than 1 while endergonic reactions have a positive ΔG and a K_{eq} less than 1. (pp. 73–74)

8 Enzymes, like all catalysts, lower the activation energy needed for all reactions. An enzyme affects only the rate of the reaction; it speeds up the reaction but does not alter the direction of the reaction, its final equilibrium, or the reaction energy involved. (pp. 75–76)

9 Most enzymes are highly specific, and each can interact only with those reactants or substrates that fit spatially and chemically into the active site of the enzyme. Anything that alters the shape of the enzyme will alter its activity. (pp. 76–78)

OBJECTIVES

After studying this chapter and reflecting on it, you should be able to carry out the following objectives.

1 Identify the various functional groups listed in Table 3.1 (p. 49). Given an unknown organic molecule, recognize and name the various functional groups and indicate whether each group is charged, polar, or nonpolar, and whether it is hydrophilic or hydrophobic. (p. 49)

2 Describe the structure of a typical monosaccharide such as glucose. Write out a condensation reaction between the two given glucose molecules, and explain what is meant by hydrolysis. (pp. 51–53)

3 Using a diagram such as Figure 3.8 (p. 53), point out the similarities and the differences between alpha and beta linkages. Explain

why the glucose molecules in starch can be used directly for food by animals whereas those in cellulose cannot. (pp. 53–54)

4 Using a diagram such as Figure 3.13 (p. 56), point out the carboxyl, or acid, group and the hydrocarbon chain. Explain the difference between a saturated fatty acid and an unsaturated fatty acid. For butter and corn oil, indicate whether each is considered a saturated or an unsaturated fat. Using a diagram such as Figure 3.12 (p. 55), explain how three fatty acids can react with glycerol to make a fat. (pp.54–56)

5 Using a diagram such as Figure 3.14 (p. 57), point out the polar and nonpolar portions of a phospholipid molecule. Specify which end of the molecule would be soluble in water. (pp. 56–57)

6 Given the structure of an amino acid, point out the α-carbon, and the carboxyl group, the amino group, and the R group that are attached to it. Be able to tell whether the R group is charged or polar, and therefore hydrophilic, or nonpolar and hydrophobic. (pp. 58–59)

7 Given the structure of two amino acids, write out a condensation reaction between the two (see Figure 3.17, p. 59) and circle the resulting peptide bond. Explain how a polypeptide chain is formed and how it is broken down by hydrolysis. (p. 59)

8 Given a picture of a protein such as the one below, do the following: point out the number of polypeptide chains it contains; point out any regions of alpha helix, label the region of the protein that would be stabilized by ionic bonds; label the region stabilized by hydrogen bonds between R groups; circle a disulfide bond, and state whether it is strong or weak. (pp. 63–68)

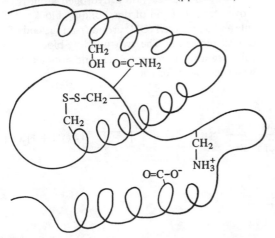

9 Differentiate among the various levels of protein structure—primary, secondary, tertiary, and quaternary. Specify the highest level of structure shown in the protein in the diagram above. Specify the level(s) of structure shown by fibrous proteins, such as hair, and the levels shown by globular proteins. Explain why proteins are so sensitive to changes in temperature and pH. (pp. 60–68)

10 Using a diagram such as Figure 3.30 (p. 70), point out an individual nucleotide and the five-carbon sugar, the phosphate group, and the nitrogen-containing base of which it is made; a base from a nucleotide on one chain bonded to a base on the opposite side; the sugar-phosphate uprights; and the hydrogen bonds between bases. (pp. 69–70)

11 Identify examples of each of the four main classes of biologically important organic molecules and the building-block units of which they are composed. (pp. 70–71)

12 Define the term free energy and explain how ΔG is derived. Give the relationship between ΔG and whether a reaction is exergonic or endergonic. (pp. 71–73)

13 Explain how the equilibrium constant (K_{eq}) is derived and state the relationship between the equilibrium constant and ΔG in describing chemical reactions. (pp. 73–74)

14 Explain what is meant by activation energy and why the activation-energy barrier provides stability for high energy molecules. Compare Figures 3.32, 3.34, and 3.35, and explain the role an enzyme plays in speeding up a chemical reaction, indicating whether the enzyme changes the ΔG of the reaction. (pp. 74–75)

15 Explain why the three-dimensional structure of an enzyme is the key to its activity. In doing so, include answers to the following questions: What is the active site of an enzyme? How are the enzyme and substrate molecules held together to form the enzyme substrate complex? What do we mean by "induced fit"? Why may changes in temperature or pH greatly reduce enzyme activity? What effect might an inhibitor molecule (both competitive and noncompetitive) have on enzyme activity? What are allosteric enzymes and how do modulators influence their activity? (pp. 76–81)

KEY TERMS

The following terms are important in this chapter; you should become familiar with them.

hydrocarbon (p. 47)
isomer (p. 48)
functional group (p. 48)
monosaccharide (p. 50)
amino group (p. 52)
disaccharide (p. 52)
condensation reaction (p. 52)
maltose (p. 52)
sucrose (p. 52)
lactose (p. 52)
hydrolysis (p. 52)
polysaccharide (p. 53)
starch (p. 53)
glycogen (p. 53)
cellulose (p. 54)
polymer (p. 55)
fatty acid (p. 55)
saturated (p. 56)
unsaturated (p. 56)
phospholipid (p. 56)
steroid (p. 56)
amino acid (p. 59)
peptide bond (p. 60)
polypeptide (p. 60)
disulfide bond (p. 60)
primary structure (p. 60)
alpha helix (p. 64)
secondary structure (p. 64)
fibrous proteins (p. 64)
keratins (p. 64)
pleated sheet (p. 65)
tertiary structure (p. 66)
quaternary structure (p. 66)
denatured proteins (p. 67)
conjugated proteins (p. 68)
prosthetic group (p. 68)
nucleic acids (p. 68)
DNA (p. 69)
adenine (p. 69)
guanine (p. 69)
cytosine (p. 69)
thymine (p. 69)
RNA (p. 70)
free energy (ΔG) (p. 71)

First Law of Thermodynamics (p. 71)
Second Law of Thermodynamics (p. 71)
exergonic (p. 72)
endergonic (p. 72)
back reaction (p. 73)
kinetic energy (p. 73)
thermal energy (p. 73)
equilibrium constant (K_{eq}) (p. 73)
activation energy (p. 74)
catalyst (p. 75)
enzyme (p. 76)
substrate (p. 76)
active site (p. 78)
induced-fit hypothesis (p. 78)
coenzymes (p. 81)
competitive inhibition (p. 81)
noncompetitive inhibition (p. 82)
allosteric inhibition (p. 82)
modulators (p. 82)
feedback inhibition (p. 82)
cooperativity (p. 82)

SUMMARY

Carbon, with its covalent bonding capacity of four, commonly forms bonds with hydrogen, nitrogen, and other carbon atoms. The carbon atoms usually join in long chains or rings, forming molecules that may be very complex. Four major classes of organic molecules found in living organisms are carbohydrates, fats, proteins, and nucleic acids.

Although these four classes of molecules differ in structure and function, they are similar in that all are built up of many simpler "building-block" molecules bonded together. In each case these building-block molecules are combined by the removal of water molecules in *condensation reactions*. In such a reaction a hydrogen atom is removed from the end of one building-block molecule and a hydroxyl (OH) group from the end of a second molecule. The two building-block molecules are now joined together, and a molecule of water has been formed:

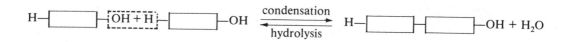

Condensation reactions are reversible; the complex organic molecules can be *hydrolyzed* into the simpler building-block molecules.

The basic building-block molecules of *carbohydrates* are the simple sugars, or *monosaccharides*. When two simple sugars are bonded together, a *disaccharide* is formed. When many simple sugars are bonded together in long chains, a *polysaccharide* is formed. Starch, glycogen, and cellulose are examples of polysaccharides. The carbohydrates are an important energy source for all organisms.

Lipids, the fats and fatlike substances, tend to be insoluble in water. *Fats* are made up of two building-block molecules: glycerol and fatty acids. *Phospholipids* are derived from the fats; they are important constituents of cell membranes.

The basic building-block molecules of the *proteins* are the *amino acids*. Amino acids are bonded together to form a protein by condensation reactions. The resulting bond is the *peptide* bond and the chains produced are *polypeptide* chains. The *primary structure* of each protein is the sequence and type of amino acids making up the polypeptide chains. Because hydrogen bonds form between one amino acid and another, the chain assumes a stable regular shape known as the *secondary structure*. These regular molecules may in turn be folded into complicated globular shapes by weak attractions between the different R groups within the chain, thus forming the *tertiary structure* of the protein. Some globular proteins are made up of two or more polypeptide chains held together by weak bonds; the way these chains fit together determines the *quaternary structure*. Because the conformation of a protein depends on weak bonds, it is easily altered, causing a change in its biological function.

The building-block unit of *nucleic acids* is the *nucleotide*, which is made up of a five-carbon sugar attached to a phosphate group and to a nitrogen-containing base. Nucleotide units are joined together through condensation reactions between the sugar of one nucleotide and the phosphate group of the next. There are four different nucleotides in each nucleic acid; it is the different sequences of the nucleotides that encode the hereditary information. The two types of nucleic acids, DNA and RNA, differ in their basic makeup, and in the number of strands in the molecule.

The course of a chemical reaction depends on whether the free energy (ΔG) in the covalent bonds in the reactants is greater or less than the free energy in the covalent bonds in the products. If a reaction results in products with less free energy in the covalent bonds than the reactants possessed (i.e., ΔG is negative), the reaction is exergonic and will proceed spontaneously. If, however, the covalent bonds of the products have more free energy than the reactants (i.e., ΔG is positive), the reaction will require a continuous input of energy and is said to be endergonic. The equilibrium constant (K_{eq}) describes the ratio of the products to the reactants at equilibrium; it depends on the ΔG of the reaction. Although exergonic reactions proceed spontaneously, initiating a reaction may require *activation energy*.

Chemical reactions can be speeded up by heat, by increasing the concentrations of the reactants, or by providing an appropriate *catalyst*. In living systems the catalysts are *enzymes*. Most enzymes are highly specific, and each can interact only with those reactants, or *substrates*, that fit spatially and chemically into the *active site* of the enzyme.

Since the formation of the enzyme-substrate complex requires the enzyme and its substrate to be complementary, anything that alters the shape of the enzyme will alter its activity. In addition to temperature and pH, many chemical substances can mask, block, or alter the shape of the enzyme and its active site.

Inhibitors decrease enzyme activity either by competing with the substrate molecules for the active site and blocking it (competitive inhibitor), or by binding to a second site on the enzyme and inducing a conformational change in the enzyme molecule (noncompetitive inhibitor).

KEY DIAGRAMS

A section of an alpha-helical structure of a protein Color each name and the appropriate structure(s) with the same color. (Reference: Figure 3.20, p. 64)

BONDS BETWEEN ATOMS:

COVALENT BONDS

HYDROGEN BONDS

COMPONENT ATOMS:

CARBON ATOMS

NITROGEN ATOMS

OXYGEN ATOMS

HYDROGEN ATOMS

R GROUPS

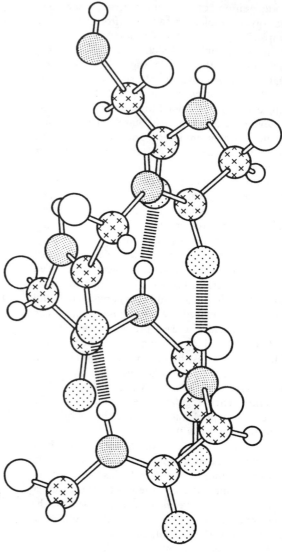

Related questions

1 List the atoms that form the backbone of the polypeptide chain:
_____

2 Which of the two types of bonds (covalent, hydrogen) are strong bonds? _____
Weak bonds? _____

3 Which of the bonds help stabilize the alpha helix? _____

Structure of a nucleotide Color each name and the appropriate structure with the same color. (Reference: Figure 3.27, p. 69)

5-CARBON SUGAR
NITROGENOUS BASE
PHOSPHATE GROUP
HYDROXYL GROUP

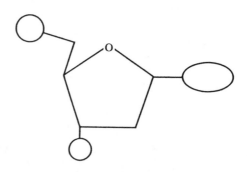

Related questions

1 If this nucleotide were joined to another nucleotide, where would it join the second?

_____

2 If a second nucleotide were to be added to the above nucleotide, where would it be added?

_____

3 What are the four nitrogenous bases in DNA? _____

QUESTIONS

Testing recall

Decide whether the following statements are true or false, and correct the false statements.

1 The bonds connecting the simple-sugar units of starch can be broken by hydrolysis. (p. 52)

2 Denaturation of a protein involves a loss of the protein's natural shape. (p. 67)

3 The hormone insulin is a steroid. (p. 61)

4 The combustion of gasoline in a car is an example of an exergonic reaction. (p. 72)

5 A chemical reaction that has a positive ΔG and a K_{eq} of 0.01 is exergonic. (p. 72)

6 Catalysts seem to work by lowering the activation energy needed for a reaction to take place. (p. 75)

7 Phospholipids are molecules with hydrophilic and hydrophobic portions. (p. 56)

8 Adding hydrogen to an unsaturated fat will make it more saturated. (p. 56)

9 Chemical reactions involving the combination of smaller building-block molecules with the removal of water molecules are hydrolysis reactions. (p. 52)

10 The bond between two adjacent amino acids in a protein molecule is a peptide bond. (p. 60)

11 Chemical reactions that require a net input of free energy to proceed are endergonic reactions. (p. 72)

12 All enzymes are proteins. (p. 76)

Questions 13–16 refer to the examples of various building-block molecules shown in the diagrams below. Before answering the questions, see if you can place each mole-cule in the class to which it belongs. The following procedures will help you with the identification. First, circle all the hydroxyl, or alcohol, groups, OH. Next, circle all the carboxyl, or acid, groups, $-C\overset{O}{\underset{OH}{\diagup}}$ *or* $-C\overset{O}{\underset{O^-}{\diagup}}$. *Now circle all the amino groups,* $NH_3{}^+$.

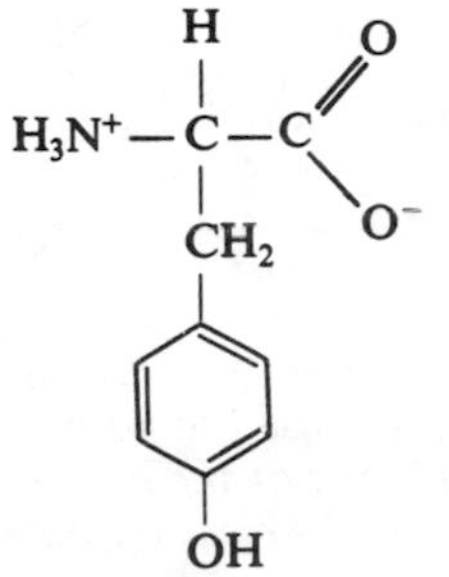

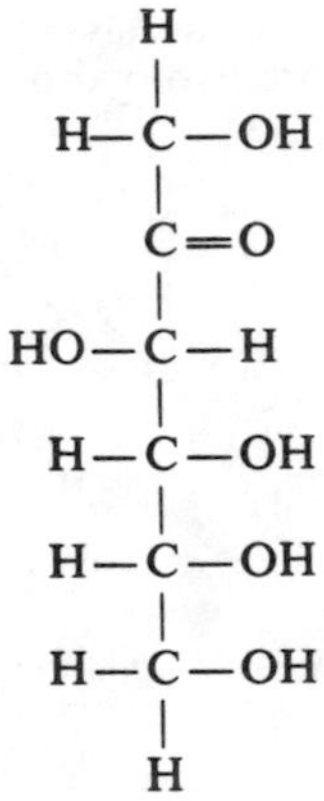

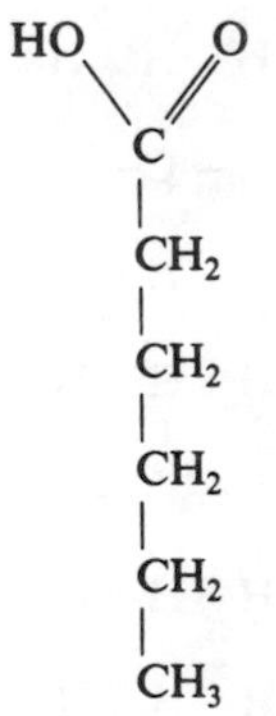

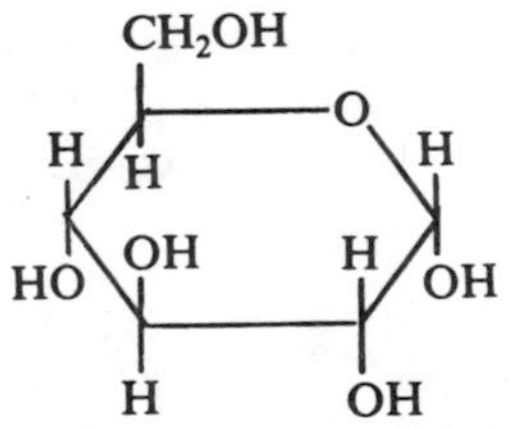

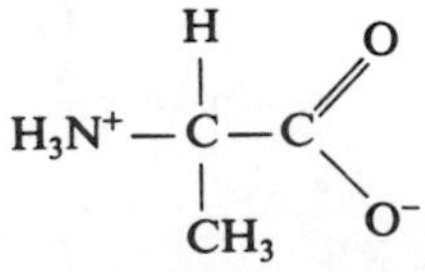

f

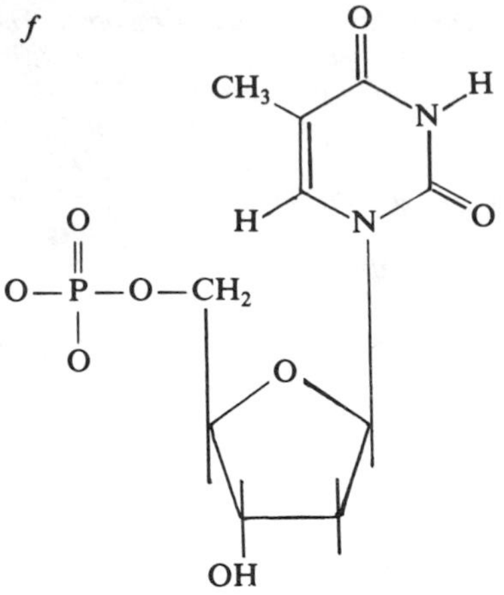

Now that you have circled the major functional groups, use the information below to assign each one of the above building-block molecules to the class of molecules to which it belongs.

13 Carbohydrates have hydrogen and oxygen atoms present in the same proportions as in water; there are two hydrogen atoms for every one oxygen. Therefore, carbohydrates have many hydroxyl groups and the grouping H — C — OH recurs frequently. Using this information, pick out any carbohydrates among the molecules above. (pp. 49–54)

14 Amino acids, as their name suggests, have both an amino and an acid group attached to the same carbon. Pick out any amino acids among the molecule(s) above. (p. 59)

15 Fatty acids have a hydrocarbon chain and an acid group. Pick out any fatty acids among the molecules above. (p. 56)

16 Nucleotides are composed of three parts: a five-carbon sugar, a phosphate group, and a nitrogenous base. Pick out the nucleotide among the molecules above, and circle the sugar, the phosphate, and the nitrogenous base. (p. 69)

Testing knowledge and understanding

17 Complete the following chemical reactions. (pp. 52–57)

a

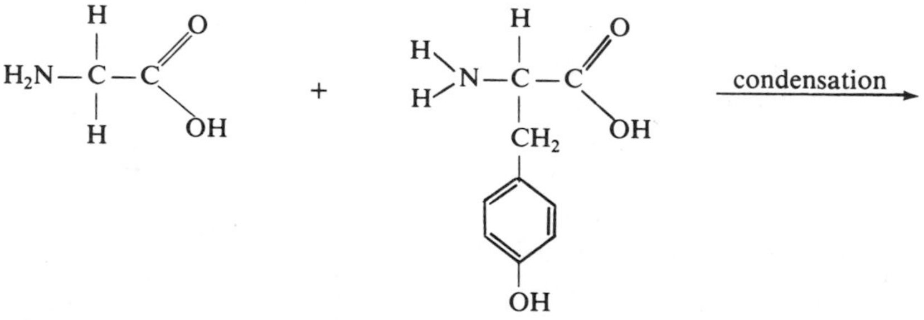

b

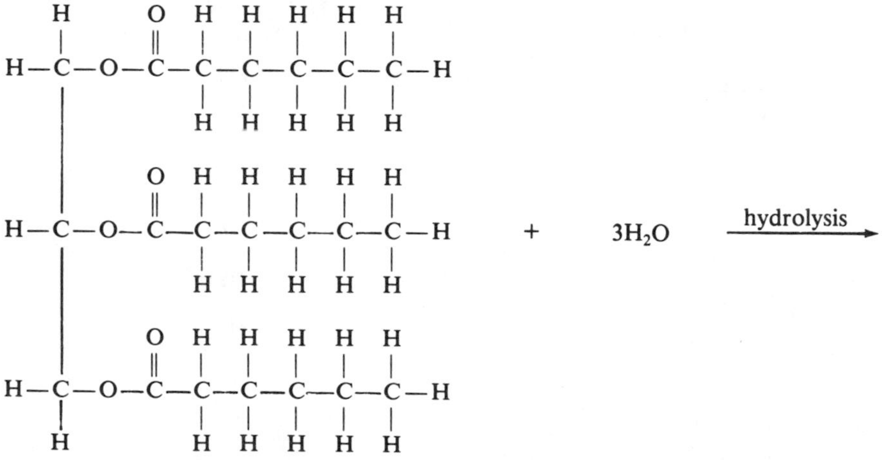

c

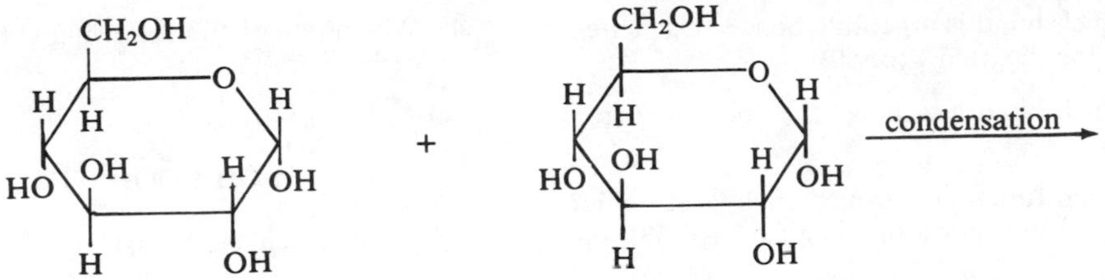

+

→ condensation →

For questions 18–23, use the structural formulas for some important organic molecules given below.

a

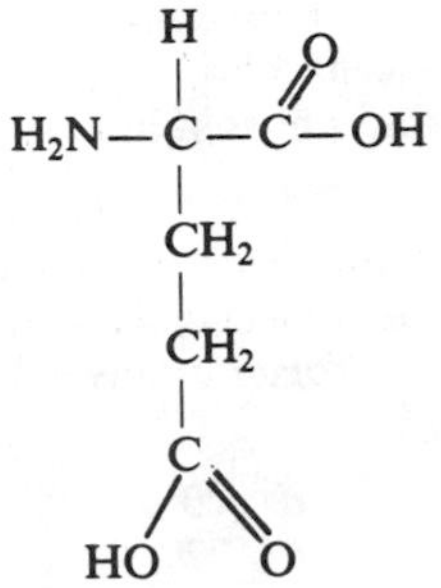

b

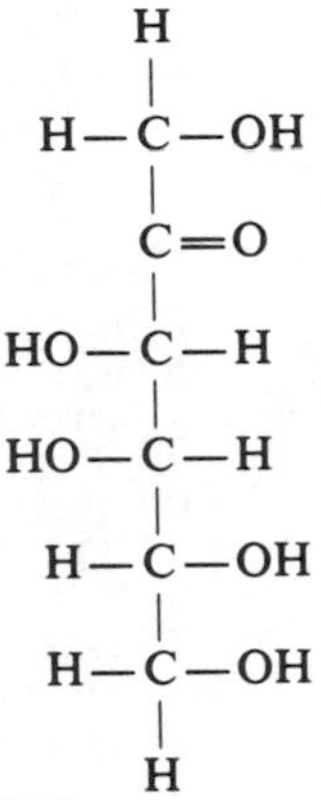

c

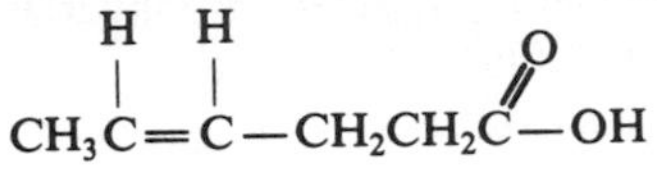

d

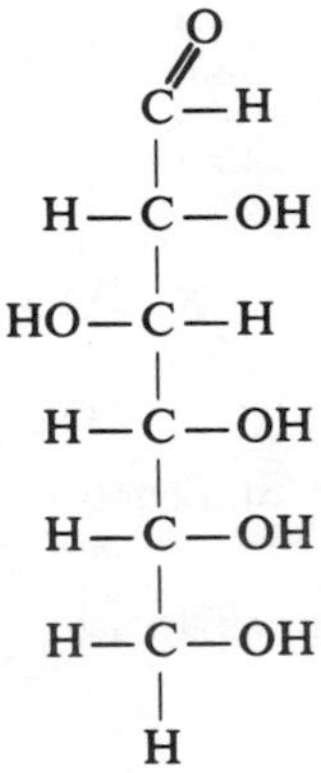

e

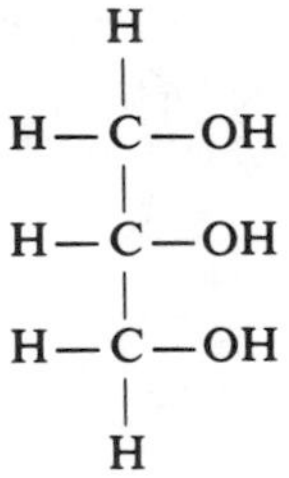

18 Which one of the above molecules is a building-block molecule for proteins? (p. 59)

19 Which one of the above molecules is glucose? (p. 51)

20 Which one of the above molecules is a fatty acid? (p. 55)

21 Which one of the above molecules would be least soluble in water? (p. 56)

22 Molecules *b* and *d* are examples of

a isomers.
b isotopes.
c organic acids.
d hydrophobic substances.
e disaccharides. (p. 46)

23 Which two of the above molecules could be the products of the hydrolysis of a fat?

a *a* and *b*   *d* *c* and *e*
b *b* and *c*   *e* *c* and *d*
c *a* and *e*   (p. 55)

For questions 24–29, use the diagram below which shows a polypeptide composed of four amino acids. Specific bonds are indicated by arrows with letters; functional groups are circled with broken lines and labeled with numbers. Use the letters and numbers on the diagram to answer the questions.

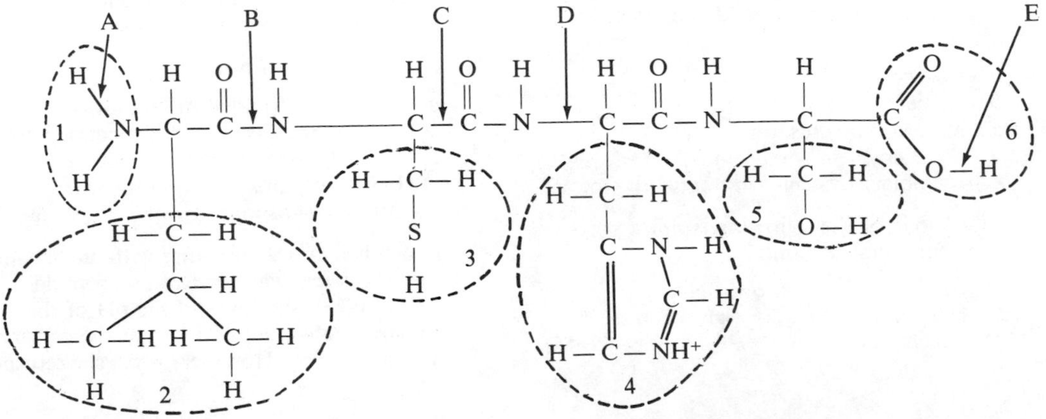

24 Which bond is a peptide bond? (Use letters on the diagram.) (p. 60)

25 Which bond has the greatest polarity? (p. 37)

26 Which functional groups are hydrophobic? (Use numbers on the diagram.) (p. 58)

27 Which functional groups are hydrophilic? (p. 58)

28 Which functional group is an amino group? (p. 59)

29 Which functional group is an acid? (p. 55)

Choose the one best answer.

30 The carbon atom can form so many different chemical compounds because

a its unstable nucleus easily gives up neutrons.
b its outer energy level contains four electrons.
c its electron shells are stable.
d it can form both ionic and covalent bonds.
e it tends to give up electrons to electron acceptors. (p. 48)

31 Below is the structural formula for a molecule composed of carbon, nitrogen, and hydrogen. The hydrogen atoms have not been drawn in. Which is the correct chemical formula?

a $C_5H_{11}N$ *d* $C_5H_{12}N$
b C_5H_9N *e* C_5H_6N
c $C_5H_{10}N$ (pp. 33, 48)

32 Compounds with the same atomic content but differing structures and properties are called

a isotopes.
b isomers.
c ionic compounds.
d polar covalent compounds.
e nonpolar covalent compounds. (p. 46)

33 Which one of the following is *not* a polymer with many subunits?

a fat   *d* DNA
b protein   *e* glycogen
c starch   (pp. 55–56)

34 Which one of the following compounds is a carbohydrate?

a $C_5H_{10}O_5$
b $C_3H_8O_3$
c $CH_3CH_2CH_2COOH$
d $C_6H_{12}O_2$
e $H_2N — CH_2COOH$ (p. 49)

35 Two classes of organic compounds typically provide energy for living systems. Representatives of these two classes are

a fats and amino acids.
b amino acids and glycogen.
c amino acids and ribose sugars.
d fats and polysaccharides.
e nucleic acids and phospholipids.
                                    (pp. 49, 55)

36 If two five-carbon sugars are combined to form a disaccharide molecule with ten carbons, how many *hydrogen* atoms will it have?

a 10   *d* 20
b 12   *e* 22
c 18   (p. 52)

37 Plants commonly store carbohydrates for an energy source as

a glycogen.   *d* sucrose.
b starch.   *e* fat.
c cellulose.   (p. 53)

38 The chemical formula

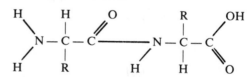

represents the product of a condensation reaction between

a a fatty acid and an amino acid.
b two fatty acids.
c two amino acids.
d an amino acid and an alcohol.
e a fatty acid and an alcohol. (p. 60)

39 Both DNA and RNA

a are single-stranded molecules.
b contain the same four nucleotide bases.
c are polymers of amino acids.
d have the same five-carbon sugar.
e contain phosphate groups. (pp. 69–70)

40 A biochemist was working with an enzyme that she knew was a *single* polypeptide chain. When she lowered the pH of the enzyme slightly she found that the enzyme's activity slowed. However, when she returned

the pH to the enzyme's normal range, the enzyme resumed normal activity and speeded up a chemical reaction. Lowering the pH probably altered the enzyme's

a   primary structure.
b   tertiary structure.
c   quaternary structure.
d   peptide bonds.   (p. 67)

41   Which of the following characteristics does not apply to a structural protein such as silk?

a   peptide bonds
b   specific primary structure
c   active site
d   hydrogen bonds between separate polypeptide chains
e   more than one kind of amino acid
                                  (pp. 64–66)

42   The conformation of a protein molecule depends on several different types of bonds and group interactions. Which of these remain intact when a protein is denatured?

a   peptide bonds
b   ionic bonds
c   hydrogen bonds
d   hydrophobic interactions
e   none of the above   (pp. 64–66)

For questions 43–47, use the diagram below of a polypeptide chain. The heavy line represents the "backbone" of the chain. Selected R groups of amino acids are shown together with various bonds and interactions that stabilize the folding of the chain. Each bond or interaction is labeled with a Roman numeral in parentheses.

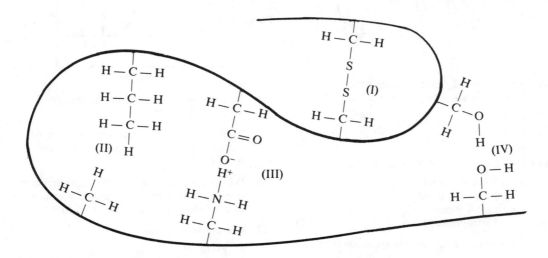

43   Which of the atoms below are bonded together linearly to form the backbone of the chain?

a   C, H, and N   d   C and O
b   C, H, and O   e   C, N, and O
c   C and N   (pp. 59–60)

44   Which of the bonds between R groups is (are) *ionic* bond(s)?

a   I   d   III and IV
b   III   e   I and II
c   IV   (p. 34)

45   Which of the R-group interactions is (are) *hydrophobic*?

a   I   d   III
b   II   e   III and IV
c   I and II   (p. 40)

46   Which of the bonds between R groups is (are) *hydrogen* bond(s)?

a   II   d   II and III
b   III   e   II and IV
c   IV   (p. 40)

47   Which of the bonds between R groups is (are) covalent bonds?

a   I   d   IV
b   II   e   I and IV
c   III   (p. 37)

For questions 48–51, use the drawing below which represents a protein molecule.

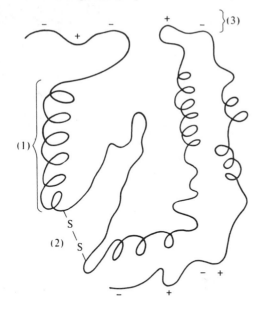

48 The region labeled (1), taken alone, illustrates which level of protein structure?

a primary   c tertiary
b secondary   d quaternary
                                (pp. 60–66)

49 In the drawing the bond labeled (2) is

a a covalent bond.
b a hydrogen bond.
c an ionic bond.
d a hydrophobic bond.   (p. 37)

50 The region of the protein molecule labeled (3) is

a hydrophobic.   c nonpolar.
b hydrophilic.   d alpha-helical.
                                (p. 40)

51 The highest level of protein structure shown in the drawing is

a primary.   c tertiary.
b secondary.   d quaternary.
                                (pp. 64–66)

52 The reaction A + B → C + D is exergonic. All of the following procedures are effective in accelerating the rate of the chemical reaction *except*

a increasing the concentration of C and D.
b heating A and B together.
c increasing the pressure on A and B.
d providing an appropriate catalyst.
                                (pp. 73–76)

53 Glucose will not burn in the air unless it is strongly heated. Why is heat required?

a Heat provides activation energy.
b Heat acts as a catalyst.
c Heat lowers the average energy content of the molecules.
d Heat increases the net amount of energy released by the reaction.
e Heat changes exergonic reactions to endergonic reactions.   (pp. 74–75)

For questions 54–58, use the diagram below, showing energy changes in the reaction X + W → Y + Z. Answers may be used once, more than once, or not at all.

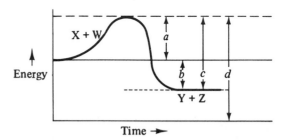

54 Which segment represents the energy of activation? (p. 75)

55 Which segment represents the amount of usable (net) energy released by this reaction? (p. 75)

56 Which segment would be the same irrespective of whether the reaction was uncatalyzed or catalyzed by an enzyme? (p. 76)

57 The above reaction is

a exergonic.
b endergonic.
c spontaneous.
d a and c are correct.   (p. 72)

58 The above reaction

a would have a negative ∆G.
b would have a positive ∆G.
c could have either a negative or positive ∆G depending on the reactants and the temperature.   (pp. 71–73)

59 If the equilibrium constant (K_{eq}) for a chemical reaction is 10,

a the reactants outnumber the products 10 to 1.
b the reaction is endergonic.
c the forward reaction predominates over the backward reaction.
d the ∆G for the reaction would be positive.
e two of the above are correct.
                                (pp. 73–74)

60 Which one of the following statements is true of enzymes?

a Enzymes lose some or all of their normal activity if their three-dimensional structure is disrupted.

b Enzymes are composed of ribose, phosphate, and a nitrogen-containing base.

c The activity of enzymes is independent of temperature and pH.

d Enzymes provide the activation energy necessary to initiate a reaction.

e An enzyme acts only once and is then destroyed. (pp. 76–78)

61 Enzymes

a impart to substrates the kinetic energy they need to react.

b may have at their active sites amino acids from widely different parts of a polypeptide chain.

c help align substrates so that the latter collide with each other in a very precise way.

d Two of the above are correct.

e a, b, and c are correct. (pp. 76–78)

Questions 62–64 refer to the following situation:

The enzyme succinic dehydrogenase normally catalyzes a reaction involving succinic acid. Another substance, malonic acid, sufficiently resembles succinic acid to form temporary complexes with the enzyme, although malonic acid itself cannot be catalyzed by succinic dehydrogenase.

62 In this example succinic acid is

a the substrate.   c an inhibitor.
b the active site.   d the product.
(pp. 76–77)

63 Malonic acid is

a the substrate.
b a competitive inhibitor.
c a positive modulator.
d a negative modulator.
e a coenzyme. (p. 81)

For question 64, also use the diagram below.

64 The solid line shows the reaction rate when a certain amount of the enzyme is added to 1 M succinic acid. Then the same amount of enzyme is added to 1 M succinic acid, but 0.5 M malonic acid is also present. Which of the dotted-line curves best shows the new rate of succinic acid catalysis?

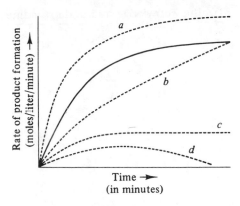

—— curve showing rate of product formation without the presence of malonic acid

---- possible curves showing rate of product formation with presence of malonic acid

ANSWERS

Testing recall

1 true
2 true
3 false—protein
4 true
5 false—endergonic
6 true
7 true
8 true
9 false—condensation reactions
10 true
11 true
12 true
13 b, d
14 a, e
15 c
16 f

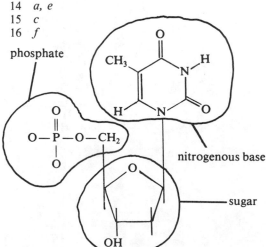

Testing knowledge and understanding

17 *a*

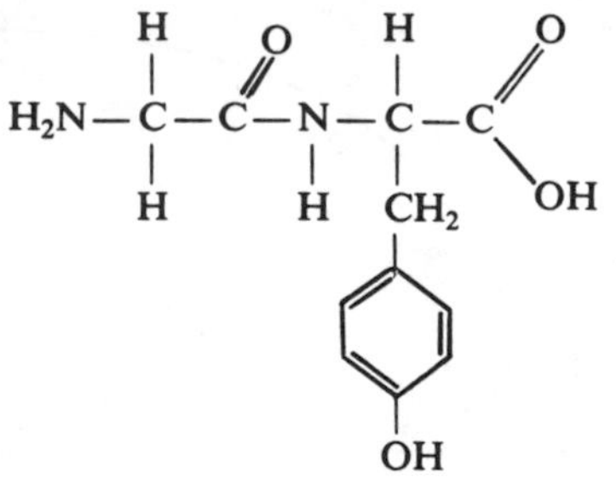

$+$ H_2O

b

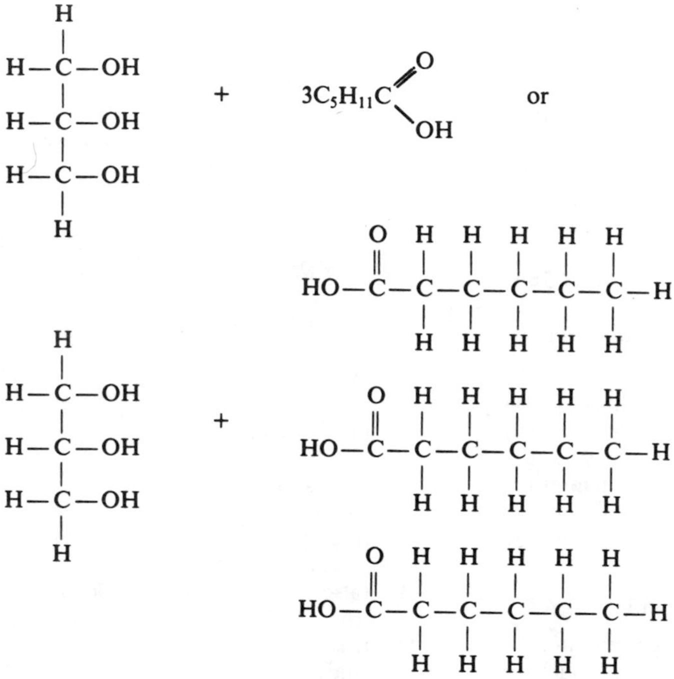

$+$ $3C_5H_{11}C$ or

c

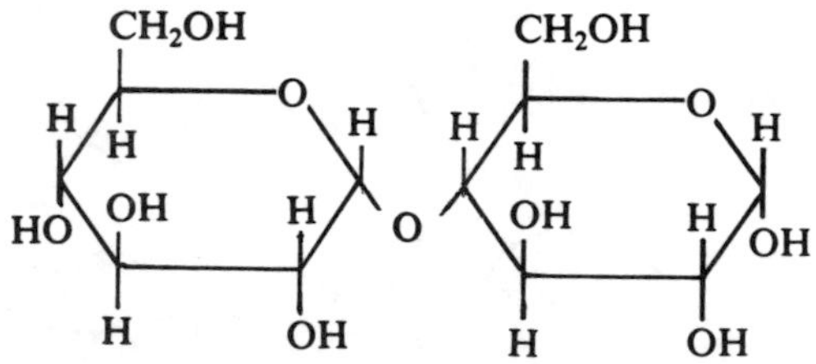

$+$ H_2O

18	*a*	21	*c*	24	*b*
19	*d*	22	*a*	25	*e*
20	*c*	23	*d*	26	2

27 1, 3, 4, 5, 6
28 1
29 6
30 *b*
31 *b*. The completed structural formula would
be

32	*b*	41	*c*	49	*a*	57	*d*
33	*a*	42	*a*	50	*b*	58	*a*
34	*a*	43	*c*	51	*c*	59	*c*
35	*d*	44	*b*	52	*a*	60	*a*
36	*c*	45	*b*	53	*a*	61	*d*
37	*b*	46	*c*	54	*a*	62	*a*
38	*c*	47	*a*	55	*b*	63	*b*
39	*e*	48	*b*	56	*b*	64	*c*
40	*b*						

AT THE BOUNDARY OF THE CELL

A GENERAL GUIDE TO THE READING

This chapter gives a brief history of our knowledge of cells and describes the microscopes used to study them. It then goes on to examine the outer boundaries of cells, notably the structure and function of the cell membrane. As you read Chapter 4 in your text, you will want to concentrate on the following topics:

1 Diffusion and osmosis. Though this material (pp. 87–94) is conceptually difficult at first, understanding it now is worth the effort. We shall be using the terms osmosis, diffusion, osmotic concentration, osmotic potential, hypertonic, hypotonic, and isotonic frequently later.

2 The fluid-mosaic model. Pay close attention to the description of this model of the cell membrane on pages 96–100 and to Figure 4.13 (p. 97). Part of the great excitement in biology in recent years results from our new understanding of membrane structure and how it relates to membrane function.

3 Membrane channels and pumps. This section is particularly important, since it introduces some of the specific mechanisms by which molecules get into and out of cells. Figure 4.18 (p. 101) is an excellent summary diagram. You will encounter more detailed information on specific channels and pumps in Chapters 7, 8, 14, 16, and 18, so you need to get a basic understanding of these phenomena here. Time spent learning the material now will save you time later on.

4 Facilitated diffusion and active transport. These concepts, explained on pages 101 and 103, will be mentioned repeatedly throughout the text.

KEY CONCEPTS

1 The fundamental organizational unit of life is the cell. All living things are composed of cells; all cells arise from pre-existing cells. (pp. 84–85)

2 A cell's interaction with its environment is crucial; materials necessary for life must be obtained from the environment, and waste products must be released into it. (pp. 93–106)

3 Membranes are barriers between cells and (as will be seen later in the text) around certain entities within cells. All substances moving into or out of a cell must pass through a membrane barrier, and the membrane of each cell can be quite specific about what is to pass through, at what rate, and in which direction. (pp. 93–106)

4 The plasma membrane consists of a bilayer of phospholipids oriented with their hydrophobic tails toward the interior of the membrane and their hydrophilic heads toward the aqueous surfaces. The proteins are distributed both on the surfaces and in the interior of the membrane. (pp. 95–100)

5 The lipid bilayer creates an effective barrier between the inside of the cell and the surrounding medium. Highly specialized channels and pumps control the passage of molecules into and out of the cell. (pp. 100–105)

OBJECTIVES

After studying this chapter and reflecting on it, you should be able to carry out the following objectives.

1 State the two components of the cell theory and be familiar with the contributions to our knowledge of cells and cell structure of each of the following scientists: Matthias Schleiden, Theodor Schwann, Rudolf Virchow, Louis Pasteur. (p. 84)

2 Describe the roles of the light microscope, transmission electron microscope, and scanning electron microscope in developing our current understanding of cell structure. (pp. 85–87)

3 Describe the process of osmosis and explain the relationship between osmosis and diffusion. In doing so, include answers to the following questions. How is the net movement of water through a membrane in osmosis related to the concentration of solute molecules? If two compounds had the same molecular weight but one ionized in solution and the other didn't, which one would have the higher osmotic concentration? Osmotic potential? (pp. 88–93)

4 Define the terms hypotonic, hypertonic, and isotonic. Explain what would happen to an animal cell placed in a hypertonic medium and to an animal cell placed in a hypotonic medium. Then explain what would happen to plant cells so placed. (p. 94)

5 Using a diagram such as Figure 4.13 (p. 97) describe the fluid-mosaic model of the cell membrane. In your description, point out the phospholipid molecules, and indicate the hydrophobic and hydrophilic portions; point out the proteins that span the interior of the membrane and those that are confined to the surface; indicate the role of the cholesterol molecules, and explain why lateral movement of molecules within the membrane is possible. List substances to which the membrane is relatively permeable, and those to which it is relatively impermeable. (pp. 96–100)

6 Describe the role that permeases play in moving material through membranes, mentioning membrane channels, gated channels, mobile carriers, and membrane pumps. Figure 4.18 (p. 101) and Figure 4.20 (p. 103) may be helpful in meeting this objective. (pp. 100–4)

7 Using a diagram such as Figure 4.19 (p. 102), explain how an electrostatic gradient can be generated across a membrane and what is meant by an electrochemical gradient. Using Figure 4.18B, show how the energy of the electrochemical gradient can be used to transport glucose into the cell. (p. 102)

8 Distinguish among simple diffusion, facilitated diffusion, and active transport. Indicate the role of these processes in the life of the cell. (pp. 101–3)

9 Using diagrams such as those in Figures 4.22–4.26 (pp. 104–7), describe the processes of endocytosis, phagocytosis, pinocytosis, and exocytosis, explaining the role of these processes in transporting substances into and out of the cell. (pp. 105–6)

10 Describe the formation and structure of the plant cell wall, including the following in your answer: primary wall, secondary wall, middle lamella, pectin, lignin, plasmodesmata, and symplast. Indicate how the cell walls of bacteria and fungi differ from those of plants. (pp. 108–9)

11 Describe the structure of the glycocalyx in animal cells and name two of its functions. (pp. 109–10)

KEY TERMS

The following terms are important in this chapter; you should become familiar with them.

cytoplasm (p. 83)
biogenesis (p. 84)
transmission electron microscope (p. 86)
scanning electron microscope (p. 87)
diffusion (p. 88)
entropy (p. 89)
osmosis (p. 90)
selectively permeable (p. 90)
osmotic concentration (p. 91)
osmotic potential (p. 91)
hypertonic (p. 94)
hypotonic (p. 94)
isotonic (p. 94)
plasma membrane (p. 94)
liposome (p. 96)
fluid-mosaic model (p. 96)
permease (p. 101)
membrane channel (p. 101)
facilitated diffusion (p. 101)
electrostatic gradient (p. 102)
electrochemical gradient (p. 102)
receptor (p. 102)
gated channel (p. 102)
membrane pump (p. 103)
active transport (p. 103)
sodium-potassium pump (p. 103)
endocytosis (p. 105)
phagocytosis (p. 105)
pinocytosis (p. 105)
pseudopodia (p. 105)
exocytosis (p. 106)
cellulose (p. 108)
primary wall (p. 108)
middle lamella (p. 108)
pectin (p. 108)
secondary wall (p. 108)
lignin (p. 108)
plasmodesmata (p. 109)
symplast (p. 109)
chitin (p. 109)
murein (p. 109)
turgor pressure (p. 109)
glycocalyx (p. 109)
contact inhibition (p. 110)

SUMMARY

The fundamental organizational unit of life is the cell. According to the cell theory, all living things are composed of cells and all cells arise from pre-existing cells. Much of our current understanding of the details of cell structure has come through use of electron microscopy.

Functioning of a membrane   The cell membrane is an active part of the cell; it regulates the movement of materials between the ordered interior of the cell and the outer environment. The general rule governing the movement of materials is that the net movement of particles of a particular substance is from regions of greater free energy (where there is an orderly, improbable arrangement) to regions of less free energy (where there is a disorderly, more probable arrangement) of that substance. This movement of particles is called *diffusion*. The plasma membrane is *selectively permeable*; it allows particles of some substances to pass through while restricting others. The movement of water through a selectively permeable membrane is called *osmosis*.

If two different solutions are separated by a selectively permeable membrane, under constant conditions of temperature and pressure the net movement of water will be from the solution with fewer dissolved particles per unit volume to the solution with more dissolved particles per unit volume; i.e., from the solution with the lower osmotic concentration to the solution with the higher osmotic concentration. *Osmotic potential* is a measure of the free energy of a solution, which under conditions of constant temperature and pressure depends on osmotic concentration (the total number of all the solute particles of all kinds —molecules or ions); water flows from regions of higher osmotic potential to regions of lower osmotic potential at a rate proportional to the difference in osmotic potential. Water will continue to move in response to this difference until an equilibrium is reached; at equilibrium the hydrostatic pressure (and hence free energy) in the solution with the higher osmotic concentration becomes so great that water molecules begin to be forced back through the membrane as fast as other water molecules are moving in. When the system reaches this dynamic equilibrium, the free energy (osmotic potential) of the solution on one side just equals the free energy (osmotic potential and hydrostatic pressure) of the solution on the other side.

The cell membrane   Because the plasma membrane is selectively permeable, the processes of osmosis and diffusion are fundamental to cell life. Cell membranes are relatively permeable to water and to certain simple sugars, amino acids, and lipid-soluble substances; they are relatively impermeable to polysaccharides, proteins, and other very large particles. Their permeability to small particles varies, but in general uncharged particles cross more rapidly than charged ones.

The cell membrane cannot completely regulate the exchange of materials. The cell in a medium that is *hypertonic* (higher osmotic concentration) relative to it tends to lose water and shrink. Conversely, a cell in a *hypotonic* medium (lower osmotic concentration relative to it) tends to gain water and swell, and may even burst. A cell in an *isotonic* medium (osmotic concentration in balance with it) neither gains nor loses appreciable water.

Cells are bounded by a *plasma membrane* composed of lipids and proteins, with many small pores. According to the *fluid-mosaic model*, the plasma membrane consists of a bilayer of phospholipids with their hydrophilic heads oriented toward the surfaces of the membrane and their hydrophobic tails toward the interior. Because the individual lipid molecules are linked only by weak bonds, many of them have lateral mobility. The proteins are distributed both on the surfaces and in the interior of the membrane. The pores are thought to be bounded by protein; the distinctive properties of these proteins make the pores selective as to what can move through them.

Enzymelike protein carriers, *permeases*, control molecular traffic into and out of the cell. Some permeases act as *membrane channels*, providing openings through which specific substances can diffuse passively across the membrane, in a process called *facilitated diffusion*. Some permeases move two specific substances cooperatively, utilizing an *electrochemical gradient*. Other permeases have a *gated channel*; a signal molecule combines with a receptor, which changes shape, opening a gate. *Mobile carrier* molecules may transport molecules across a membrane. Still other permeases, known as *pumps*, carry on *active transport*, using energy to move substances against their concentration gradients.

Sometimes substances are taken into the cell by an active process called *endocytosis*, in which a substance is enclosed in a membrane-bound vesicle pinched off from the cell membrane. The reverse sequence, in which materials within vesicles are conveyed to the surface of the cell and discharged, is called *exocytosis*.

Cell walls and coats The plant *cell wall* is located outside the membrane and is composed mainly of cellulose. The *primary wall*, the first portion of the wall laid down, consists of a loose network of fibrils. Many plant cells add further layers, forming a thicker, more compact *secondary wall*. Adjacent cells are bound together by the *middle lamella*.

Fungi and bacteria have cell walls made not of cellulose but of other complex polysaccharide molecules. The presence of the cell wall enables the cells of plants, fungi, and bacteria to exist in hypotonic media without bursting.

Most animal cells have a cell coat, or *glycocalyx*, composed of carbohydrates covalently bonded to protein or lipid molecules in the plasma membrane. The glycocalyx functions in cell recognition and in contact inhibition, and has recognition sites for interaction between the cell and important molecules.

KEY DIAGRAM

The coloring exercise on the following page is designed to help you learn the basic features of membrane structure. You will need seven different colors for this picture. Note the number and arrangement of the different titles. Using the seven colors and the information in Figure 4.18 (p. 97), color the titles and the corresponding structures, using a different color for each structure.

Ask yourself:

1 Are extrinsic proteins found only on the outer surfaces of membranes, or can they be found on the inner surface as well?

2 Do intrinsic proteins too have hydrophilic and hydrophobic sections? If so, where are they located?

3 What factors alter the fluidity of membranes?

QUESTIONS

Testing recall

Match the names of the following scientists with their contributions.

a Pasteur
b Schleiden and Schwann
c Singer and Nicolson
d Virchow
e Danielli and Davson

1 formulated the cell theory (p. 84)

2 advanced the theory of biogenesis (p. 84)

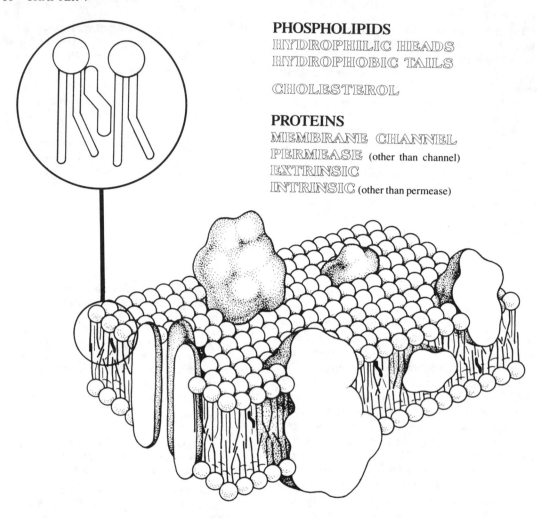

PHOSPHOLIPIDS
HYDROPHILIC HEADS
HYDROPHOBIC TAILS

CHOLESTEROL

PROTEINS
MEMBRANE CHANNEL
PERMEASE (other than channel)
EXTRINSIC
INTRINSIC (other than permease)

3   disproved the theory of spontaneous generation (p. 84)

4   formulated the fluid-mosaic model (p. 96)

It is important not to confuse the following pairs of terms. In one sentence, distinguish between the words in each pair.

5   osmosis–diffusion (pp. 89–90)

6   hypertonic–hypotonic (p. 94)

7   osmotic concentration–osmotic potential (p. 91)

8   facilitated diffusion–active transport (pp. 101, 103)

9   endocytosis–exocytosis (pp. 105–6)

10   pinocytosis–phagocytosis (p. 105)

11   cell wall–cell coat (pp. 108–9)

12   primary cell wall–secondary cell wall (p. 108)

Testing knowledge and understanding

Questions 13–22 refer to the following situation.

The solutions in the two arms of the U-tube are separated at the bottom of the tube by a selectively permeable membrane. At the beginning of the experiment the volumes in both arms are the same, and the level of the liquid is therefore at the same height. The membrane is permeable to water and to sodium chloride, but *not* to glucose. The apparatus is allowed to stand for three days.

Initial set-up

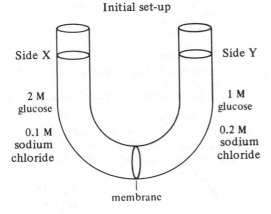

membrane

For each of the next ten questions, select the most appropriate phrase using the following key:

a Both the *statement* and the *reason* are correct.
b The *statement* is correct, but the *reason* is incorrect.
c The *statement* is incorrect, but the *reason* is a fact or principle.
d Both the *statement* and the *reason* are incorrect.

13 The sodium chloride solution on side X will become more concentrated and that on side Y less concentrated *because* a substance tends to diffuse from regions of lower concentration to regions of higher concentrations of that substance. (p. 88)

14 The concentrations of the glucose solutions on sides X and Y will remain unchanged *because* the membrane is impermeable to glucose and it cannot diffuse from one side to the other. (pp. 90–91)

15 The concentration of sodium chloride on side X will eventually equal that on side Y *because* sodium and chloride ions will move by diffusion from one side to the other until a uniform density is reached. (p. 88)

16 The concentration of glucose on side X will decrease and that on side Y increase *because* water molecules will diffuse through the membrane from side Y to side X by osmosis, thus lowering the glucose concentration on side X. (pp. 90–91)

17 The fluid level will increase on side Y and decrease on side X *because* water molecules will move through the membrane from regions of higher to regions of lower concentration of water molecules. (pp. 90–91)

18 The fluid level on side X will rise *because* the water molecules on that side at the beginning of the experiment have more free energy than those on side Y. (pp. 90–91)

19 The net movement of water molecules will be from side X to side Y *because* water molecules will move from the solution with the lower osmotic potential to the solution with the higher osmotic potential when the two are separated by a selectively permeable membrane. (pp. 90–91)

20 Water molecules will move only from side Y to side X and not from side X to side Y *because* water molecules move only from regions of higher to regions of lower concentration. (pp. 91–93)

21 The fluid level on side X will rise *because* the solution in side X had lower osmotic potential than the solution in side Y. (pp. 90–91)

22 Water molecules will tend to move from side Y to side X *because* the net movement of water molecules will be from the solution with the lower to the solution with the higher osmotic concentration. (pp. 90–91)

Choose the one best answer.

Questions 23–25 refer to the following situation.

Two beakers are connected by a tube partitioned by a membrane permeable to water but not to protein. Beaker A contains a 2 percent protein solution and beaker B contains a 4 percent protein solution.

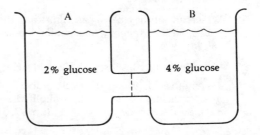

23 Assuming uniform temperature and pressure, which one of the following statements best describes what will happen in this system?

a Water will move from A to B.
b Water will move from B to A.
c Water will move equally in both directions, so that there will be no net change in the system.
d Water will move in both directions, but the net flow will be from A to B.
e Water will move in both directions, but the net flow will be from B to A.
(pp. 90–93)

24 Side A is _____ compared to side B.

a hypertonic
b hypotonic
c isotonic (p. 94)

25 Suppose that instead of protein, we dissolve 1,000 molecules of NaCl in beaker A and 1,000 molecules of sucrose in beaker B. Suppose further that the membrane in the connecting tube is impermeable to both NaCl and sucrose. What will happen in the system? (Select one of the answers listed under question 23.) (pp. 90–93)

Questions 26–30 refer to the following diagram of the plasma membrane.

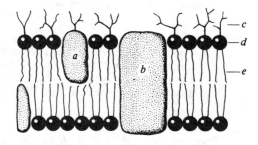

26 Of the five labeled parts of the diagram (a, b, c, d, and e), which one is now thought to play a critical role in cell recognition? (p. 110)

27 Which one of the labeled parts illustrates a globular protein that probably acts as a permease? (p. 101)

28 Which one of the labeled parts illustrates the hydrophilic portion of a phospholipid molecule? (p. 97)

29 When hydrolyzed, substance b would yield

a amino acids. d nucleotides.
b fatty acids. e phospholipids.
c monosaccharides. (pp. 97, 59)

30 Which one of the labeled parts shows the most hydrophobic region of the membrane? (p. 97)

31 Molecules move by diffusion. Diffusion is

a faster in a liquid than in a gas.
b faster in a solid than in a liquid.
c from a region of high concentration to a region of low concentration.
d from a region of low concentration to a region of high concentration. (p. 88)

32 Facilitated diffusion

a involves a permease and requires energy.
b involves a permease but does not require energy.
c requires energy but does not involve a permease.
d moves substances from a region of lower concentration to a region of higher concentration.
e does not require energy or a permease. (p. 101)

33 Which of the following molecules would cross a cell membrane most easily? Assume there is no active transport or facilitated diffusion.

a amino acid d lipid-soluble
b starch substance
c protein e nucleotide
(pp. 90, 100)

34 The concentration of potassium in a red blood corpuscle is much higher than it is in the surrounding blood plasma, yet potassium continues to move into the cell. The process by which potassium moves into the cell is called

a osmosis.
b simple diffusion.
c facilitated diffusion.
d active transport.
e pinocytosis. (p. 103)

35 A dehydrated plant cell is placed in pure water. Which of the following statements best describes what will happen?

a Water will enter the cell, but the cell will be prevented from bursting by the cell wall.
b Water will enter the cell, and the cell will burst.
c Water will be drawn out of the cell until the cell and the water are at equilibrium.
d Water will be drawn out of the cell until the cell dies.
e Water will not move either into or out of the cell. (pp. 90, 109)

36 The only way in which a very large molecule such as a protein could cross a cell membrane is by

   a active transport.
   b endocytosis.
   c simple diffusion.
   d facilitated diffusion.
   e osmosis.   (p. 105)

37 According to the fluid-mosaic model of the cell membrane, the proteins are located

   a in a continuous layer over the outer surface of the membrane.
   b in a continuous layer over the inner surface of the membrane.
   c in a continuous layer over both the outer and inner surfaces of the membrane.
   d in the middle of the membrane, between the lipid layers.
   e in discontinuous arrangements, both on the surface and in the interior of the membrane.   (pp. 97–98)

38 Among the cell structures and materials listed below, which one would be *inner*most if they were arranged in proper order?

   a glycocalyx
   b plasma membrane
   c primary cell wall
   d secondary cell wall
   e pectin   (p. 108)

39 You are studying the transport of a certain type of molecule into cells. You find that transport slows down when the cells are poisoned with a chemical that inhibits energy production. Under normal circumstances the molecule you are studying is probably transported into the cell by

   a simple diffusion.
   b facilitated diffusion.
   c active transport.
   d osmosis.
   e exocytosis.   (p. 103)

40 Cell biologists have recently discovered special regions between epithelial cells, called gap junctions, where tiny channels run from the cytoplasm of one cell to the cytoplasm of the adjacent cell. The equivalent structures between plant cells are called

   a pectins.   d primary walls.
   b middle lamellae. e permeases.
   c plasmodesmata.   (p. 109)

For further thought

1 Distinguish between osmosis and diffusion and give a physical explanation for each. What is osmotic pressure? Discuss the implications of these processes for both an animal cell and a plant cell living in a hypotonic medium, a hypertonic medium, and an isotonic medium. (pp. 88–94)

2 Briefly describe the fluid-mosaic model of membrane structure. Explain why membranes are generally impermeable to charged molecules and to highly polar molecules such as sugars and amino acids. Discuss the importance of this impermeability to (a) osmosis and (b) the establishment of an electrochemical gradient across a membrane. (pp. 96–104)

3 A 5 percent glucose solution is frequently administered intravenously to persons who have undergone surgery. Would you expect the 5 percent glucose solution to be hypertonic, hypotonic, or approximately isotonic to the blood? Give reasons for your answer. (p. 94)

ANSWERS

Testing recall

1 b   3 a
2 d   4 c

5 *Osmosis* is the movement of water through a selectively permeable membrane; *diffusion* is the movement of any particles and may or may not be through a membrane.

6 A *hypertonic* solution has a relatively higher concentration of solute and will gain water by osmosis; a *hypotonic* solution has a relatively lower concentration of solute and will lose water.

7 *Osmotic potential* is a measure of the tendency of water to diffuse across a membrane; it varies inversely with the concentration of osmotically active particles in the solution, the *osmotic concentration*.

8 *Facilitated diffusion* is passive movement with the concentration gradient; *active transport* is generally movement against the concentration gradient and requires energy.

9 *Endocytosis* is a process by which a substance is drawn into a cell in a membrane-
bounded vesicle. *Exocytosis* is essentially
the reverse of endocytosis.

10 *Pinocytosis* and *phagocytosis* differ only in
the size of the particles taken in; if the particles are large the process is termed phagocytosis; if liquid or very small, pinocytosis.

11 The *cell wall* is entirely separate from the
plasma membrane, but the molecules of the
cell coat attach directly to the molecules of
the plasma membrane.

12 The *primary cell wall* is found in all plant
cells, whereas the thicker *secondary wall* is
laid down inside the primary wall only in
certain plant tissues.

Testing knowledge and understanding

13	b	20	d	27	b	34	d
14	c	21	a	28	d	35	a
15	a	22	a	29	a	36	b
16	a	23	d	30	e	37	e
17	c	24	b	31	c	38	b
18	b	25	e	32	b	39	c
19	d	26	c	33	d	40	c

INSIDE THE CELL

A GENERAL GUIDE TO THE READING

This chapter focuses on the internal structure of cells, presenting basic information on the various organelles. As you read Chapter 5, you will want to concentrate on the following topics.

1 Organelles. You will want to learn the basic structure and function of the various organelles since we shall be referring to them frequently throughout the book.

2 The differences between procaryotic and eucaryotic cells. These differences, summarized on pages 130 and 134 and in Table 5.1 (p. 135), will be referred to again in later chapters. The differences are quite simple to understand and should be mastered now.

KEY CONCEPTS

1 A living cell is an extraordinarily complex unit with an intricate internal structure; its activities are precisely integrated and controlled.

2 Eucaryotic cells have a membrane-enclosed nucleus whereas procaryotic cells lack a membrane-enclosed nucleus and internal membranous organelles. (pp. 111–15, p. 134)

OBJECTIVES

After studying this chapter and reflecting on it, you should be able to carry out the following objectives.

1 For each of the organelles listed below, describe its structure; give a major function; indicate whether it is surrounded by a single membrane, a double membrane, or no membrane; and state whether it is found in plant, animal, or bacterial cells. (pp. 111–30)

nucleus	vacuole
nucleolus	microtubules
endoplasmic reticulum	microfilaments
ribosome	microtrabecular
Golgi apparatus	lattice
mitochondrion	centriole
lysosome	basal bodies
plastids (chromoplasts and leucoplasts)	cilla and flagella

2 List four differences between procaryotic and eucaryotic cells. (pp. 130–135)

3 Describe the Margulis endosymbiont hypothesis for the origin of the eucaryotic cell and give three lines of evidence for the validity of this hypothesis. (pp. 134–36)

KEY TERMS

The following terms are important in this chapter; you should become familiar with them.

organelles (p. 111)
nucleus (p. 111)
procaryotic cells (p. 112)
eucaryotic cells (p. 112)
chromosomes (p. 112)
genes (p. 113)
nucleoli (p. 113)
nuclear membrane (p. 113)
endoplasmic reticulum (rough and smooth) (p. 116)
cytosol (p. 116)
ribosomes (p. 116)
Golgi apparatus (p. 118)
lysosomes (p. 120)
peroxisomes (p. 121)
mitochondria (p. 122)
plastids (p. 122)
chromoplasts (p. 122)
chloroplasts (p. 122)
leucoplasts (p. 122)
chlorophyll (p. 122)
carotinoids (p. 122)
stroma (p. 123)

thylakoids (p. 123)
grana (p. 123)
vacuoles (p. 124)
betacyanin (p. 125)
anthocyanin (p. 125)
microfilaments (p. 125)
actin (p. 126)
myosin (p. 126)
cytoskeleton (p. 126)
microtubules (p. 126)
microtrabecular lattice (p. 127)
centrioles (p. 128)
basal bodies (p. 129)
flagella (p. 129)
cilia (p. 130)
endosymbiotic hypothesis (p. 134)

SUMMARY

Eucaryotic cells have a membrane-bound nucleus, whereas *procaryotic cells* (bacteria and cyanobacteria) lack a membrane-bound nucleus. The following discussion concerns only eucaryotic cells.

The *nucleus* contains the *chromosomes*, which contain the genes. It can therefore direct the cell's life processes. The nucleus also contains one or more *nucleoli*, where ribosomal *RNA* is synthesized and combined with proteins before moving into the cytoplasm to become part of the ribosomes. Separating the nucleus from the cytoplasm is a double *nuclear membrane* interrupted by pores. The nuclear membrane is continuous at places with the endoplasmic reticulum.

The *endoplasmic reticulum* (ER) forms a system of interconnected membrane-enclosed spaces. Sometimes the membranes of the ER are "rough," with *ribosomes* on their outer surfaces; when no ribosomes are present, the ER is "smooth." Ribosomes are sites of protein synthesis. The ER functions both as a passageway for intracellular transport and as a manufacturing surface.

The *Golgi apparatus* consists of stacks of membrane-enclosed vesicles that function in the storage, modification, and packaging of secretory products.

Located within the cytoplasm are many other organelles. The *mitochondria* are the powerhouses of the cell; chemical reactions within the mitochondria provide energy for the activities of the cell. *Lysosomes* are membranous sacs that function as storage vesicles for powerful digestive enzymes; they may act as the cell's digestive system, hydrolyzing materials taken in by endocytosis.

Peroxisomes are also membranous sacs of enzymes; these are oxidative rather than digestive. Most plant cells have large membranous organelles called *plastids*. The two principal categories of plastids are the colored *chromoplasts* and the colorless *leucoplasts*. Chloroplasts are chromoplasts that contain the green pigment chlorophyll; they capture the energy of sunlight and utilize it in the manufacture of organic compounds. The leucoplasts' primary function is storage of starch, oils, or protein granules. Membrane-enclosed, fluid-filled spaces termed *vacuoles* are found in many cell types, and perform a variety of functions. Most mature plant cells have a large central vacuole occupying much of the volume of the cell; it plays an important role in maintaining the turgidity of the cell and in the storage of important substances, including wastes.

Microtubules and *microfilaments* appear to function in intracellular movement and to support the cell. Microtubules also form the spindle of dividing cells and are essential components of centrioles, cilia, and flagella. The *centrioles* are hollow cylindrical bodies located just outside the nucleus of most animal cells; they help organize the microtubular spindle for cell division. *Cilia* and *flagella* are hairlike projections from the cell's surface that move the cell or move materials across the cell's surface. Inside the stalk of cilia and flagella is a 9 + 2 arrangement of microtubules. At the base of the stalk is the *basal body*, whose structure is the same as that of the centriole.

Procaryotic cells lack the internal membranous organelles discussed above, but do have ribosomes and a *nucleoid* containing a circular chromosome of DNA. Some cells have flagella but these lack microtubules.

KEY DIAGRAM

In the diagram below, color each name and the appropriate structure(s) with the same color.

CELL MEMBRANE
MICROVILLI
NUCLEAR
 MEMBRANE
NUCLEUS
NUCLEOLUS
CYTOSOL
ENDOPLASMIC
 RETICULUM
RIBOSOMES
GOLGI COMPLEX
MITOCHONDRIA
VACUOLES
PINOCYTOTIC
 VESICLES
LYSOSOME
CENTRIOLES

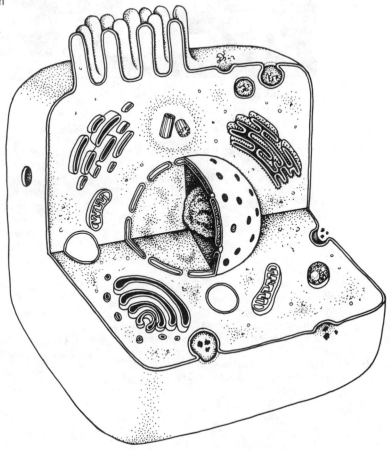

QUESTIONS

Testing recall

Do the following crossword puzzle.

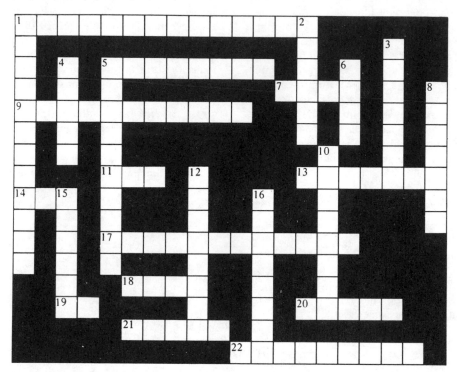

Across

1 long threadlike structures found in muscle fibers (p. 125)
5 protein carrier molecules in the membrane (p. 101)
7 opening in the nuclear envelope (p. 114)
9 chromoplast containing chlorophyll (p. 122)
11 prefix meaning "same as" (p. 94)
13 control center of cell (p. 111)
14 contains the hereditary information (p. 113)
17 hollow cylindrical structures found in spindle fibers and centrioles (p. 126)
18 prefix meaning "less than" (p. 94)
19 abbreviation for the organelle involved in intracellular transport (p. 115)
20 have a 9 + 2 arrangement of microtubules (p. 130)
21 prefix meaning "greater than" (p. 94)
22 focus of the microtubular spindle in animal cells (p. 128)

Down

1 sites of chemical reactions that provide energy for cell's activities (p. 122)
2 type of endoplasmic reticulum (p. 116)
3 occupies most of the volume of most mature plant cells (p. 124)
4 membrane-bound vesicles involved in packaging secretory products (p. 118)
5 contains oxidative enzymes (p. 121)
6 fundamental organizational unit of life (p. 84)
8 the fluid part of the cytoplasm (p. 116)
10 dark oval bodies found in nucleus (p. 113)
12 cell's digestive system (p. 120)
15 carrier-mediated transport that requires energy (p. 103)
16 site of protein synthesis (p. 116)

Listed below are ten substances or organelles found in many cells. For each, write E if the item is found only in eucaryotic cells, P if it is found only in procaryotic cells, and B if it is found in both. (Table 5.1, p. 135)

1 nuclear membrane

2 ribosome

3 flagella with 9 + 2 structure

4 DNA in chromosome

5 mitochondrion

6 lysosome

7 plasma membrane

8 chlorophyll

9 endoplasmic reticulum

10 chloroplast

Testing knowledge and understanding

11 A particular cell steadily secretes a protein into the surrounding medium. The secretion is released from the cell by exocytosis from membranous vesicles derived from the Golgi apparatus. The cell also functions as a storage depot for glycogen. The cell can be identified as

   a a bacterium.
   b a cyanobacterium.
   c a cell from a terrestrial plant.
   d a cell from an animal.   (pp. 53, 118)

12 The antibiotic streptomycin is thought to combine with the ribosomes in bacteria and thus disrupt their normal functioning. In other words, this antibiotic destroys bacteria by

   a preventing the synthesis of proteins.
   b interfering with normal cell reproduction.
   c slowing energy production within the cell.
   d preventing transport within the endoplasmic reticulum.
   e interfering with materials entering and leaving the cell.   (p. 116)

13 An organelle surrounded by a double membrane is the

   a lysosome.   *d* centriole.
   b Golgi apparatus.   *e* ribosome.
   c chloroplast.   (p. 123)

14 Membranes are found as part of all of the following subcellular structures *except* the

   a endoplasmic reticulum.
   b ribosomes.
   c Golgi apparatus.
   d mitochondria.
   e lysosomes.   (pp. 115–22)

15 Tubulin

   a is a major component of the flagella of eucaryotic cells.
   b assembles into microfilaments.
   c is a major component of centrioles.
   d forms the major portion of the microtrabecular lattice.
   e Two of the above are correct.   (p. 126)

16 All of the following structures are found in both procaryotic and eucaryotic cells *except*

   a plasma membrane.   *d* cell wall.
   b Golgi apparatus.   *e* ribosomes.
   c chromosomes.   (p. 135)

17 Which one of the following is *false* concerning lysosomes?

   a They act as the digestive system of the cell.
   b They function as storage vesicles for oxidative enzymes which catalyze certain condensation reactions.
   c Lipid-digesting lysosomes sometimes lack a particular enzyme, resulting in Tay-Sachs disease.
   d They are produced by budding from the Golgi apparatus.
   e They possess a selectively permeable membrane that allows certain substances to pass through but is impermeable to the enzymes stored within it.   (p. 120)

18 The relatively homogeneous internal proteinaceous portion of the chloroplast is called the

   a leucoplast.   *d* grana.
   b stroma.   *e* vacuole.
   c thylakoid.   (p. 123)

19 A structure found in both plant and animal cells but which has its greatest development in plant cells is a(n)

   a mitochondrion.   *c* vacuole.
   b lysosome.   *d* Golgi apparatus.
   (p. 124)

20 Which of the following is *not* found in procaryotic cells?

   a mitochondrion   *d* flagella
   b plasma membrane   *e* ribosome
   c DNA   (p. 135)

21 In the following comparison of procaryotic and eucaryotic cells, which item is *incorrect*?

	Characteristic	Procaryotes	Eucaryotes
a	nuclear membrane	absent	present
b	chromosomes	DNA	DNA and protein
c	mitochondria	present	present
d	ribosomes	small	large
e	flagella	lack 9 + 2 structure	have 9 + 2 structure (p. 135)

22 All of the following organelles have chromosomes *except*

a ribosomes.   c chloroplasts.
b nucleus.   d mitochondria.
(pp. 113, 135)

23 It has been proposed that mitochondria and chloroplasts (and flagella) are modern descendants of primitive forms of procaryotic cells that took up residence within primitive cells and evolved independently there. Which one of the following is *not* evidence of this view?

a Like procaryotic cells, mitochondria and chloroplasts have DNA that is not wound on protein spools.

b The ribosomes in procaryotic cells, mitochondria, and chloroplasts are very similar, and differ from the cytoplasmic ribosomes of eucaryotes.

c Mitochondria and chloroplasts are known to have their own genes and ribosomes and to conduct protein synthesis.

d Mitochondria and chloroplasts build their own membranes.

e The chromosomes of mitochondria and chloroplasts have genes that code for *all* their enzymes; the organelles are completely independent of the host cell's genes. (p. 136)

ANSWERS

Testing recall

1	E	6	E
2	B	7	B
3	E	8	B
4	B	9	E
5	E	10	E

Testing knowledge and understanding

11	d	14	b	17	b	20	a	22	a
12	a	15	e	18	b	21	c	23	e
13	c	16	b	19	c				

MULTICELLULAR ORGANIZATION AND THE DIVERSITY OF ORGANISMS

A GENERAL GUIDE TO THE READING

This chapter introduces basic information about plant and animal tissues and the variety of living things. This information will serve as a foundation for material presented in other chapters. The chapter is well illustrated; you will find learning this material easier if you pay close attention to the photographs and drawings. As you read Chapter 6 in your text, you will want to concentrate on the following topics.

1 Plant tissues. The section on these tissues (pp. 140–45) provides an introduction for Chapters 8, 9, and 12. Time spent learning the material now will save you time later on.

2 Epithelium and connective tissue. Read carefully the material on these tissues (pp. 146–48), since they will not be discussed further. The subsequent material on cartilage, bone, muscle, and nerves (pp. 150–51) serves primarily as an introduction; these tissues will be discussed in more detail in later chapters.

3 The five kingdoms of organisms. These kingdoms, and the major groups of organisms in each, are described briefly on pages 152–66. This introduction to the diversity of life is provided here to familiarize you with the major groups that will be mentioned in later chapters.

KEY CONCEPTS

1 The bodies of most multicellular organisms are organized on the basis of specialized tissues, organs, and systems. (pp. 137–38)

2 All plant tissues can be divided into two major categories: meristematic tissue and permanent tissue. (pp. 140–45)

3 All animal tissues are divided into four categories: epithelium, connective tissue, muscle, and nerve. Each of these is a diverse assemblage of different subtypes. (pp. 145–51)

4 Living things are classified on the basis of the evolutionary relationships thought to exist among them. (pp. 152–66)

OBJECTIVES

After studying this chapter and reflecting on it, you should be able to carry out the following objectives.

1 Describe meristematic cells and indicate where in the plant body such cells are found. Differentiate between the apical and lateral meristems and give the function of each. (p. 141)

2 List the three subcategories of permanent tissue and give examples of each. (pp. 141–45)

3 Give the characteristics of epidermal cells; indicate where they are found in the plant body; specify the function of the cuticle; and describe the periderm. (pp. 141–42)

4 Describe parenchyma cells; specify their location in the plant body; and list some of their functions. Contrast collenchyma cells (Fig. 6.4, p. 143) and sclerenchyma cells (Fig. 6.5, p. 143) with respect to function, the thickness of the cell walls, and whether or not the cells are living or dead at maturity. Explain where the endodermis is located and describe the Casparian strip. (pp. 142–44)

5 Give the function of xylem; indicate what types of cells are present in it; and explain how a complex tissue such as xylem differs from simple tissues such as parenchyma and collenchyma. Give the function of phloem, and indicate what types of cells are present in it. (pp. 144–45)

6 Using a diagram of a flowering-plant body, such as Figure 6.7 (p. 145), identify the root, shoot, flower, and apical meristems.

7 Explain where epithelial tissues are found in the body, and give their functions. Differentiate among squamous, cuboidal, and columnar cells. Explain how you would recognize simple epithelium, stratified epi-

thelium, and pseudostratified epithelium. (pp. 146–48)

8 List the four main types of connective tissue. Explain how you would recognize dense fibrous connective tissue and loose connective tissue, and specify where in your body you might expect to find each. Describe cartilage and bone, and give the function of each; in doing so, show how the matrix of cartilage differs from that of bone, and how that difference is reflected in the respective functions of cartilage and bone. (pp. 148–51)

9 Give the function of muscle cells, and list the three principal types. (p. 151)

10 Give the function of a nerve cell and explain how its structure, as shown in Figure 6.15 (p. 151), is adapted to its function.

11 Using a diagram such as Figure 6.16 (p. 153), point out the different tissue types and explain how the cells are integrated to form an organ. (pp. 151–52)

12 Name the five kingdoms of organisms and give the distinguishing characteristics of each. (pp. 152–66)

13 Give one important characteristic of each of the following groups of plants: green algae, brown algae, red algae, mosses, vascular plants. (pp. 157–60)

14 Give one important characteristic and one example of each of the following groups of animals: coelenterates, flatworms, molluscs, annelids, arthropods, echinoderms, chordates. (pp. 161–66)

KEY TERMS

The following terms are important in this chapter; you should become familiar with them.

tissue (p. 137)
organ (p. 138)
system (p. 138)
fibroblast (p. 138)
intercellular matrix (p. 138)
spot desmosomes (p. 139)
belt desmosomes (p. 139)
tight junctions (p. 139)
gap junctions (p. 140)
meristematic tissue (p. 140)
permanent tissue (p. 140)
apical meristem (p. 141)
lateral meristem (p. 141)

SUMMARY

The bodies of multicellular organisms are organized on the basis of *tissues*, organs, and systems. Aggregations of cells are either held together by connective tissue or by specific cell-to-cell recognition and adherence. Cells may be joined together by spot desmosomes, belt desmosomes, tight junctions, or gap junctions.

All plant tissues can be divided into meristematic tissue (growth tissue—undifferentiated cells capable of dividing) and permanent tissue (mature differentiated cells). Regions of *meristematic tissue* are found at the growing tips of roots and stems (*apical meristems*) and, in many plants, in areas toward the periphery of the roots and stems (*lateral meristems*). The permanent tissues fall into three subcategories: surface tissues, fundamental tissues, and vascular tissues. The surface tissues (*epidermis, periderm*) form the protective outer covering of the plant body. There are four types of fundamental tissues: *parenchyma* (thin-walled, loosely packed cells found throughout the plant body), *collenchyma* (supportive tissue whose cell walls are irregularly thickened), *sclerenchyma* (supportive tissue with very thick cell walls), and *endodermis* (a single layer of cells surrounding the vascular cylinder of roots and occasionally of stems). The vascular, or conductive, tissue is characteristic of higher plants and consists of two principal types of complex tissue: xylem and phloem. *Xylem* supports the plant and transports water and dissolved minerals upward. *Phloem* conducts organic materials up and down the plant body. The body of higher land plants is divided into two major parts: the root and the shoot.

Animal tissues are divided into epithelium, connective tissue, muscle, and nerve. *Epithelial tissue* covers or lines the internal and external surfaces of all free body surfaces. *Connective tissue*, composed of cells embedded in an extensive extracellular matrix, connects, supports, or surrounds other tissues or organs; examples of connective tissue are blood and lymph, connective tissue proper, cartilage, and bone. The three types of *muscle tissue* (skeletal or striated, smooth, and cardiac) consist of cells specialized for contraction and are responsible for most movement in higher animals. *Nervous tissue* is highly specialized for the ability to respond to stimuli.

The process of evolution, repeated over billions of years, has given rise to the various groups of organisms seen today. The various organisms can be classified into five kingdoms: Monera, Protista, Plantae, Fungi, and Animalia. The organisms belonging to the Monera and Protista are predominantly unicellular; the others are all multicellular.

The kingdom *Monera* includes two groups of unicellular organisms: the Eubacteria (including the bacteria and the Cyanobacteria) and the Archaebacteria. Moneran cells lack the membrane-enclosed nucleus and other membranous structures present in the cells of other organisms.

The kingdom *Protista* includes a wide variety of organisms, mostly unicellular. Two major groups of Protista are the plantlike protists and the animal-like protists (*Protozoa*).

The organisms belonging to the kingdom *Plantae* all have cells with rigid cell walls and chloroplasts. Thus they are photosynthetic and can synthesize their own food. The red, brown, and green algae are aquatic plant groups that show little tissue differentiation. The land-plant groups (mosses, liverworts, and the vascular plants) probably evolved from ancestral green algae.

Organisms of the kingdom *Fungi* also have cell walls, but they lack chlorophyll and cannot manufacture their own food; they must obtain their nutrients already synthesized. Fungi depend entirely on absorption of nutrient molecules.

Members of the kingdom *Animalia* differ from plants and fungi in the lack of rigid cell walls, and in their mode of nutrition; animals ingest large particles of food. Seven animal groups (coelenterates, flatworms, molluscs, annelids, arthropods, echinoderms, and chlordates) will be frequently referred to throughout the book.

QUESTIONS

Testing recall

Match each of the following plant and animal tissues with its characteristics.

a	collenchyma	g	muscle
b	connective	h	nervous
c	endodermis	i	parenchyma
d	epidermis	j	phloem
e	epithelium	k	sclerenchyma
f	meristematic	l	xylem

1 thin-walled, loosely packed, unspecialized plant tissue (p. 142)

2 protective outer covering of the plant body (p. 142)

3 blood and bones are examples (pp. 148–51)

4 undifferentiated plant cells capable of dividing (p. 141)

5 conducts organic materials up and down the plant body (p. 145)

6 specialized for contraction (p. 151)

7 uniformly thick-walled supportive tissue (p. 143)

8 conducts water and dissolved materials upward in the plant body (p. 144)

9 specialized for stimulus reception and conduction (p. 151)

10 lines the inner and outer surfaces of the animal body (pp. 146–48)

Below are listed the five kingdoms of organisms used in the classification system presented in your textbook. Answer questions 11–17 with the appropriate letter or letters.

a	Animalia	d	Plantae
b	Fungi	e	Protista
c	Monera		

11 Which kingdoms have only unicellular organisms? (p. 154)

12 Which kingdoms have multicellular organisms? (pp. 156–61)

13 Which kingdoms include organisms whose cells have cell walls? (pp. 159–60)

14 Which kingdoms include at least some organisms that can synthesize their own food? (pp. 154–60)

15 Which kingdoms include organisms that ingest particulate food? (pp. 156, 161)

16 In which kingdom are nearly all the organisms photosynthetic? (p. 157)

17 In which kingdom are all the organisms totally dependent on absorption to obtain their high-energy nutrients? (p. 160)

Testing knowledge and understanding

Choose the one best answer.

18 Which one of the following cell types would you *not* expect to find in a leaf?

a	epidermal cell	d	tracheid
b	parenchyma cell	e	collenchyma cell
c	periderm		(pp. 140–45)

19 The region of tissue at the growing tips of roots and stems would be classified as

   a surface tissue.   *d* periderm.
   b apical meristem. *e* parenchyma.
   c lateral meristem.   (p. 141)

20 The corky outer bark on trees is classified as

   a epidermis.   *d* sclerenchyma.
   b cuticle.   *e* periderm.
   c endodermis.   (p. 142)

21 All of the following are simple tissues *except*

   a parenchyma.   *d* endodermis.
   b collenchyma. *e* phloem.
   c sclerenchyma.   (pp. 144–45)

22 A tissue is made up of several varieties of cells that are embedded in an extensive matrix composed of ground substance and many irregularly arranged fibers. This tissue would be classified as

   a lymph.
   b loose connective tissue.
   c dense connective tissue.
   d cartilage.
   e bone.   (pp. 148–51)

23 Which one of the following is *not* an example of connective tissue?

   a blood   *d* cartilage
   b bone   *e* muscle
   c lymph   (pp. 148–51)

24 Skin is an animal organ. Of which of the following tissue types is it composed?

   a connective tissue *d* nerve
   b epithelium   *e* all of the above
   c muscle   (pp. 151–52)

Below are listed some of the important animal groups that you are expected to become familiar with. Match each of the characteristics in questions 25–31 with the appropriate group or groups of animals.

   a annelids   *e* echinoderms
   b arthropods   *f* flatworms
   c chordates   *g* molluscs
   d coelenterates

25 digestive cavity with one opening (pp. 161–62)

26 two distinct tissue layers (p. 161–62)

27 jointed legs and hard outer covering (p. 165)

28 radial symmetry (pp. 161, 165)

29 shells (p. 162)

30 internal hard skeleton (p. 166)

31 segmented worms (p. 163)

Referring to the list above, match each animal in questions 32–44 with the group it belongs to.

32 grasshopper (p. 165)

33 clam (p. 162)

34 lobster (p. 165)

35 planarian (p. 162)

36 hydra (pp. 161–62)

37 seastar (p. 165)

38 fish (p. 166)

39 crayfish (p. 165)

40 jellyfish (pp. 161–62)

41 snake (p. 166)

42 frog (p. 166)

43 human being (p. 166)

44 earthworm (p. 163)

For questions 45–52, use the following key.

   a algae   *c* mosses
   b fungi   *d* vascular plants

45 Which group consists of the dominant land plants? (pp. 159–60)

46 Which group obtains its nutrients only by absorption? (p. 160)

47 Which group includes angiosperms and gymnosperms? (pp. 159–60)

48 Which group includes aquatic plants that show little tissue differentiation? (pp. 157–58)

49 Which group shows the greatest tissue specialization? (pp. 159–60)

50 Which group includes the flowering plants? (pp. 159–60)

51 Which group lacks chlorophyll (p. 160)

52 Which group possesses an effective transport system? (pp. 159–60)

ANSWERS

Testing recall

1	*i*	5	*j*	8	*l*
2	*d*	6	*g*	9	*h*
3	*b*	7	*k*	10	*e*
4	*f*				

11 *c,e*
12 *a,b,d*
13 *b,c,d,e*
14 *c,d,e*
15 *a,e*
16 *d*
17 *b*

Testing knowledge and understanding

18	*c*	21	*e*	23	*e*
19	*b*	22	*b*	24	*e*
20	*e*				

25 *d,f*
26 *d*
27 *b*
28 *d,e*

29	*g*	35	*f*	41	*c*	47	*d*
30	*c*	36	*d*	42	*c*	48	*a*
31	*a*	37	*e*	43	*c*	49	*d*
32	*b*	38	*c*	44	*a*	50	*d*
33	*g*	39	*b*	45	*d*	51	*b*
34	*b*	40	*d*	46	*b*	52	*d*

Chapter 7

ENERGY TRANSFORMATIONS: RESPIRATION AND OTHER CATABOLIC PATHWAYS

A GENERAL GUIDE TO THE READING

This chapter discusses the process by which the energy stored in complex organic compounds is released and converted into a form that can be used by the cell to do work. Because the chapter involves much organic chemistry, your first impression may be that the material is hopelessly complex. However, if you focus your attention on the following aspects of the process, you will find that the material is readily understandable. In addition, the concepts and objectives listed in the next sections of this *Study Guide* provide a valuable tool for learning this complex information.

1 Oxidation and reduction. These terms, defined on pages 169–70, will be used frequently in this chapter and the next; you will want to understand the paired concepts of oxidation and reduction.

2 The compound adenosine triphosphate (ATP). The synthesis and functioning of ATP, the universal energy currency of living things, is first discussed on pages 170–71. You need not learn the structure of the molecule, but you will want to be familiar with the symbols ATP, ADP, AMP, $\textcircled{P}$, and P_i and with the relationships among the substances they represent.

3 Stages in the breakdown of glucose. Notice that the complete breakdown of glucose involves five stages, discussed in successive sections of the text: glycolysis (Stage I), oxidation of pyruvic acid to acetyl-CoA (Stage II), the Krebs citric acid cycle (Stage III), the respiratory electron-transport chain (Stage IV), and the chemiosmotic synthesis of ATP (Stage V). (pp. 172–79)

4 Glycolysis (Stage I). This stage, described on pages 172–76, is extremely important; it occurs in all living organisms and is therefore a fundamental characteristic of life. Focus your attention on Figure 7.3 (p. 172) and on the summary at the bottom of pages 175–76. Although Figure 7.3 is an accurate depiction of the entire reaction series for glycolysis, you are not expected to memorize these details. Rather, the diagrams are provided merely to stress that each step involves a specific enzyme which rearranges specific atomic bonds in the substrate. Often it is easier to learn the basic concepts when you can see what is happening from step to step. The summary diagram shown in Figure 7.7 (p. 182) may also be helpful.

5 Fermentation. The role of this process is summed up in the first paragraph in the section on fermentation (p. 177). Reread this paragraph until it is clear to you. In particular, you need to understand that fermentation allows a cell to continue making ATP when oxygen is absent; without fermentation all glycolysis would cease under anaerobic conditions.

6 Oxidation of pyruvic acid to acetyl-CoA (Stage II). Notice that this stage is summed up in a single reaction (p. 179).

7 The Krebs citric acid cycle (Stage III) is complex, but if you study Figure 7.5 (p. 180) and 7.6 (p. 181) carefully you will understand the basic idea of the cycle.

8 The respiratory electron-transport chain (Stage IV). This stage, described on pages 182–84, is probably the most difficult for most students to grasp. Study carefully Figure 7.8 and the accompanying caption.

9 The anatomy of respiration and chemiosmotic synthesis of ATP (pp. 184–87). You should learn where each of the five stages takes place, and their relationships to one another. Figure 7.10 (p. 186) is particularly helpful in visualizing how the stages fit together. You should also understand how a chemiosmotic gradient is built up across the inner membrane of the mitochondrion, and the gradient's relationship to ATP production.

10 Summary of respiration energetics (pp. 187–89). You should learn the ATP yield from the various stages. Figure 7.11 (p. 187) summarizes the essential information for you. Concentrate on ATP production—on where and how much ATP is produced at each stage.

11 Respiration of fats and proteins (pp. 189–90). Students often make the mistake of thinking that glycolysis and the Krebs cycle occur only in carbohydrate metabolism; actually, fats and proteins, using various metabolic pathways, also feed into the Krebs cycle, as Figure 7.7 (p. 182) makes clear.

12 Body temperature and metabolic rate (pp. 190–94). You should understand the relationship between temperature and metabolic rate, and become familiar with the terms "ectothermic" and "endothermic" since these terms will be used later in the book.

KEY CONCEPTS

1 Within a living cell, a constant supply of energy is required to drive the various chemical reactions that maintain life. (p. 167)

2 The ultimate energy source for most organisms is sunlight; green plants transform light energy into energy-rich compounds like glucose, which can be used directly or passed to other organisms. (p. 168)

3 The energy stored in complex organic molecules must be transformed into the energy of ATP—the universal energy currency of living organisms—in order to be used by the cell; this transformation must occur in every living cell. (pp. 167–71)

4 The energy stored in complex organic molecules is not liberated through a single large reaction; rather, the universal catabolic process by which the molecules are broken down occurs as a series of small reactions, each catalyzed by its own specific enzyme. (p. 187)

5 All living cells break down sugars by the process of glycolysis. The glycolytic pathway consists of a series of coupled reactions in which the product of one reaction becomes the substrate for the next. In such reactions, the exergonic steps push or pull endergonic steps, with the favorable *net* free-energy change of the steps taken together enabling the sequence of reactions to proceed. (pp. 172–76)

6 Fermentation enables a cell to continue the reactions of glycolysis in the absence of oxygen, by providing reactions in which NAD_{ox} is regenerated from NAD_{re}. (pp. 177–78)

7 Considerably more energy can be extracted from glucose if oxygen is present than if it is absent; consequently a plentiful supply of oxygen is essential for most organisms if their energy demands are to be met. (p. 179)

8 Cells living under aerobic conditions obtain most of their energy from cellular respiration, a process in which pyruvic acid is oxidized to acetyl coenzyme A and then acetyl-CoA is oxidized in the Krebs citric acid cycle. (pp. 179−82)

9 The NAD_{re} and FAD_{re} synthesized during Stages I, II, and III pass their electrons to oxygen indirectly by way of a series of electron carrier molecules. The energy thus liberated is used to pump H^+ ions from the matrix into the outer compartment of the mitochondrion, creating a chemiosmotic gradient across the inner membrane. ATP is produced when H^+ ions move back across the membrane. (pp. 182−87)

10 The metabolic breakdown of high-energy compounds is an inefficient process; more than half the available energy is lost as heat. Some animals have evolved mechanisms for retaining this heat and can thus maintain a uniformly high body temperature and metabolic rate. (pp. 190−94)

OBJECTIVES

After studying this chapter and reflecting on it, you should be able to carry out the following objectives.

1 Define oxidation and reduction in terms of gain and loss of electrons, gain and loss of oxygen, and gain and loss of hydrogen. Indicate whether reducing a substance stores or releases energy in that substance and do the same for oxidizing a substance. (pp. 169−70)

2 Using a diagram that shows the structure of an ATP molecule, such as Figure 7.2 (p. 171), identify the adenine and the ribose portions, and the parts constituting adenosine monophosphate, adenosine diphosphate, and adenosine triphosphate, respectively. Explain how ATP is formed from ADP and inorganic phosphate, and state whether the reaction involved is exergonic or endergonic. Describe the role ATP plays in the transfer of energy. (pp. 170−71)

3 Using a diagram such as Figure 7.3 (p. 172), point out and name the starting product and the end products of glycolysis; point out the reactions in which phosphates from ATP are transferred to glucose; point out the oxidation-reduction reaction, and point out the two reactions in which molecules of ATP are synthesized. Summarize the ATP production of glycolysis, specifying the number of ATP molecules used in the preparatory reactions, the number of ATP molecules synthesized, and the net gain. (pp. 172−76)

4 Explain what is meant by *coupled reactions*, and describe how exergonic reactions can be used to push or pull endergonic reactions in order to get them to proceed. (p. 173)

5 Explain why NAD_{ox} must be regenerated from NAD_{re} in order for glycolysis to continue. Next describe how NAD_{ox} is regenerated in the absence of oxygen. Specify what the end product of fermentation would be in your body cells in the absence of oxygen, and what it would be in most plant cells. (pp. 177−78)

6 Summarize in an equation the conversion of pyruvic acid into acetyl-CoA. Notice that two of the six carbon atoms in the original glucose molecule have been released as CO_2. Indicate whether or not this reaction is a redox reaction. (p. 179)

7 Using a diagram such as Figure 7.5 (p. 180), point out the reactions in which CO_2 is produced, specify how many CO_2 are produced in each "turn" of the Krebs cycle, and indicate how many turns of the cycle are necessary to oxidize the four carbons remaining from the original molecule of glucose; point out the redox reactions, and point out the reaction in which ATP is produced. (pp. 180−82)

8 List the products of the oxidation of two acetyl-CoA molecules in the Krebs cycle, as shown in a diagram such as Figure 7.11 (p. 187).

9 Using diagrams such as Figures 7.8 (p. 184) and 7.10 (p. 186), trace the pathway the electrons follow as they move down the respiratory electron-transport chain and indicate the sites where hydrogen ions are pumped into the outer compartment. Indicate what is meant by a chemiosmotic gradient and explain why this gradient is important. State the number of ATP molecules formed per molecule of glucose resulting from electron transport and chemiosmotic synthesis of ATP. (pp. 184−87)

10 Sketch a mitochondrion, labeling the outer membrane, the inner membrane, the matrix, the inner compartment, and the outer compartment. Indicate where in the mitochondrion the Krebs cycle takes place, where the electron transport enzymes are located, where the concentration of H^+ would be highest, and where the concentration would be lowest. (pp. 184–85)

11 Using a diagram such as Figure 7.11 (p. 187), summarize the ATP yield from the complete breakdown of glucose to carbon dioxide and water. Indicate how many net ATP are formed in glycolysis, how many via the Krebs cycle, and how many are formed by chemiosmotic synthesis. Compare the number of ATP produced from the metabolism of one glucose molecule under *anaerobic* conditions with the number produced under *aerobic* conditions. (pp. 187–89)

12 Using a diagram such as Figure 7.12 (p. 189), explain how fats and proteins can be metabolized to yield energy in the form of ATP. (pp. 189–90)

13 State the relationship between body temperature and metabolic rate and indicate what is meant by the term Q_{10}. Cite differences between ectothermic (poikilothermic) and endothermic (homeothermic) animals; explain how these differences relate to their activity level. (pp. 190–91)

14 State the relationship between body size and metabolic rate, and explain its implications for small animals. (pp. 192–94)

KEY TERMS

The following terms are important in this chapter; you should become familiar with them.

photosynthesis (p. 167)
metabolism (p. 168)
anabolism (p. 168)
catabolism (p. 168)
chemosynthesis (p. 168)
reduction (p. 170)
oxidation (p. 170)
redox reactions (p. 170)
adenosine triphosphate (ATP) (p. 170)
adenosine diphosphate (ADP) (p. 171)
adenosine monophosphate (AMP) (p. 171)
inorganic phosphate (P_i) (p. 171)
anaerobic (p. 172)

glycolysis (p. 172)
coupled reactions (p. 173)
PGAL (p. 174)
NAD (p. 174)
pyruvic acid (p. 175)
fermentation (p. 178)
lactic acid (p. 178)
ethanol (p. 178)
aerobic (p. 179)
cellular respiration (p. 179)
acetyl-CoA (p. 179)
Krebs citric acid cycle (p. 181)
FAD (p. 181)
respiratory electron-transport chain (p. 182)
chemiosmotic gradient (p. 186)
poikilothermic (p. 190)
ectothermic (p. 190)
homeothermic (p. 191)
endothermic (p. 191)
Q_{10} (p. 191)
hibernate (p. 194)

SUMMARY

Within a living cell, a constant supply of energy is required to drive the various chemical reactions that maintain life. The first living organisms arose about 3.5 billion years ago; they metabolized abiotically synthesized organic molecules. Later, *chemosynthetic* organisms evolved that obtained their energy from inorganic materials. About 3 billion years ago, certain organisms acquired the ability to use the energy of light to synthesize energy-rich compounds like glucose in the process called *photosynthesis*. Life today depends almost entirely on this process. The energy stored in glucose is released through the process of *respiration* in which oxygen and glucose are combined to yield CO_2 and water. The released energy can be used to run all the cell's *metabolic* reactions, which include two phases: *anabolism*—the assembling of organic molecules—and *catabolism*, the breaking down of organic molecules to yield energy.

The initial storage of energy in compounds like glucose and the release of energy in respiration involve oxidation and reduction (*redox*) reactions. *Reduction* is the addition of one or more electrons to an atom or molecule, while *oxidation* is the removal of electrons. Reduction stores energy in the reduced compound; oxidation liberates energy from the oxidized substance. During photosynthesis, CO_2 is reduced to form energy-rich glucose; the energy stored in glucose is released when glucose is oxidized during respiration.

ATP is a universal energy currency used by cells to do all manner of work. It is composed of adenosine bonded to three phosphate groups in sequence: adenosine − ⓟ ~ ⓟ ~ ⓟ. If the terminal phosphate group is removed by hydrolysis, energy is released and the compounds *ADP* and inorganic phosphate are left. New ATP can be synthesized from ADP and inorganic phosphate in an energy-demanding process called *phosphorylation*.

Cellular metabolism Before the energy stored in lipids and carbohydrates can be used by the cell to do work, the molecules must be broken down in a series of chemical reactions and the energy transferred to ATP. The complete degradation of an energy-rich compound such as glucose to carbon dioxide and water involves many enzymatically controlled reactions.

Anaerobic metabolism The first series of reactions in the degradation of glucose is termed *glycolysis*; it is the breakdown of glucose to two molecules of pyruvic acid, with the production of two molecules of NAD_{re} and a net gain of two ATP molecules. The glycolytic pathway is a series of *coupled reactions*, in which exergonic steps push or pull endergonic steps, with the favorable *net* free-energy change of the steps taken together enabling the sequence of reactions to proceed. This process, which is common to all living cells, is *anaerobic*; i.e., it does not require molecular oxygen.

The fate of the pyruvic acid depends on the oxygen supply. In the absence of sufficient O_2, the pyruvic acid may be reduced by NAD_{re} to form CO_2 and ethyl alcohol in some kinds of organisms, or lactic acid in others. The NAD_{ox} molecules formed in this reaction are then available to be reused in glycolysis. The process whereby the glycolytic pathway leads to the production of alcohol or lactic acid from pyruvic acid is called *fermentation*; it enables the cell to continue synthesizing ATP by the breakdown of nutrients under anaerobic conditions.

Aerobic metabolism Under *aerobic* conditions—when molecular oxygen is available—the pyruvic acid can be further oxidized, with the accompanying synthesis of ATP; this process is called *cellular respiration*. The process begins with the breakdown of the two pyruvic acid molecules to form two molecules each of acetyl-CoA, CO_2, and NAD_{re}. The two-carbon acetyl-CoA is fed into a complex circular series of reactions, the *Krebs citric acid cycle*. Here the acetyl-CoA combines with a four-carbon compound to form the six-carbon molecule citric acid. In subsequent reactions two carbons are lost as CO_2, leaving a four-carbon molecule that can combine with another acetyl-CoA and start the cycle over again. In the course of the cycle, a molecule of ATP is synthesized, and eight hydrogens are removed and picked up by carrier compounds (NAD_{ox} and FAD_{ox}). Since one molecule of glucose gives rise to two molecules of acetyl-CoA, two turns of the cycle occur for each molecule of glucose oxidized.

The next stage of respiration involves the passage of the electrons from the carrier molecules NAD_{re} and FAD_{re} down a "respiratory chain" of electron transport molecules to oxygen, with which hydrogen ions combine to form water. The electron transport chain molecules are found in the inner membrane of the mitochondrion. The transfer of electrons along the electron transport chain results in the pumping of H^+ ions from the matrix through the inner mitochondrial membrane into the outer compartment of the mitochondrion. As a result, the H^+ concentration increases in the outer compartment and an electrostatic and osmotic concentration gradient is built up across the inner membrane. The flow of H^+ ions down this *chemiosmotic gradient* is used to make ATP from ADP.

Anaerobic metabolism takes place in the cytoplasm, while aerobic respiration is confined to the mitochondrion.

The total number of new ATP molecules produced by the complete metabolic breakdown of glucose is 36, two from glycolysis, two from the Krebs cycle, and 32 from chemiosmotic synthesis in the mitochondrion.

Cells can also extract energy in the form of ATP from fats and proteins. The fats and proteins are hydrolyzed to form products that can be fed into the carbohydrate metabolic pathways.

Body temperature and metabolic rate Cellular respiration captures about 39 percent of the energy of glucose and converts it into ATP, the rest of the energy is released, mostly as heat. All plants and the vast majority of animals—termed cold-blooded or *poikilothermic* ("of variable temperature") or *ectothermic* ("externally heated")—promptly lose most of this heat to their environment. The body temperature of such animals is largely determined by the environmental temperature. As their body temperature increases, their metabolic rate also increases (within limits); their activity is thus correlated with the environmental temperature.

A few animals, notably birds and mammals, maintain a constant high body temperature. Such animals are said to be warm-blooded or *homeothermic* ("of uniform body temperature") or *endothermic* ("internally heated"). Their metabolic rate can accordingly be maintained at a uniformly high level, and they can remain active in cold weather. In both ectothermic and endothermic animals, as well as in plants, the normal metabolic rate is inversely related to body size; the smaller the organism, the higher the metabolic rate.

KEY DIAGRAM

The following coloring exercise is designed to help you understand and learn the processes involved in ATP synthesis in the mitochondrion. The diagram is more complicated than any you have colored so far, so you will want to follow the directions closely and color the structures in the order given. First color the title and then the appropriate structure, using a different color for each structure. (Reference: Figure 7.10, p. 186)

Outer Compartment of Mitochondrion

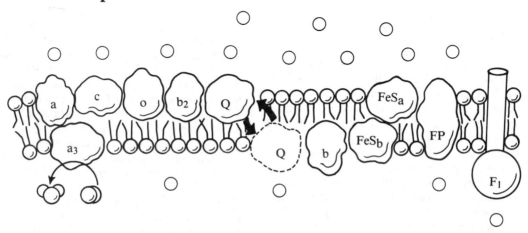

Inner Compartment of Mitochondrion

HYDROGEN IONS (H$^+$)
ATP SYNTHESIS COMPLEX
H$_2$O
O$_2$

H$^+$ PUMP COMPLEXES

Color the three respiratory enzyme complexes where H$^+$ ions are pumped from the inner compartment to the outer compartment (respiratory enzymes labeled FP, Q, and cytochromes a – a$_3$). Note that enzyme Q is a shuttle molecule that accepts electrons and hydrogen ions on the inner surface and transports them to the outer surface where the electrons are passed to cytochrome b$_2$ and the H$^+$ is released into the outer compartment.

ELECTRON TRANSPORT MOLECULES

Color the remaining electron transport molecules (labeled FeS$_a$, FeS$_b$, and cytochromes b$_2$, o, and c). Note that we have colored in two types of carrier complexes, those that transport electrons and H$^+$ ions, and those that transport only electrons. The fact that some of the carrier complexes accept only electrons means that the H$^+$ ions will be left in the outer compartment. Do you see why this is vital for the functioning of the pump?

Draw in NAD_{re} *and the pathway that the* electrons *follow from* NAD_{re} *to* O_2.

Draw in the pathways that hydrogen ions *follow when being pumped from the inner compartment to the outer compartment.*

Ask yourself

1   In what part of the mitochondrion is the pH the lowest?

2   What would happen if the inner mitochondrial membrane was freely permeable to hydrogen ions? Could ATP be synthesized?

QUESTIONS

Testing recall

1   Complete the following chart; it will be a useful study aid.

	Starting Compound(s)	Products
Stage I (pp. 172–76)		
Fermentation (pp. 177–79))		
Stage II (pp. 179–81)		
Stage III (pp. 181–82))		
Stages IV and V (pp. 182–87)		

For Stages IV and V:

_____ NAD_{re} from Stage I

_____ NAD_{re} from Stage II

_____ NAD_{re} from Stage III

_____ FAD_{re} from Stage III

$\Bigg\} \rightarrow$ _____ ATP

Decide whether the following statements are true or false, and correct the false statements.

2   The end product of glycolysis is lactic acid. (p. 175)

3   Alcoholic fermentation produces CO_2 and ethyl alcohol in most plant cells and in many microorganisms. (p. 178)

4   The reactions of the Krebs cycle take place within the mitochondrion. (p. 185)

5   Most of the ATP yield from the complete oxidation of glucose results from the chemiosmotic gradient created by the electron-transport chain. (p. 185)

6   Reactant A in the reaction A + H → AH is oxidized. (pp. 169–70)

7   The reaction ADP + P_i → ATP releases energy. (p. 171)

8   Animals whose body temperature varies with the environmental temperature are said to be homeothermic. (p. 190)

Questions 9–13 refer to the following diagram of the mitochondrion.

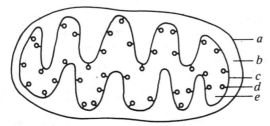

9   Where is ATP synthesized? (p. 185)

10   Where is the concentration of H^+ ions highest? (p. 185)

11   Where do the reactions of the Krebs cycle take place? (p. 185)

12   Where are the molecules of the respiratory electron-transport chain located? (p. 185)

13   Where is the matrix of the mitochondrion? (p. 185)

Testing knowledge and understanding

Choose the one best answer.

14   In the reaction NADH + H^+ + FAD → NAD^+ + $FADH_2$, which reactant will be oxidized as the reaction proceeds to the right?

a   NADH
b   FAD
c H^+
d NAD^+
e $FADH_2$
        (pp. 169–70)

15   The glycolytic pathway from glucose to pyruvic acid involves a lengthy series of different chemical reactions. Each individual reaction requires

a   a molecule of ATP.
b   a molecule of NAD.
c   a molecule of ADP.
d   a molecule of a specific enzyme.
e   a molecule of NADP.   (p. 172)

Questions 16–20 refer to the following diagram.

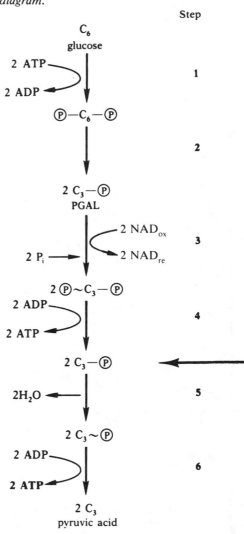

16 The diagram shows a series of reactions involved in glycolysis. How many *net* molecules of ATP can be produced from the products of the breakdown of one glucose molecule by the time it is degraded to $2C_3 - P$ (note arrow)? Assume that *no oxygen is present.*

a 2ATP
b 4ATP
c 6ATP
d 8ATP
e no net gain or loss of ATP   (p. 176)

17 The reactions shown in the diagram take place in the

a chloroplast.   d nucleus.
b cytoplasm.   e ribosomes.
c mitochondria.   (p. 185)

18 What is the *net gain* in ATP molecules produced during the reactions of glycolysis under anaerobic conditions?

a 2   d 8
b 4   e 10
c 6   (p. 176)

19 Which of the numbered steps in the diagram show(s) an oxidation—reduction reaction?

a 1, 2   d 1, 2, 3, 4, 5, 6
b 1, 2, 3   e 3 only
c 3, 5   (pp. 169—70)

20 The reaction shown in step 3 has a ΔG of +3.0 whereas the reaction shown in step 4 has a ΔG of −9.0. Which of the following statements is (are) correct about these two reactions?

a The reaction in step 3 can take place spontaneously because it is an exergonic reaction.
b The energy liberated from the reaction in step 3 can drive the reaction in step 4.
c The reactions in steps 3 and 4, when taken together, are strongly exergonic so these two coupled reactions can proceed.
d The reaction in step 4 is so unfavorable energetically that it will proceed only if subsequent reactions liberate enough free energy to pull the reactants past this step.
e Two of the above are correct.   (p. 176)

21 When a muscle cell is metabolizing glucose in the complete absence of molecular oxygen, which one of the following substances is *not* produced?

a PGAL   d lactic acid
b ATP   e acetyl-CoA
c pyruvic acid   (p. 178)

For questions 22–26 use the following three answers. An answer may be used once, more than once, or not at all.

a Process would be inhibited under anaerobic conditions.
b Process would be promoted under anaerobic conditions.
c Process would occur at the same rate under aerobic and anaerobic conditions.

22 Active transport of glucose into cells. (p. 172)

23 The formation of ethyl alcohol in certain microorganisms. (p. 178)

24 The accumulation of NAD_{re} in cells. (p. 178)

25 The Krebs citric acid cycle. (p. 181)

26 Chemiosmotic synthesis of ATP. (pp. 182—84)

27 An important function of fermentation is to

a regenerate NAD_{ox}.
b produce alcohol as a nutrient source.
c prevent oxidative phosphorylation.
d produce NAD_{re}.
e synthesize glucose.   (p. 178)

28 Which one of the following statements concerning glycolysis is *false*?

a It proceeds in a step-by-step series of chemical reactions, each catalyzed by an enzyme.
b Phosphorylation occurs during the process.
c Oxygen is not required for the process to occur.
d The end products are carbon dioxide and water.
e ATP is formed.   (pp. 175—76)

29 How much molecular oxygen is required in the *fermentation* of one molecule of glucose?

a none   d 36 molecules
b 1 molecule   e 38 molecules
c 24 molecules   (p. 178)

30 Which metabolic pathway is a common pathway to both anaerobic and aerobic metabolism?

a the electron-transport chain
b the citric acid cycle
c the oxidation of pyruvic acid
d glycolysis
e none of the above   (p. 172)

Questions 31–34 refer to the following diagram of the Krebs citric acid cycle.

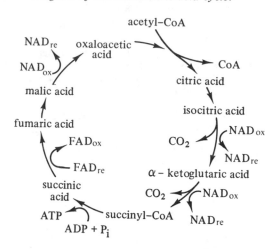

31 If two acetyl-CoA molecules are fed into the cycle, how many ATP molecules are synthesized directly in the cycle?

a 1   d 12
b 2   e more than 20
c 4   (pp. 180–82)

32 If two acetyl-CoA molecules are fed into the cycle, how many NAD_{re} molecules are produced?

a 0   d 12
b 3   e 32
c 6   (pp. 180–82)

33 If citric acid has six carbon atoms, how many carbon atoms does succinic acid have?

a 1   d 6
b 4   e 12
c 5   (p. 180)

34 Besides supplying the cell with ATP and NAD_{re}, the Krebs citric acid cycle

a breaks down glucose to CO_2 and water.
b produces ATP from the NAD_{re}.
c provides intermediates from which various other compounds can be made.
d utilizes oxygen in its stepwise reactions.
e converts lactic acid to pyruvic acid.
     (pp. 180–82)

35 The internal membranes of an organelle taken from a eucaryotic cell are found to contain enzymes and coenzymes that carry out phosphorylation of ADP via a series of step-by-step reactions. The organelle is a

a chloroplast.   d Golgi apparatus.
b lysosome.   e mitochondrion.
c nucleus.   (pp. 184–85)

36 The final electron acceptor in respiratory electron transport is

a O_2.   d NAD.
b H_2O.   e FAD.
c CO_2.   (p. 183)

37 Cyanide blocks the respiratory electron-transport chain. As a result

a the Krebs cycle speeds up.
b electrons and hydrogens cannot flow from NAD_{re} to oxygen.
c three ATPs are produced for every pair of electrons.
d production of water increases.
e glycolysis is inhibited.   (pp. 182–84)

38 According to the chemiosmotic hypothesis,

a H^+ ions diffuse out of the mitochondrion; this movement releases energy that can be used to make ATP.
b H^+ ions are actively transported from the cytoplasm into the inner compartment of the mitochondrion.
c the hydrogen reservoir in the mitochondrion accumulates in the mitochondrial matrix.
d the electron-transport molecules synthesize ATP.
e ATP is produced in the mitochondrion when the hydrogen ions stored in the outer compartment flow back into the matrix.   (pp. 185–86)

39 Which one of the following summary reactions (each a part of the process of metabolic breakdown of glucose) ultimately yields the most new ATP molecules, under aerobic conditions? Assume that all electrons complete their trip down the respiratory electron chain.

a 2 pyruvic acid → 2 acetyl-CoA
b glucose → 2 pyruvic acid
c 2 PGAL → 2 pyruvic acid
d 2 acetyl-CoA → 4 carbon dioxide + water
e 2 acetyl-CoA → 2 citric acid
     (pp. 172–73, 179–80)

40 For a living animal, which of the following compounds has the greatest amount of energy per molecule?

a ATP   d NAD_{re}
b ADP   e pyruvic acid
c H_2O (p. 182)

41 When a molecule of glucose is completely broken down in a cell to water and carbon dioxide, some ATP molecules are synthesized directly and some by chemiosmotic phosphorylation via the electron-transport

system. What percentage of the total number of ATP molecules formed comes from the latter process?

a 94 percent   d 78 percent
b 89 percent   e 6 percent
c 83 percent   (p. 189)

42 When glucose is broken down to carbon dioxide and water during aerobic respiration, more than 60 percent of its energy is released as

a oxygen.   d ATP.
b carbon dioxide.   e NAD.
c heat.   (p. 189)

43 Which one of the following is *not* associated with cold-blooded animals?

a body temperature close to environmental temperature
b low metabolic rate
c insulating hair or feathers
d metabolic rate that varies with environmental temperatures
e sluggish behavior in cold temperature
(p. 191)

For further thought

1 How does ATP store and release energy? What do we do with all our ATP?

2 Compare the energy levels of glucose, pyruvic acid, NAD_{re}, ATP, CO_2, and H_2O.

3 Give an account of the formation of ATP, referring to each of the following molecules: H_2O, CO_2, O_2, 6-carbon carbohydrate, 3-carbon carbohydrate, 2-carbon molecule, 4-carbon molecule, 5-carbon molecule, 6-carbon molecule.

4 Why must glycolysis take place in all cells?

5 Summarize the complete respiratory breakdown of one molecule of glucose, explaining (a) the main stages in the process, (b) the role of oxygen in the process, (c) the significance of the electron-carrier system, and (d) the fate of the energy contained in the glucose.

6 What is the relative efficiency of glycolysis and cellular respiration?

7 Explain how the energy stored in NAD_{re} is converted into energy stored in ATP.

ANSWERS

Testing recall

1

	Starting Compound(s)	*Products*
Stage I	Glucose 2 ATP 2 NAD_{ox}	2 pyruvic acid 2 NAD_{re} 4 ATP (2 net ATP)
Fermentation	2 pyruvic acid 2 NAD_{re}	2 ethyl alcohol + 2 CO_2 + 2 NAD_{ox} or 2 lactic acid + 2 NAD_{ox}
Stage II	2 pyruvic acid 2 CoA 2 NAD_{ox}	2 acetyl-CoA 2 CO_2 2 NAD_{re}
Stage III	2 acetyl-CoA 2 four-carbon compounds (oxaloacetic acid)	2 CoA 4 CO_2 6 NAD_{re} 2 FAD_{re} 2 ATP 2 four-carbon compound (oxaloacetic acid)
Stages IV and V		

$$\left. \begin{array}{l} \underline{\quad 2 \quad} \; NAD_{re} \text{ from Stage I} \\ \underline{\quad 2 \quad} \; NAD_{re} \text{ from Stage II} \\ \underline{\quad 6 \quad} \; NAD_{re} \text{ from Stage III} \\ \underline{\quad 2 \quad} \; FAD_{re} \text{ from Stage III} \end{array} \right\} \rightarrow \underline{\quad 32 \quad} \; ATP$$

2 false—pyruvic acid or glycolysis to fermentation
3 true
4 true
5 true
6 false—reduced
7 false—stores
8 false—poikilothermic or ectothermic
9 *d*
10 *b*
11 *e*
12 *c*
13 *e*

Testing knowledge and understanding

14	*a*	22	*a*	30	*d*	37	*b*
15	*d*	23	*b*	31	*b*	38	*e*
16	*e*	24	*b*	32	*c*	39	*d*
17	*b*	25	*a*	33	*b*	40	*e*
18	*a*	26	*a*	34	*c*	41	*b*
19	*e*	27	*a*	35	*e*	42	*c*
20	*c*	28	*d*	36	*a*	43	*c*
21	*e*	29	*a*				

ENERGY TRANSFORMATIONS: PHOTOSYNTHESIS

A GENERAL GUIDE TO THE READING

The focus of this chapter is photosynthesis—the process by which green plants capture the sun's radiant energy and use it to synthesize complex organic molecules. As you read Chapter 8 in your text, special attention to the following topics will help you concentrate on the primary aspects of the process.

1 Photosynthesis: the light reactions. Be sure you understand the organization of the photosynthetic unit (p. 201) and the effect of light on chlorophyll (p. 202) before you go on to read the sections on cyclic and noncyclic photophosphorylation.

2 Cyclic and noncyclic photophosphorylation. You will want to read these sections slowly since important conceptual material is presented. Figures 8.9 (p. 203) and 8.10 (p. 205) are crucial to understanding the processes.

You will need to remember that the products of the light reactions provide the energy for the synthesis of carbohydrate in the dark reactions.

3 The anatomy of photophosphorylation. It is important that you understand the relationships among the flow of electrons along the electron-transport carriers, the movement of H^+ ions, and the synthesis of ATP. Study Figure 8.12 (p. 208) carefully since it shows how the electrons from water are moved along the two electron-transport chains, and how this leads to the accumulation of H^+ ions inside the thylakoid and the establishment of an electrochemical gradient. A comparison of Figures 8.12 (p. 208) and 7.10 (p. 186) will help you understand the similarities between H^+ movement and ATP synthesis in the mitochondrion and the chloroplast.

4 Photosynthesis: the dark reactions. The Calvin cycle is discussed on pages 210–12. Figure 8.14 (p. 211) is particularly helpful in understanding this process.

5 Photorespiration. Take time to read carefully the material on this subject (pp. 212–13) since a knowledge of this topic is prerequisite to understanding the importance of C_4 photosynthesis.

6 Leaf anatomy is described on pages 213–14; try to relate each part of the leaf to its function in enabling the plant to carry out photosynthesis.

7 C_4 photosynthesis. You will find this material easy if you go over the anatomy of C_3 and C_4 leaves (see Figure 8.16, p. 213) and study carefully Figure 8.19 (p. 216).

6 A cyclic series of reactions, the Calvin cycle, is used to reduce CO_2 to form carbohydrate. (pp. 210–12)

7 The hydrogen and electrons necessary for the reduction of CO_2 come from water. When water is split, the oxygen is not needed and is released as a gas. (p. 206)

8 Under certain conditions, the RuBP necessary for the Calvin cycle is oxidized in the process called photorespiration. The RuBP is thus unavailable for the Calvin cycle. (pp. 213–14)

9 Some plants have evolved structural and biochemical adaptations that enable the plants to fix CO_2 in a different manner, circumventing the photorespiration process. (pp. 214–17)

KEY CONCEPTS

1 The ultimate energy source for most organisms is sunlight; green plants transform light energy into chemical energy, which can be used directly or passed to other organisms. (pp. 195–96)

2 When a photon of light is absorbed by a chlorophyll molecule, the photon's energy raises an electron to a higher energy level and the excited state is passed through the photosystem. The energy released in the transfer of excited electrons from one acceptor to the next is converted into a form that can be used by the cell. (pp. 199–207)

3 During cyclic photophosphorylation, light energy is used to move electrons from chlorophyll, through a series of electron acceptor molecules, and back to the chlorophyll molecule. The free energy released in the electron transfer is indirectly converted into ATP. (pp. 203–4)

4 During noncyclic photophosphorylation, light energy is used to pull electrons and hydrogen away from water. The electrons are passed through two electron-transport chains, eventually reaching the final acceptor. (pp. 204–7)

5 The energy released as the electrons flow along the electron-transport chains is used to move H^+ ions into the inside of the thylakoid. ATP is synthesized when the H^+ ions flow back through special channels in the membrane, down the concentration gradient. (pp. 207–10)

OBJECTIVES

After studying this chapter and reflecting on it, you should be able to carry out the following objectives.

1 Write a summary equation for the reactions of photosynthesis, with glucose as the end product. (p. 198)

2 Explain why almost all organisms depend directly or indirectly on photosynthesis to satisfy their energy needs. Mention the terms "autotroph" and "heterotroph" in your answer. (pp. 195–96)

3 Describe the general organization of a photosynthetic unit. Explain why most of the chlorophyll molecules in a photosynthetic unit are referred to as "antenna" molecules and what is meant by a reaction-center molecule. (pp. 201–3)

4 Using a diagram such as Figure 8.9 (p. 203), trace the flow of electrons during cyclic photophosphorylation; point out the portion of the pathway that is indirectly linked to ATP synthesis, and explain what cyclic photophosphorylation accomplishes. (pp. 203–4)

5 Using a diagram such as Figure 8.10 (p. 205), trace the process that results in the flow of electrons from water to NADP. In doing so, start with the light being absorbed by Photosystem I and trace the flow of electrons activated by that event, specifying what the final electron acceptor molecule in that system is called; explain how the electron holes in Photosystem I are filled by means of another light event, how the resulting electron holes

in Photosystem II are filled, and what happens to the released oxygen when, during this process, water is split. Point out the portion of the pathway that is indirectly linked to ATP synthesis. List the products of noncyclic photophosphorylation. (pp. 204–7)

6 Using a diagram such as Figure 8.12 (p. 208), trace the flow of electrons from water through the various electron carriers to NADP, indicating how a H^+ ion concentration gradient is established within the thylakoid. Point out the enzyme complex that can use the energy of the gradient of phosphorylate ADP. (pp. 207–10)

7 Explain how the products of the light reactions are used to reduce CO_2 in the Calvin cycle to form PGAL, and describe the fate of this PGAL. Figure 8.14 (p. 211) may be helpful. (pp. 210–12)

8 Describe the process of photorespiration and indicate why it appears to be a wasteful process. Discuss the relationship between photosynthesis and photorespiration. (pp. 212–13)

9 Sketch a chloroplast, including in your diagram the thylakoids, grana, and stroma. Indicate where the various photosynthetic reactions take place. (p. 208)

10 Describe the structural differences between the leaves of C_3 and C_4 plants; explain the functional significance of the Kranz anatomy. Using Figure 8.19 (p. 216), discuss the biochemical adaptation that allows C_4 plants to use the Calvin cycle under conditions that would normally lead to photorespiration. (pp. 213–14)

KEY TERMS

The following terms are important in this chapter; you should become familiar with them.

photon (p. 195)
autotroph (p. 196)
heterotroph (p. 196)
chlorophyll (p. 198)
photophosphorylation (p. 199)
photosynthetic unit (p. 201)
reaction-center molecule (p. 203)
P700 (p. 203)
cyclic photophosphorylation (p. 204)
NADP (p. 204)
Photosystem I (p. 204)
carbon fixation (p. 205)
Photosystem II (p. 205)

P680 (p. 205)
noncyclic photophosphorylation (p. 206)
thylakoid (p. 208)
grana (p. 208)
stroma (p. 208)
Calvin cycle (p. 211)
RuBP (p. 211)
C_3 photosynthesis (p. 212)
photorespiration (p. 212)
parts of the leaf: petiole, blade, cuticle, epidermis, mesophyll, stomata, guard cells, bundle sheath (pp. 213–14)
Kranz anatomy (p. 214)
C_4 photosynthesis (p. 215)
PEP (p. 216)

SUMMARY

The ultimate energy source for most living things is sunlight, transformed by green plants into chemical energy in the process called *photosynthesis*. The green plants utilize the energy of light to remove the electrons and hydrogen from water and use them to reduce carbon dioxide to organic material. Oxygen is formed as a by-product. The summary equation for the photosynthetic process is

$$6\ CO_2 + 12\ H_2O + light \xrightarrow{\text{chlorophyll}}$$

$$6\ O_2 + C_6H_{12}O_6 + 6\ H_2O$$

The reactions of photosynthesis can be divided into two parts: the light reactions, in which light energy is trapped and stored in specialized energy-transfer molecules, and the dark reactions, in which the stored energy is used to convert carbon dioxide into carbohydrates like glucose.

Different wavelengths of light, especially red and blue light, are trapped by various pigment molecules organized in two photosynthetic units within the chloroplasts. When a photon of light is absorbed by a chlorophyll molecule, the photon's energy raises an electron to a higher energy level and the excited state is passed by pigment molecule to pigment molecule, eventually reaching a specialized reaction-center molecule where it is trapped. The energized electron is then passed through a series of acceptors, releasing energy at each step. This energy is used for generating an energy-storage compound.

Cyclic photophosphorylation In cyclic photophosphorylation, light energy is trapped by pigment molecules in the P700 antenna system and the excited state eventually reaches the reaction-center molecule P700. An energized electron from P700 then begins passage from one acceptor molecule to the next, releasing free energy at each step and eventually returning to the P700 molecule. Some of the energy released as the electron is eased down the energy gradient is used by the cell indirectly to phosphorylate ADP into ATP. Because the electrons are returned to the chlorophyll molecules from which they originated, this process is termed *cyclic photophosphorylation.*

Noncyclic photophosphorylation Like the cyclic process previously described, noncyclic photophosphorylation begins when light energy excites electrons in the pigments of the P700 antenna system and the energized electrons are passed to P700. However, in noncyclic photophosphorylation the electrons are passed to a different series of electron-transport molecules, eventually reaching the final acceptor, *NADP*, which retains two electrons and their associated protons. The antenna molecules, the P700 reaction center, and the electron-transport chain constitute *Photosystem I.*

Photosystem I is now short of electrons; it has electron "holes." Another light event occurs in a different photosynthetic unit; light energy is trapped by pigment molecules in the P680 antenna system and the excited state eventually reaches the reaction-center molecule P680. Next the excited electrons are passed to an electron acceptor Q, which in turn passes them, via a series of transport molecules, step by step down an energy gradient to the electron holes in the P700 system. As the electrons move down the transport chain, some of the energy released along the way is used by the cell indirectly to synthesize ATP. The antenna molecules and the P680 reaction center, plus its special set of electron-transport molecules, constitute *Photosystem II.*

Photosystem II is now short of electrons; the deficit is filled by electrons pulled from water. The splitting of water also produces free protons and molecular oxygen.

$$2 H_2O \rightarrow 4 e^- + 4 H^+ + O_2$$

Since the electrons are not passed in a circular chain in this process (some leave the system via $NADP_{re}$ and others enter from water), this series of reactions is termed *noncyclic photophosphorylation*; $NADP_{re}$ and O_2 are the end products.

The antenna pigments, reaction centers, and the electron-transport chain molecules are precisely arranged within the membranes of flattened sacs called *thylakoids.* Disc-shaped thylakoids often lie close together in stacks called *grana.* The thylakoid membrane serves as a barrier between the interior of the thylakoid and the interior of the chloroplast, which is known as the *stroma.* Like the inner mitochondrial membrane, the thylakoid membrane makes possible an electrochemical gradient, which supplies energy for the synthesis of ATP. During the light reactions of photosynthesis, water is split and the resulting H^+ ions are deposited in the interior of the thylakoid, thereby increasing the H^+ concentration. In addition, as electrons flow through Photosystems I and II, some of the energy is used to move H^+ ions from the stroma to the thylakoid interior, adding to the H^+ concentration. When the electrons reach $NADP_{ox}$ they reduce it, which removes H^+ ions from the stroma. The net effect is that H^+ ions accumulate in the interior of the thylakoids, while the outer compartment, the stroma, becomes negatively charged. The membrane of the thylakoid contains numerous enzyme complexes that can utilize the energy of the gradient to phosphorylate ADP to ATP.

The dark reactions The ATP and $NADP_{re}$ produced in the light reactions are used to reduce carbon dioxide to carbohydrate in a series of reactions called the *Calvin cycle.* First, a five-carbon sugar, ribulose biphosphate (RuBP), is combined with CO_2 to form an unstable six-carbon molecule that splits into two three-carbon molecules. These are then phosphorylated by ATP and reduced by $NADP_{re}$ to form PGAL, a three-carbon sugar. Five-sixths of the PGAL produced is used to synthesize more RuBP, the remaining one-sixth may be used for synthesis of glucose. The reactions of the Calvin cycle occur in the *stroma* of the chloroplast.

Under certain conditions, however, RuBP is oxidized by means of the very same enzyme that under more favorable conditions would facilitate its carboxylation. This breakdown of RuBP is called *photorespiration*—a seemingly wasteful process, because no ATP is formed. Photorespiration predominates over photosynthesis when CO_2 levels are low and O_2 levels are high, but even at normal levels most photosynthetic production is undone by concurrent photorespiration.

Structure of leaves and chloroplasts The leaves of higher green plants are the principal organs of photosynthesis. The outer surfaces of a leaf are made up of epidermal tissue, covered with a waxy *cuticle* impermeable to water. The region between the upper and lower epidermis is filled with paren-chyma cells making up the *mesophyll* region. The mesophyll cells have many chloroplasts and, being loosely packed, leave air spaces that communicate with the outside by tiny holes in the epidermis called *stomata*. Veins, which contain the xylem and phloem cells for transport, branch profusely within the mesophyll. The veins are usually sur-rounded by tightly packed parenchyma cells mak-ing up a *bundle sheath*.

C_4 photosynthesis Some angiosperm plants of tropical origin have a distinct leaf structure known as *Kranz* anatomy. In Kranz plants the bundle-sheath cells have numerous chloroplasts (those of other plants usually do not), and mesophyll cells are clustered in a ring-like arrangement aound the bundle sheath. Such plants can carry out photo-synthesis under conditions of high temperature and intense light, when loss of water induces clo-sure of the stomata; with the stomata closed, con-centrations of CO_2 in the air spaces inside the leaf fall and those of O_2 rise. Under these conditions most plants (C_3 plants) have a net loss of CO_2 because of photorespiration, but Kranz plants (C_4 plants) do not, thanks to a special way of fixing

CO_2 initially. They combine CO_2 with a three-carbon compound in the mesophyll cells to form a four-carbon compound that passes into the bundle-sheath cells, where the CO_2 is regenerated; they can thus maintain a CO_2 level in the bundle-sheath cells that allows carboxylation of RuBP in the Calvin cycle to predominate over its oxidation in photorespiration.

KEY DIAGRAM

The following coloring exercise is designed to help you understand and learn the processes involved in ATP synthesis in the chloroplast. The diagram is a companion to the key diagram in Chapter 7 of this Study Guide *(see p. 52). You will notice that you are asked to color many of the same struc-tures. You will find it easier to compare the two processes if you use the same colors as you did in Chapter 7. Follow the directions closely and color the structures in the order given. First color the title and then the appropriate structure, using a different color for each structure. (Reference: Figures 7.10, p. 186; 8.10, p. 205; and 8.12, p. 208)*

Stroma (PH2)

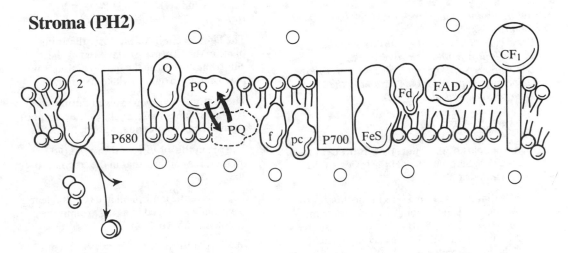

Thylakoid Interior (PH4)

HYDROGEN IONS (H^+)
ATP SYNTHESIS COMPLEX (CF_1)
H_2O
O_2
PHOTONS OF LIGHT
REACTION CENTER MOLECULES
(P680 and P700)
H^+ PUMP COMPLEX **(PQ)**

Color the enzyme complex (labeled PQ) where H+ ions are pumped from the outer surface into the inner compartment of the thylakoid. Note that enzyme PQ is a shuttle molecule that accepts electrons and hydrogen ions on the outer surface and transports them to the inner surface where the electrons are passed to cytochrome f and the H+ is released into the interior of the thylakoid.

ELECTRON TRANSPORT MOLECULES

Color the remaining electron-transport molecules (labeled Z, Q, f, PC, FeS, Fd, and FAD). Note that we have colored in two types of carrier complexes: those that transport electrons and H+ ions, and those that transport only electrons. The fact that some of the carrier complexes accept only electrons means that the H+ ions will be left in the inner compartment. Do you see why this is vital for the functioning of the pump?

PHOTOSYSTEM I

Draw an outline around each molecule that is part of Photosystem I.

PHOTOSYSTEM II

Draw an outline around each molecule that is part of Photosystem II.
Draw in $NADP_{re}$ and the pathway that the electrons follow from water to $NADP_{re}$.
Draw in the pathway that hydrogen ions follow when being pumped from the stroma to the interior of the thylakoid.

Ask yourself

1   In what part of the chloroplast is the pH the lowest? Where was the pH the lowest in the mitochondrion?

2   Note that the hydrogen ions in the interior of the thylakoid come from two sources; name the two sources. Contrast this with the source of hydrogen ions in the outer compartment of the mitochondrion.

3   Compare ATP generation in the mitochondrion with that in the chloroplast, showing how hydrogen ion movement is coupled to electron transport and how ATP is produced.

QUESTIONS

Testing recall

Select the correct term or terms to complete each supplement.

1   Reactant A in the reaction $A + H \rightarrow AH$ is (oxidized, reduced). (p. 170)

2   The reaction $ADP + P_i \rightarrow ATP$ (stores, releases) energy. (p. 171)

3   The wavelengths of light most effective in driving photosynthesis are those in the (red, yellow, blue-violet, green) part of the visible spectrum. (p. 201)

4   The indirect product of cyclic photophosphorylation is (O_2, NADP, ATP, PGAL, glucose). (pp. 203–4)

5   The direct products of noncyclic photophosphorylation are (O_2, PGAL, ADP, ATP, $NADP_{re}$). (p. 205)

6   The products of the light reactions necessary to drive the dark reactions are (ATP, $NADP_{re}$, O_2, CO_2). (p. 210)

7   Products of the Calvin cycle, or dark reactions, include (O_2, PGAL, ADP, $NADP_{ox}$). (pp. 211–12)

8   The light reactions of photosynthesis take place in the (stroma, thylakoids) of the chloroplast whereas the dark reactions take place in the (stroma, thylakoids). (p. 208)

9   The molecular oxygen evolved during photosynthesis comes from (cyclic photophosphorylation, carbon dioxide, water). (p. 206)

10   The pH in the interior of the thylakoid is (higher, lower) than that in the stroma. (p. 208)

11   The cells in the mesophyll in a (C_3, C_4) leaf are usually arranged in a ring around the bundle sheath. (p. 214)

12   In a C_4 plant, carbon dioxide is first combined with a compound called (RuBP, PEP). (p. 216)

13   Under conditions of high light intensity, intense heat, and dryness, a C_4 plant would (photorespire, synthesize carbohydrate, fix carbon dioxide into PEP). (p. 217)

Of the basic processes of photosynthesis

a cyclic photophosphorylation
b noncyclic photophosphorylation
c both light reactions
d Calvin cycle
e both light and dark reactions

which process (or group of processes) involves

14 utilization of CO_2? (p. 210)

15 oxidation–reduction reactions? (pp. 203, 205, 211)

16 light-energized electrons? (pp. 203–5)

17 production of $NADP_{re}$? (p. 205)

18 synthesis of PGAL? (p. 211)

19 production of O_2? (p. 206)

20 occurrence in stroma of chloroplasts? (p. 208)

21 splitting of water? (p. 206)

22 chlorophyll as both the initial electron donor and the ultimate electron acceptor? (p. 203)

Match each item below with the associated part of the leaf shown in cross section in the drawing. Answers may be used once, more than once, or not at all.

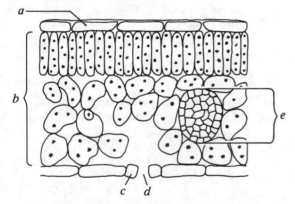

23 cells containing large numbers of chloroplasts that are active in photosynthesis (p. 213)

24 system that delivers water and minerals to cells of the leaf (p. 213)

25 place where CO_2 enters the leaf (p. 213)

26 structure containing xylem and phloem (p. 213)

27 layer that protects internal tissues of the leaf (p. 213)

28 mesophyll tissues (p. 213)

29 cell that regulates the size of the stomatal opening (p. 213)

Testing knowledge and understanding

30 Suppose you are studying a plant that is an unusual blue color. If you extracted its photopigments, which of the three photopigments whose absorption spectra are shown below would you predict would be present?

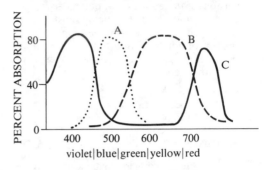

a pigment A
b pigment B
c pigments B and C
d pigments A and C
e pigments A and B   (p. 201)

31 In the *reaction centers* during photophosphorylation,

a water is oxidized.
b NADP is reduced.
c light energy is converted into chemical energy.
d ATP is synthesized from ADP and P_i.
e an electrochemical gradient is established.
                                                (p. 203)

32 During cyclic photophosphorylation,

a electron flow causes H^+ ions to be transported into the thylakoid.
b $NADP_{re}$ is produced.
c water is split.
d both photosystems I and II are involved.
e oxygen is generated.   (pp. 203–4)

33 The light reactions of photosynthesis take place within the

a plasma membrane of the cell.
b membranes of the mitochondria.
c membranes of the thylakoids.
d membranes surrounding the chloroplast.
e stroma of the chloroplast.   (p. 208)

34 The dark reactions of photosynthesis (Calvin cycle) take place within the

a membranes surrounding the chloroplast.
b thylakoids of the chloroplast.
c cytoplasm outside the chloroplast.
d stroma of the chloroplast.
e vacuole.   (p. 208)

35 Which one of the following is *not required* for photosynthesis to proceed?

a oxygen   d light
b carbon dioxide   e chlorophyll
c water   (pp. 198)

36 In photosynthesis the reduction of CO_2 to carbohydrate requires that both energy currency and a strong reducing substance (hydrogen donor) be in ample supply. The process of noncyclic photophosphorylation provides these requisites, using light energy to synthesize

a ADP and ATP.
b ATP and P700.
c ATP and $NADP_{re}$.
d ADP and $NADP_{ox}$.
e P700 and P680.   (p. 205)

37 If photosynthesizing green algae are provided with CO_2 synthesized with heavy oxygen (^{18}O), later analysis will show that all but one of the following compounds produced by the algae contain the ^{18}O label. That one exception is

a PGA.   d RuDP.
b PGAL.   e O_2.
c glucose.   (p. 211)

38 Which one of the following statements concerning noncyclic photophosphorylation is *false*?

a Two different light-driven events are necessary if electrons are to be moved all the way from H_2O to NADP.
b The pigment molecules that trap light energy are built into the thylakoid membranes.
c There are at least two different places in the overall noncyclic pathway where energized electrons are passed energetically downhill via a series of electron-carrier substances.
d One of the products of noncyclic photophosphorylation that help make possible the dark reactions of the Calvin cycle is $NADP_{re}$.
e Some of the energy released during electron transport is used to hydrolyze ATP to ADP and inorganic phosphate.   (pp. 204–8, 210)

39 The "first step" in photosynthesis is the

a formation of ATP.
b energizing of an electron of chlorophyll by a photon of light.
c splitting of water into H and O components.
d addition of CO_2 to a five-carbon sugar.
e combining of two molecules of PGAL to form a molecule of glucose.   (p. 204)

40 The light reactions of photosynthesis

a provide CO_2 for the dark reactions.
b produce carbohydrate.
c provide the energy required for the dark reactions.
d use O_2 in the production of ATP.
e include two of the above.   (p. 210)

41 What is the source of the electrons that reduce $NADP_{ox}$ during photosynthesis?

a O_2 d ATP
b water   e PGAL
c light   (p. 206)

42 In photosynthesis, water molecules must be continuously split to

a provide the O_2 needed for photophosphorylation.
b provide the electrons needed to reduce $NADP_{ox}$.
c provide the electrons needed for cyclic photophosphorylation.
d provide the energy needed to oxidize P_{680} and P_{700}.
e provide the energy for ATP synthesis.   (p. 206)

43 In biology a "limiting factor" is a condition or substance that, by its absence or short supply, limits the rate at which a biological process can proceed. Which one of the following would be *least* likely to be a limiting factor for photosynthesis?

a oxygen   d light
b carbon dioxide   e chlorophyll
c water   (p. 206)

44 Which of the following occurs in noncyclic but *not* in cyclic photophosphorylation?

a flow of electrons
b synthesis of ATP through an H^+ gradient
c synthesis of $NADP_{re}$
d absorption of light by chlorophyll   (p. 198)

45 The oxygen in our atmosphere is a product of

a the splitting of CO_2 during photosynthesis.
b cyclic photophosphorylation.
c noncyclic photophosphorylation.
d both cyclic and noncyclic photophosphorylation.
e the Calvin cycle.   (p. 206)

46 In an oak tree, most photophosphorylation takes place in

a the parenchyma cells of the roots.
b the xylem cells of the stem.
c the epidermal cells of the leaves.
d the mesophyll cells of the leaves.   (p. 203)

47   The electron-transport molecules of photo-phosphorylation are

a   built into the thylakoid membrane.
b   built into the outer membrane of the chloroplast.
c   located in the interior of the thylakoid.
d   located in the inner membrane of the mitochondrion.
e   located in the stroma.   (p. 208)

48   In the dark reactions of photosynthesis,

a   PGAL is synthesized.
b   oxygen is produced.
c   water is split.
d   ATP is synthesized.
e   electrons are returned to the chlorophyll molecule.   (p. 211)

49   Which of the following compounds is involved in *both* the light and dark reactions of photosynthesis?

a O_2
b $NADP_{re}$
c CO_2
d   chlorophyll
e   PGAL   (p. 210)

50   Two different principal mechanisms of CO_2 fixation have been found in green plants—the Calvin cycle (or C_3 pathway) and the C_4 pathway. Which one of the following statements concerning these is *false*?

a   The C_4 pathway is more common in tropical plants than in temperate-zone plants.
b   The C_3 pathway is more common in temperate-zone plants than in tropical plants.
c C_4 plants are more efficient than C_3 plants at fixing CO_2 when the concentration of available CO_2 is low and oxygen high.
d   The C_4 pathway requires two different light-driven events, whereas the C_3 pathway requires only one.
e   Some C_4 plants are now very important crop plants in the Middle West.   (pp. 215–17)

51   All of the following take place in both photosynthesis and respiration *except*

a   electron flow.
b   splitting of water molecules.
c   synthesis of ATP.
d   transfer of electrons to carrier molecules.
e   establishment of an H^+ ion gradient.   (pp. 186, 208)

52   Which one of the following statements *best* describes the relationship between photosynthesis and respiration?

a   Respiration is the exact reversal of the biochemical pathways of photosynthesis.
b   Photosynthesis stores energy in complex organic molecules, and respiration releases it.
c   Photosynthesis takes place only in the light, and respiration takes place only in the dark.
d   Photosynthesis occurs only in plants and respiration only in animals.
e   ATP molecules are produced in photosynthesis and used up in respiration.   (pp. 195–96)

For further thought

1   Describe the principal events in both cyclic and noncyclic photophosphorylation, and contrast these two processes. (pp. 203–7)

2   Describe the Calvin cycle, by which carbohydrates are synthesized from CO_2 in the dark reactions. Indicate the starting materials and the end products, and describe the relationship between photophosphorylation and carbon fixation. (pp. 210–12)

3   Discuss structural features of a typical leaf that are important in making it an efficient organ for carrying out photosynthesis. (pp. 213–14)

4   Compare ATP generation in the mitochondria with that in the chloroplast. In your answer: a) show how H^+ movement is coupled to electron transport; b) indicate the sources of the electrons that are passed along the electron-transport chains; c) indicate which part of the mitochondrion and thylakoid have the highest H^+ ion concentration; and d) explain where ATP is produced. (pp. 186, 208)

ANSWERS

Testing recall

1 reduced
2 stores
3 red, blue-violet
4 ATP
5 O_2, $NADP_{re}$, ATP
6 ATP, $NADP_{re}$
7 PGAL, ADP, $NADP_{ox}$
8 thylakoids, stroma
9 water
10 lower
11 C_4
12 PEP
13 synthesize carbohydrate, fix CO_2 into PEP

14	d	18	d	22	a	26	e
15	e	19	b	23	b	27	a
16	c	20	d	24	e	28	b
17	b	21	b	25	d	29	c

Testing knowledge and understanding

30	c	36	c	42	b	48	a
31	c	37	e	43	a	49	b
32	a	38	e	44	c	50	d
33	c	39	b	45	c	51	b
34	d	40	c	46	d	52	b
35	a	41	b	47	a		

NUTRIENT PROCUREMENT AND PROCESSING BY PLANTS AND OTHER AUTOTROPHS

A GENERAL GUIDE TO THE READING

This chapter begins the second unit of the text, "The Biology of Organisms." The chapter describes the processes by which plants obtain the nutrients they need in order to synthesize their own organic compounds. As you read Chapter 9 in your text, you will find it useful to focus on the following sections:

1 The introduction, which contrasts autotrophic and heterotrophic modes of nutrition. You should learn the terms autotrophic and heterotrophic (p. 221), since they are used frequently.

2 The section "Roots as Organs of Procurement." This section (pp. 227–34) requires careful reading, with particular attention to Figures 9.3–9.8 (pp. 227–31). Make sure you understand the terms "apoplast" and "symplast," and the importance of the endodermis with its Casparian strip.

KEY CONCEPTS

1 Autotrophic organisms manufacture their own organic compounds from inorganic raw materials absorbed directly from the environment. (pp. 221–33)

2 In order to carry out their life processes, all organisms require prefabricated high-energy organic compounds or the raw materials from which these compounds can be synthesized. (p. 221)

3 Because of the presence of the Casparian strip, all materials absorbed by the root must pass through the living endodermal cells to reach the stele, and the plant can exercise some control over the movement of substances into the vascular tissue. (pp. 229–32)

OBJECTIVES

After studying this chapter and reflecting on it, you should be able to carry out the following objectives.

1 Contrast the nutrient requirements of autotrophic and heterotrophic organisms. (p. 221)

2 For each of the nutrients listed below, indicate whether it is required by plants, and if so whether it is a macronutrient, needed in relatively large amounts, or a micronutrient, needed in minute amounts; then give one function of each nutrient.

nitrogen	potassium
protein	magnesium
carbon dioxide	water
copper	vitamins
phosphorus	fats
glucose	calcium

(pp. 223–26)

3 Explain why most of the mass of a plant's body comes from air, not from the solid earth in which it grows. (p. 223)

4 Explain how the root system is adapted to provide the extensive absorptive surface a plant needs if it is to obtain sufficient nutrients to support its large volume. (p. 227)

5 Using a cross section of a root like those in Figures 9.4 (p. 228) and 9.7 (p. 230), identify the epidermis, root hairs, cortex, endodermis, pericycle, stele, xylem and phloem, and vascular tissues. (pp. 228–31)

6 Explain how water can move through the cortex of the plant root by flowing through the symplast or apoplast. (p. 232)

7 Draw several endodermal cells with their Casparian strips and relate their structure to the role of the endodermis in water and mineral absorption. (p. 229)

8 Describe the mutualistic relationship between mycorrhizae and plant roots. (p. 234)

9 Describe the process of nitrogen fixation in *Rhizobium* and certain cyanobacteria, and explain why nitrogen fixation is a useful process. (p. 234)

10 Explain how insectivorous plants supplement their diet and name the most beneficial nutrient that is obtained from this process. (p. 236)

KEY TERMS

The following terms are important in this chapter; you should become familiar with them.

autotrophic (p. 221)
heterotrophic (p. 221)
epiphytes (p. 226)
primary root (p. 227)
secondary root (p. 227)
fibrous root system (p. 227)
taproot system (p. 227)
adventitious roots (p. 227)
root hairs (p. 227)
epidermis (p. 229)
cortex (p. 229)
endodermis (p. 229)
Casparian strip (p. 229)
stele (p. 231)
pericycle (p. 231)
xylem (p. 231)
phloem (p. 231)
pith (p. 231)
symplast (p. 232)
apoplast (p. 232)
nitrogen fixation (p. 234)
mycorrhizae (p. 234)
mutualism (p. 234)

SUMMARY

All organisms require high-energy organic compounds to carry out their life functions. Organisms capable of manufacturing their own organic nutrients from inorganic raw materials are *autotrophic*; those that cannot synthesize organic materials from inorganic raw materials and require prefabricated complex organic materials from the environment are *heterotrophic*. These two groups differ in both their nutrient requirements and the problems associated with nutrient procurement.

Most autotrophs are *photosynthetic*; they use light energy to drive their synthesis of organic compounds. Green plants, the largest group of photosynthetic organisms, require carbon dioxide and water as raw materials. In addition, they need certain mineral nutrients. Some minerals, required in substantial amounts (*macronutrients*), function as structural components of complex organic molecules; others, required in minute amounts (*micronutrients*), function as parts of enzymes or coenzymes.

The CO_2 required by the plant is absorbed by diffusion into cells in the interior of the leaf. Tiny openings or *stomata* in the epidermis allow the CO_2 to penetrate to the interior of the leaf, where it can circulate in the intercellular spaces and be available to individual cells.

Most higher land plants take in water and mineral nutrients from the soil through their roots. To absorb enough of these needed materials, plants require a sufficiently large surface area. As an organism gets bigger, its volume increases much faster than its surface area. Therefore, a large multicellular plant must have an enormous absorptive surface area to support its large volume. A typical root is extensively branched and subdivided and has many *root hairs*, tiny hairlike extensions of the epidermal cells, which vastly increase the total absorptive surface. Roots also store organic compounds and anchor the plant to the substrate.

Water from the soil may be absorbed directly into the epidermal cells and move from cell to cell through the *cortex* to the *stele* by flowing through the *symplast*. Alternatively, it may move across the epidermis and cortex without entering a cell by flowing along the cell walls and intercellular spaces that make up the *apoplast*. However, the *endodermis*, which separates the cortex from the vascular cylinder, presents a barrier to movement. The side and end walls of the endodermal cells contain a *Casparian strip* that forms a waterproof seal between the endodermal cells. Consequently all materials must pass through the living endodermal cells to reach the stele, and the plant can exercise some control over the movement of substances into the vascular tissue.

Minerals are absorbed in ionic form from the soil water. The availability of inorganic ions for absorption depends on the fertility, physical structure, and pH of the soil. Each mineral is absorbed at its own rate, independent of the rate of absorption of the other minerals and also independent of the absorption of water. Simple diffusion, facilitated diffusion, and active transport are all involved in the absorption process. Some plant roots and fungi form mutualistic associations called mycorrhizae, which greatly facilitate mineral uptake.

Plants of the legume group form a mutualistic association with certain bacteria that live in the plant's root nodules. Such bacteria convert (fix) molecular nitrogen into ammonia, which can be used by the plant.

A few photosynthetic plants supplement their inorganic diet with organic compounds obtained by trapping and digesting insects and other small animals. The nitrogenous compounds appear to be most beneficial to the plant.

QUESTIONS

Testing recall

Match each item below with the associated part of the root shown in cross section in the drawing. Answers may be used once, more than once, or not at all.

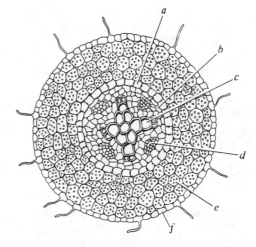

1 region that functions in carbohydrate storage (p. 229)

2 outer boundary of the stele (p. 229)

3 part of enormous surface area where much water absorption occurs (p. 229)

4 cells with a waterproof band along side and end walls (p. 229)

5 cells that conduct water and dissolved minerals to other parts of the plant body (p. 231)

6 area where water and dissolved minerals must enter a living cell to reach the vascular tissue (p. 229)

7 area that contains meristematic cells that may give rise to lateral roots (p. 231)

8 region where adjacent plant cells are interconnected, forming a symplast through which water can flow (p. 232)

Testing knowledge and understanding

Choose the one best answer.

9 Autotrophic organisms

    a must digest their nutrients before taking them into the cell.
    b synthesize their own organic materials from inorganic materials in the environment.
    c require complex organic molecules already synthesized by other organisms.
    d require no external energy source since they synthesize their own high-energy compounds. (p. 221)

10 Three nutrients needed by green plants and included in considerable quantity in most fertilizers are

    a calcium, boron, and lead.
    b carbon dioxide, water, and nitrogen.
    c copper, zinc, and sodium.
    d glucose, PGAL, and sucrose.
    e nitrogen, phosphorus, and potassium. (p. 225)

11 Nitrogen is needed for the formation of

    a sugars.   *d* proteins.
    b fats.   *e* starches.
    c cellulose.   (p. 224)

12 Of the four most abundant elements in most plants (C, H, O, and N), which does a terrestrial green plant procure mainly through its roots from the soil?

    a H and O   *d* H and N
    b C and O   *e* C and N
    c O and N   (p. 223)

13 Plants rely on membrane selectivity to control which substances will enter the xylem for transport. The cell layer most responsible for this regulation is the

    a cortex.   *d* xylem.
    b endodermis.   *e* pith.
    c pericycle.   (p. 227)

14 Most of the mineral nutrients required by plants are absorbed by the root cells by the process of

    a osmosis.   *d* phagocytosis.
    b passive diffusion. *e* pinocytosis.
    c active transport.   (p. 232)

15 The nutrient from which 93 percent of the total weight of glucose is synthesized is

    a water.   *d* nitrogen.
    b carbon dioxide. *e* phosphorus.
    c potassium.   (p. 223)

16 Below are shown endodermal cells with their Casparian strips shaded in. Which one of the arrows represents a correct pathway for water? (p. 227)

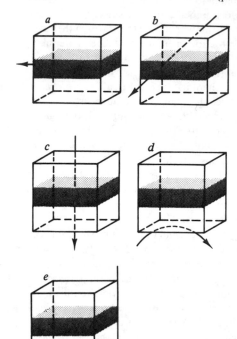

17 The nitrogen source used by bacteria of the genus *Rhizobium* living in the root nodules of legume plants is

    a N_2.   *d* NH_3.
    b NO_2^-.   *e* NH_4^+.
    c NO_3^-.   (p. 234)

18 Insectivorous green plants supplement their diet by trapping and digesting insects. The most beneficial nutrient obtained from this process is

    a nitrogen.   *d* phosphorus.
    b glucose.   *e* sulfur.
    c calcium.   (p. 236)

For further thought

1 Contrast the nutrient requirements of animals and green plants. Why are animals ultimately dependent on green plants? (p. 221)

2 Describe the processes involved in the uptake of water and minerals by a plant root. Explain why such factors as soil acidity may affect the absorption of ions. (pp. 223–34)

ANSWERS

Testing recall

1	e	5	c
2	a	6	a
3	f	7	b
4	a	8	e

Testing knowledge and understanding

9	b	13	b	16	c
10	e	14	c	17	a
11	d	15	b	18	a
12	d				

NUTRIENT PROCUREMENT AND PROCESSING BY ANIMALS AND OTHER HETEROTROPHS

A GENERAL GUIDE TO THE READING

This chapter describes the way in which hetero-trophs—bacteria, fungi, and animals—procure the complex organic nutrients necessary for their activities. The following are some of the high-lights of Chapter 10.

1 Definitions of saprophytic, parasitic, herbi-vore, carnivore, and omnivore. You will want to learn these terms, introduced on pages 238–39, as they are used frequently.

2 Nutrient requirements. The material on essential amino acids (pp. 239–40) is particu-larly relevant, since many college students are vegetarians.

3. Nutrient procurement by protozoans and animals. When you read this material (pp. 248–58) pay special attention to the differ-

ences between intracellular and extracellular digestion. Figures 10.7–10.12 (pp. 250–54) are particularly helpful in learning this material.

4 The digestive system of vertebrates. You will want to learn thoroughly this material (pp. 258–65), which focuses on the human diges-tive tract.

5 Enzymatic digestion in humans. Although this section seems complicated at first glance, it really is not difficult if you focus on the main aspects of the process. Notice that in most cases digestion is a two-step process: one type of enzyme hydrolyzes the polymer into smaller units by breaking internal bonds in the molecule, and a second type of enzyme completes the hydrolysis to the building-block units by removing the units one by one from the ends of the molecule. The summary

diagrams on p. 78 of this guide provide a helpful and easily learned summary of enzymatic digestion. (pp. 266–71)

KEY CONCEPTS

1 Heterotrophic organisms must obtain high-energy organic compounds already synthesized; ultimately heterotrophs depend on nutrients synthesized by autotrophs. (pp. 238–39)

2 Much of the diversity among living things is a result of adaptations toward one of the three major nutritive modes: photosynthetic, absorptive, and ingestive. (p. 239)

3 Heterotrophs must digest their food, breaking it down into smaller molecules, before their cells can absorb it. (p. 238)

4 Extracellular digestion is an adaptation for utilizing larger pieces of food than could be eaten with only intracellular digestion; it is the general rule in multicellular animals. (pp. 247–65)

5 The presence in a complete digestive tract of specialized sections for different functions produces an efficient digestive system. (pp. 254–65)

OBJECTIVES

After studying this chapter and reflecting on it, you should be able to carry out the following objectives.

1 Define these terms: saprophyte, parasite, herbivore, carnivore, and omnivore. Differentiate between absorptive and ingestive heterotrophs and give an example of each. (pp. 238–39)

2 List the nutrients required by heterotrophs, specifying for each nutrient at least one of its roles in metabolism or physiology. (pp. 239–47)

3 Define "essential amino acid" and explain why it is important to eat several different types of protein at each meal. (pp. 239–40)

4 Discuss the role of the various vitamins in the human body and indicate why a vitamin deficiency may lead to symptoms of disease. (pp. 241–45)

5 Explain what digestion is and why heterotrophs need to digest complex foods before they can be absorbed. (pp. 247–48)

6 Briefly describe how fungi procure nutrients, indicating whether their digestion is intracellular or extracellular. Describe a way in which fungi are similar to animals, and a major way in which they are different. (pp. 247–48)

7 For *Paramecium*, hydra, planaria, and earthworm, explain how each organism procures its food, indicate for each whether digestion is primarily intracellular or extracellular, and explain how each digests and absorbs its food. (pp. 249–55)

8 Explain what is meant by a complete digestive tract and describe some of its advantages. In doing so, be sure to discuss the importance of the mechanical breakup of bulk food in animals, giving three examples of adaptations for this process; also, explain the adaptive significance of a storage chamber in the digestive tract. (pp. 254–65)

9 Trace the route food follows from the time it enters your mouth until the undigestible residue leaves the anus, indicating where most digestion and absorption occur. Explain how the lining of the small intestine is adapted to increase the absorptive surface area. Compare in length the intestine of a herbivore and the intestine of a carnivore. Give three functions of your large intestine. (pp. 258–64)

10 Explain how organisms such as cows, rabbits, and horses can derive some nutrients from the cellulose they eat, even though mammals cannot digest cellulose. (pp. 264–65)

11 Assume you have just eaten a hamburger and bun. Describe the steps in your digestion that will break down the starch in the bun and the protein and fat in the meat. In doing so, use a diagram such as Figure 10.28 (p. 266) to show how starch is hydrolyzed to maltose and how maltose is hydrolyzed to glucose, specifying what enzymes are involved in these reactions and where they are produced; explain how proteins are broken down to amino acids, specifying what enzymes are involved in these reactions and where they are produced (Figure 10.29, p. 267, may be helpful); explain how fat can be digested into fatty acids and glycerol, specifying the role bile plays in fat digestion, and indicating whether fat must be hydrolyzed prior to absorption. Finally, indicate how the ingested nucleic acids are digested to their building-block molecules, naming the enzymes involved. (pp. 265–71)

KEY TERMS

The following terms are important in this chapter; you should become familiar with them.

absorptive heterotroph (p. 238)
ingestive heterotroph (p. 238)
saprophytic (p. 238)
parasitic (p. 238)
herbivore (p. 238)
carnivore (p. 238)
omnivore (p. 239)
essential amino acid (p. 239)
kwashiorkor (p. 240)
essential fatty acid (p. 240)
vitamin (p. 241)
scurvy (p. 241)
beriberi (p. 241)
pellagra (p. 243)
pernicious anemia (p. 243)
xerophthalmia (p. 244)
rickets (p. 244)
digestion (p. 247)
extracellular digestion (p. 247)
rhizoids (p. 247)
haustoria (p. 247)
intracellular digestion (p. 249)
food vacuole (p. 249)
oral groove (p. 250)
cytopharynx (p. 250)
gastrovascular cavity (p. 251)
nematocyst (p. 252)
complete digestive tract (p. 254)
anus (p. 254)
pharynx (p. 254)
esophagus (p. 254)
crop (p. 254)
gizzard (p. 254)
typhlosole (p. 255)
filter feeder (p. 256)
incisor
canine
premolar
molar
} teeth (p. 258)
peristalsis (p. 261)
sphincter (p. 261)
mucosa (p. 261)
gastric juice (p. 261)
pyloric sphincter (p. 261)
duodenum (p. 261)
villi (p. 262)
microvilli (p. 262)
colon (p. 263)
caecum (p. 263)
appendix (p. 263)
rumen (p. 264)
rectum (p. 265)
amylase (p. 265)
pepsin (p. 266)
zymogens (p. 267)

pancreas (p. 268)
intestinal glands (p. 268)
trypsin (p. 268)
chymotrypsin (p. 268)
endopeptidase (p. 269)
exopeptidase (p. 269)
maltase (p. 269)
sucrase (p. 269)
lactase (p. 269)
nucleases (p. 269)
phosphatases (p. 269)
N-glycolases (p. 269)
liver (p. 269)
gall bladder (p. 271)
bile (p. 271)
lipase (p. 271)

SUMMARY

All organisms require high-energy organic compounds to carry out their life functions. Since many of the organic molecules found in nature are too large to be absorbed unaltered through cell membranes, they must first be hydrolyzed (i.e. digested) by enzymes into their constituent building-block molecules.

There are four main groups of heterotrophic organisms: bacteria, fungi, protozoans, and animals. Bacteria and fungi are *absorptive heterotrophs*; they lack internal digestive systems and depend mainly on absorption as their mode of feeding. In contrast, protozoans and animals are *ingestive heterotrophs*; they take in particulate or bulk food and digest it inside their body.

Many bacteria and fungi thrive on a diet containing only carbohydrates because they can synthesize other organic compounds from carbohydrates. However, most heterotrophic organisms require carbohydrates, fats, and proteins in bulk. In addition, they require certain vitamins and minerals in small quantities.

Nine *amino acids* are *essential* in the diet of most animals because the animals cannot synthesize them. Since amino acids cannot be stored in the body, all the essential amino acids must be ingested simultaneously and in the correct proportions if effective protein synthesis is to take place.

Some animals require no fat in their diet; others cannot synthesize enough of certain fatty acids for their needs, so these *essential fatty acids* must be included in their diet.

Vitamins are organic compounds necessary in small quantities to given organisms that cannot synthesize them. Most vitamins function as coenzymes or parts of coenzymes. A prolonged vitamin deficiency impairs metabolic processes within the cell, often producing symptoms of a deficiency disease in the organism. Vitamins are classified into two groups on the basis of solubility: those of the B complex and vitamin C are water soluble, while vitamins A, D, E, and K are fat soluble.

Digestion in fungi is *extracellular*; digestive enzymes are secreted directly into the food supply and the products of digestion are then absorbed. Many fungi are *saprophytic* (living on dead material) while others are *parasitic* (living on or in other living organisms). A few fungi supplement their diets by trapping and digesting small animals.

Like fungi, animals must digest their food before it can cross their cell membranes. In protozoans and some animals, digestion is *intracellular*; the food is ingested directly into the cell by endocytosis, then hydrolyzed in a food vacuole, and the products of digestion are absorbed. In others, digestion is *extracellular*, either in the environment or in a specialized digestive structure.

The unicellular protozoans carry on only intracellular digestion. The radially symmetrical coelenterates have a saclike body containing a *gastrovascular cavity*. The specialized cells lining the cavity carry on both extracellular and intracellular digestion. Extracellular digestion allows the organism to eat larger pieces of food than it could handle using only intracellular digestion; it is the general rule in multicellular animals.

The free-living flatworms are bilaterally symmetrical, elongate animals with a gastrovascular cavity. They carry out some extracellular digestion, but most of the food is digested intracellularly.

Animals above the level of coelenterates and flatworms have a *complete digestive tract*, each section of which can be specialized for a different function—mechanical breakup of bulk food, temporary storage, enzymatic digestion, absorption of products, reabsorption of water, storage of wastes, etc. For example, the earthworm has a muscular *pharynx* for sucking in food, a *crop* for temporary storage, a *gizzard* for mechanical breakup, and an *intestine* with a large surface for extracellular enzymatic digestion and absorption.

The digestive tract of mammals consists of a series of chambers specialized for different functions. The first chamber is the *oral cavity*, where the teeth break up the food mechanically. The teeth of different vertebrates are specialized in a variety of ways and may differ in number, structure, and arrangement. Sharp pointed teeth, poorly adapted for chewing, characterize *carnivores* (meat eaters) whereas broad flat teeth, well adapted for chewing, characterize *herbivores* (animals that eat plant materials). Humans, who eat both plant and animal material, are *omnivores*; their teeth are rather unspecialized.

The muscular tongue manipulates the food during chewing and mixes it with *saliva*. The lubricated food is pushed backward through the *pharynx* into the *esophagus*. The food moves quickly through the long esophagus, pushed along by waves of muscular contraction called *peristalsis*, into the *stomach*, where it is stored and further broken up by the squeezing and churning action of the stomach muscles. Pepsin secreted in the *gastric juice* begins protein digestion. The food moves from the stomach into the *duodenum*, where ducts carrying secretions from the liver and pancreas enter. The rest of the *small intestine* is long and coiled. Most of the digestion and the absorption of the products of digestion occurs in the small intestine, where several structural adaptations increase the absorptive surface area. The length of the small intestine in various animal species is proportional to the amount of plant material in their diet; herbivores ordinarily have a longer small intestine than carnivores, because the cellulose cell walls of plants are difficult to break up and tend to interfere with digestion and absorption.

Indigestible material, water, and unabsorbed substances move on into the *large intestine*, or *colon*. A *caecum* projects from the junction of the small and large intestine. Two important functions of the large intestine are the reabsorption of water and the excretion of certain salts when their concentration in the blood is too high. The last portion of the large intestine, the *rectum*, functions as a storage chamber for the feces until release.

Enzymatic digestion, the complete hydrolysis of organic nutrients into their building-block molecules, occurs mainly in the mouth, stomach, and small intestine. It is generally a two-step process. First the complex nutrients are hydrolyzed enzymatically into smaller fragments; then other enzymes complete the hydrolysis into the building-block compounds. Absorption of the simple sugars and amino acids into the blood involves active transport. A summary of the action of the various enzymes (with their sites of production shown in parentheses) is shown below.

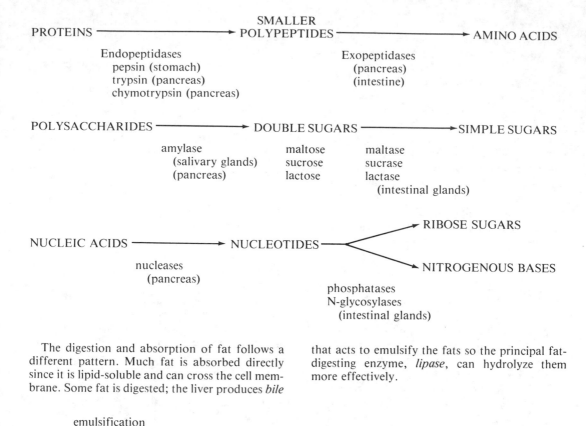

PROTEINS ⟶ SMALLER POLYPEPTIDES ⟶ AMINO ACIDS

Endopeptidases
pepsin (stomach)
trypsin (pancreas)
chymotrypsin (pancreas)

Exopeptidases
(pancreas)
(intestine)

POLYSACCHARIDES ⟶ DOUBLE SUGARS ⟶ SIMPLE SUGARS

amylase
(salivary glands)
(pancreas)

maltose
sucrose
lactose

maltase
sucrase
lactase
(intestinal glands)

NUCLEIC ACIDS ⟶ NUCLEOTIDES ⟶ RIBOSE SUGARS
⟶ NITROGENOUS BASES

nucleases
(pancreas)

phosphatases
N-glycosylases
(intestinal glands)

The digestion and absorption of fat follows a different pattern. Much fat is absorbed directly since it is lipid-soluble and can cross the cell membrane. Some fat is digested; the liver produces *bile* that acts to emulsify the fats so the principal fat-digesting enzyme, *lipase*, can hydrolyze them more effectively.

emulsification

FATS ⟶ EMULSIFIED FAT ⟶ FATTY ACIDS AND GLYCEROL

bile salts (liver)

lipase (pancreas)

QUESTIONS

Testing recall

1 Which of the following substances are required by animals?

a	carbon dioxide	*f*	fats
b	amino acids	*g*	water
c	vitamins	*h*	sugars
d	polysaccharides	*i*	proteins
e	phosphates	*j*	nitrates

(pp. 239–47)

Match each statement below with the associated part of the digestive system in the drawing on the next page. Answers may be used once, more than once, or not at all.

2 Carbohydrate digestion begins here. (p. 265)

3 Protein digestion begins here. (p. 266)

4 Enzymes that digest fat, protein, and carbohydrates are produced here. (p. 269)

5 Most water reabsorption occurs here. (p. 265)

6 Most fat digestion occurs here. (p. 271)

7 Most products of digestion are absorbed here. (p. 262)

8 Fat-digesting enzymes are synthesized here. (p. 271)

9 This organ produces a substance that emulsifies fat. (p. 269)

10 This organ secretes enzymes active at a low pH. (p. 266)

11 This structure is lined by villi. (p. 262)

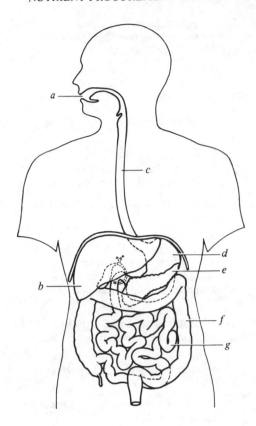

Testing knowledge and understanding

Choose the one best answer.

12 Nutritionally, a lion could best be described as

   a an autotroph.
   b an absorptive heterotroph.
   c an ingestive heterotroph.
   d a herbivore.
   e an omnivore.   (p. 238)

13 Which one of the following statements most accurately describes a vitamin?

   a It can be obtained in the diet only from animal products, such as meat, eggs, milk, and cheese.
   b It is essential for all animals but can be synthesized only by green plants.
   c Though it is a useful addition to the diet, it is not essential.
   d It is a universal building block for the body's proteins.
   e It is an organic compound which is required in small quantities in the diet and which cannot be synthesized by the body.
                                      (p. 241)

14 Essential amino acids

   a can be synthesized from other amino acids within the body.
   b can only be obtained from animal protein.
   c must all be present at the same time in the proper proportions for proper utilization.
   d can be supplied by eating large quantities of a single plant protein.   (pp. 239–40)

15 Which one of the following statements most accurately describes an essential amino acid?

   a It can be obtained in the diet only from animal products such as meat, eggs, milk, and cheese.
   b It is a universal building block for the body's proteins, found in all the proteins our bodies synthesize.
   c Though it is necessary for life, the organism in question cannot synthesize it and hence must obtain it in the diet.
   d It functions as a coenzyme in a biochemical pathway that is essential for life.
   e It is essential for all animals but can be synthesized only by green plants.
                                      (pp. 239–40)

16 A deficiency of vitamin C (ascorbic acid) in the human diet results in the disease called scurvy, characterized by bleeding gums, loosening of teeth, delayed healing of wounds, and painful and swollen joints. These deficiency symptoms are understandable because ascorbic acid plays a major role in

   a forming red blood cells.
   b maintaining good night vision.
   c facilitating calcium absorption.
   d promoting blood clotting.
   e forming collagen fibers in connective tissue.   (p. 242)

17 Beriberi can be prevented or cured if the diet contains sufficient

   a vitamin A.   d vitamin D.
   b vitamin B_1.   e vitamin K.
   c vitamin C.   (p. 241)

18 Which one of the following organisms obtains its nourishment *only* by extracellular digestion?

   a hydra   d fungus
   b amoeba   e green plant
   c planaria   (p. 247)

19 Gastrovascular cavities

   a are found only in unicellular organisms such as *Paramecium*.
   b have only one opening to the exterior, which functions as both mouth and anus.

c generally include a gizzard instead of teeth to break up bulk food.
d are commonly found in herbivorous animals such as sheep and goats.
e usually have several chambers, each specialized for a different function.
(p. 251)

20 All of the following are functions of the stomach in humans *except*

a temporary food storage.
b mechanical breakup of food.
c digestion of proteins.
d fermentation by microorganisms.
e secretion of acid. (p. 261)

21 All of the following are adaptations to a herbivorous diet *except*

a shearing teeth.
b pouches, such as the rumen, for microbial digestion.
c a caecum.
d a long small intestine.
e partial dependence on bacterial synthesis and digestion. (pp. 259, 261–65)

22 Which one of the following animals has the least subdivision of its digestive system into separate chambers specialized for different operations on food?

a earthworm *d* hydra
b cow *e* chicken
c human being (pp. 251–52)

23 Which one of the following animals would you expect to have the most developed grinding surface on the molar teeth?

a dog *d* sheep
b cat *e* lion
c human (p. 258)

24 Which one of the following statements is *false* concerning the digestive system of most higher land animals?

a It begins with a mouth and ends with an anus.
b It frequently depends on microorganisms to perform vital functions.
c The surface area is reduced as much as possible to permit easy passage of food.
d It has a mechanism for chewing and grinding the food.
e It includes a chamber for food storage. (pp. 258–65)

25 Which one of the following digestive enzymes would be most likely to catalyze the reaction $C_{12}H_{22}O_{11} + H_2O \rightarrow C_6H_{12}O_6 + C_6H_{12}O_6$

a amylase *d* endopeptidase
b sucrase *e* exopeptidase
c lipase (p. 266)

26 Which one of the following mammals would you expect to have the longest small intestine relative to its body size?

a human *d* seal
b dog *e* sheep
c cat (p. 262)

Questions 27–29 refer to the following situation.

A hungry student eats a ham and cheese sandwich and drinks a glass of milk for lunch. He has a piece of apple pie for dessert.

27 Digestion of the starch in this meal is carried out in the

a oral cavity and stomach.
b stomach and small intestine.
c small intestine and large intestine.
d oral cavity and small intestine.
e stomach and large intestine. (pp. 265, 268)

28 The products of digestion of the protein in the student's meal are absorbed in the

a stomach. *d* pancreas.
b small intestine. *e* large intestine.
c liver. (p. 262)

29 Enzymatic hydrolysis in the mouth will primarily affect the

a ham. *d* milk sugar.
b cheese. *e* milk fats.
c bread. (p. 265)

30 The digestion of fats takes place almost entirely in the

a mouth. *d* small intestine.
b esophagus. *e* large intestine.
c stomach. (p. 271)

31 The enzyme that catalyzes hydrolysis of fat is

a amylase. *d* trypsin.
b pepsin. *e* bile.
c lipase. (p. 271)

32 Bile aids in fat digestion by

a hydrolyzing the bonds between glycerol and fatty acids.
b breaking peptide bonds.
c converting unsaturated fats to saturated fats.
d emulsifying fat droplets so that more surface area is exposed to the action of digestive enzymes.
e stimulating increased release of fat-digesting enzymes from the pancreas. (p. 271)

33 Which one of the following organs is part of the digestive system but does *not* secrete digestive enzymes?

a pancreas
b large intestine
c small intestine
d stomach
e salivary gland
(p. 265)

34 Which one of the following can be absorbed in the digestive tract without any hydrolysis?

a fat molecule
b sucrose
c polypeptide
d starch
e dipeptide
(p. 271)

35 Which of the following pairings of enzyme and function is *incorrect*?

a salivary amylase—starches hydrolyzed to disaccharides
b pepsin—peptides hydrolyzed to amino acids
c lipase—fats hydrolyzed to glycerol and fatty acids
d maltase—disaccharides hydrolyzed to monosaccharides
e trypsin—internal peptide bonds hydrolyzed
(pp. 265–71)

36 Which one of the following statements is *true* concerning lactose digestion?

a Adults of all mammals other than human beings have high levels of lactase.
b Lactose is synthesized in muscles by lactase.
c The undigested lactose in the colon of intolerant individuals is fermented by bacteria, resulting in diarrhea.
d There is no evidence for a genetic transmission of lactose tolerance.
e Adult humans in most parts of the world are lactose-tolerant; hence powdered milk is an ideal food for nutritional supplements.
(p. 270)

For further thought

1 Define "vitamin" and make a general statement of how vitamins function. Choose one vitamin to discuss, including characteristic symptoms of its deficiency, foods in which it is found, etc. (pp. 241–44)

2 What is the general function of the digestive enzymes? Name some factors in the digestive tract that influence the function of the enzymes and discuss their effect. Describe some specific ways enzymes may "cut up" a protein molecule. (pp. 265–71)

3 Discuss and compare adaptations for nutrient procurement and processing in hydra, earthworm, cow, and human being. (pp. 251–71)

4 Follow a hamburger with roll through the digestive tract. Name the important enzymes involved, their source, and their actions. What happens to the end products? (pp. 258–71)

ANSWERS

Testing recall

1	b, c, d, e, f, g, h, i	7	g
2	a	8	e
3	d	9	b
4	e	10	d
5	f	11	g
6	g		

Testing knowledge and understanding

12	c	19	b	25	b	31	c
13	e	20	d	26	e	32	d
14	c	21	a	27	d	33	b
15	c	22	d	28	b	34	a
16	e	23	d	29	c	35	b
17	b	24	c	30	d	36	c
18	d						

GAS EXCHANGE

A GENERAL GUIDE TO THE READING

This chapter is designed to give you an understanding of the problem of gas exchange and of how various organisms, both plant and animal, have solved it. As you read Chapter 11 in your text, you will want to concentrate on the following aspects of the topic.

1   The nature of the problem of gas exchange. Factors that complicate the maintenance of an adequate respiratory surface in various organisms are discussed on pages 273–74. You will find the subsequent material easier if you first learn the four basic requirements for gas exchange-systems (p. 274). Then, when reading the remainder of the chapter, try to understand how each gas-exchange mechanism meets these four requirements.

2   Solutions in terrestrial plants. Remember that the gas-exchange problem in plants (pp. 274–80) is different from that in animals. Like animal cells, all living plant cells require

O_2 for cellular respiration, but those cells carrying on photosynthesis also require CO_2. Transpiration, a process resulting in the loss of water in plants, is mentioned in this chapter; it will be discussed in more detail in Chapter 12.

3   The gills of a fish. These are described in some detail (pp. 281–84) to give you a good understanding of the problems associated with gas exchange in water and of how fish have solved these problems. Figures 11.13 (p. 283) and 11.14 (p. 284) will help you understand how the gills of a fish function.

4   Human lungs. The human gas-exchange system is discussed on pages 286–89. Make sure you understand how negative-pressure breathing works and how it differs from positive-pressure breathing. (p. 290).

5   Respiratory cycle of birds. Note that the respiratory cycle of birds differs fundamentally from that of mammals. Figure 11.21 (p. 289) is helpful in understanding how the system works.

KEY CONCEPTS

1   A basic problem for most living organisms, both plant and animal, is procuring oxygen for respiration and eliminating carbon dioxide. (pp. 272–74)

2   Gas exchange between a living cell and its environment always takes place by diffusion across a thin, moist cell membrane. (pp. 273–74)

3   All organisms require a protected respiratory surface of adequate dimensions relative to their volume. The larger the organism, the more surface it requires. (p. 274)

4   Many organisms need a transport system to carry the gases from the respiratory surface to the internal cells. (p. 274)

OBJECTIVES

After studying this chapter and reflecting on it, you should be able to carry out the following objectives.

1   Discuss the problem of gas exchange faced by nearly all living organisms. In doing so, consider the reason why most organisms require molecular oxygen, and in particular why plants require a supply of oxygen even though they produce it during photosynthesis. (pp. 273–80)

2   Explain why an organism such as a goldfish has evolved a special respiratory surface whereas the large algae have no special gas-exchange mechanism. In your answer indicate how gas exchange is affected by the relationship of surface to volume, the shape of an organism, and the permeability of the outer covering of an organism. (pp. 274–75, 281–84)

3   List the four basic requirements for gas-exchange systems. (p. 274)

4   Indicate how the structure of leaves of a typical terrestrial plant meets the four basic requirements of gas-exchange systems, and indicate where the actual gas-exchange process takes place. In doing so, specify what two gases pass in and out the stomata. (pp. 274–75)

5   Using Figures 11.4 (p. 276) and 11.5 (p. 277), explain the structure and function of guard cells. Discuss the mechanisms involved in changing the turgidity of guard cells, and the effect this has on opening and closing the stomata.

6   Discuss the problem of gas exchange in stems and roots. In doing so, answer the following questions. What are lenticels, and what is their function? Do roots have specialized structures for gas exchange? How do gases enter and leave the roots? Do plants, like animals, have specialized structures for transporting gases? How are gases distributed effectively in plants? Why wouldn't the same method work in animals? (pp. 278–79)

7   Give three reasons why obtaining oxygen is a greater problem for aquatic organisms than for air breathers. (p. 284)

8   Explain how the sea star, the segmented marine worm, the squid, and the fish meet the four basic requirements for gas-exchange systems. Indicate what features the gas-exchange mechanisms of these organisms have in common. (pp. 281–84)

9   Using diagrams such as Figures 11.13 (p. 283) and 11.14 (p. 284), show how the flow of water across the gills of a fish maximizes the amount of oxygen picked up. Indicate what is meant by a countercurrent exchange system. (pp. 282–84)

10 Distinguish between an invaginated and an evaginated respiratory surface, and give an example of each. Give three reasons why most land animals have evolved invaginated rather than evaginated respiratory systems. (p. 285)

11 Trace the route a molecule of air follows in your body from inhalation to your lung. Then, explain how the structure of your lung satisfies the four basic requirements for a respiratory system. (pp. 286–87)

12 Describe the process of breathing in your body, and explain how this differs from the breathing process in a frog. (pp. 288–90)

13 Give reasons why birds and mammals have such high oxygen requirements, and explain how birds obtain oxygen from the air more efficiently than mammals. (pp. 191, 288–90)

14 Outline the evolution of the swim bladder of modern fish, and explain how the swim bladder allows a fish to swim at different depths. (p. 291)

15 Describe the tracheal system of an insect, and identify two fundamental ways in which this invaginated system differs from the respiratory system of the land vertebrates. Indicate whether a tracheal system would be sufficient for larger animals. (pp. 293–94)

KEY TERMS

The following terms are important in this chapter; you should become familiar with them.

stomata (p. 274)
guard cell (p. 276)
transpiration (p. 277)
lenticel (p. 278)
gills (p. 281)
countercurrent exchange system (p. 282)
lung (p. 285)
external nares ⎫
nasal cavity ⎪
pharynx ⎪ parts of
glottis ⎬ the human
epiglottis ⎪ respiratory
trachea ⎪ system
bronchus ⎪ (pp. 286–87)
alveolus ⎭
breathing (p. 288)
diaphragm (p. 288)
rib cage (p. 288)
vital capacity (p. 289)
negative-pressure breathing (p. 290)
positive-pressure breathing (p. 290)
swim bladder (p. 291)
tracheae (p. 293)
spiracle (p. 293)
tracheal gills (p. 294)

SUMMARY

Since aerobic respiration is the chief method of respiration in both plants and animals, the vast majority of organisms must have some method of obtaining oxygen and getting rid of waste carbon dioxide.

Gas exchange between a living cell and its environment always takes place by *diffusion* across a moist cell membrane. The cells of most small aquatic organisms are in close contact with the surrounding medium and carry out gas exchange across their whole body surface; no special respiratory devices are necessary. With the evolution of large three-dimensional organisms came the necessity for a corresponding evolution of specialized respiratory surfaces to meet the increased oxygen requirements. A wide variety of structures for gas exchange have evolved, but each respiratory system must meet four basic requirements:

1 A respiratory surface of adequate dimensions
2 A moist exchange surface
3 A method of transporting the gases to the cells
4 Protection for the fragile respiratory surface

In plants, gas exchange in the leaves takes place at a high rate. The loosely packed mesophyll cells in the interior of the leaf provide an adequate surface area, and the epidermis with its waxy cuticle protects the mesophyll from mechanical damage and water loss. Tiny openings in the epidermis (*stomata*) allow the gases to penetrate to the interior, where they circulate through the numerous intercellular spaces.

The size of each *stoma* in the epidermis is regulated by two *guard cells*. When the guard cells are turgid, the stoma is open; when the guard cells lose water and are flaccid, the stoma is closed. When the stomata are open, gases can move into the interior of the leaf, but at the same time the plant loses water by *transpiration*.

Root cells obtain oxygen by simple diffusion across the moist membranes of the individual cells; air reaches the interior cells via the intercellular spaces. Gas exchange in stems takes place through *lenticels* connected with the extensive apoplast system.

Most large multicellular animals have evolved some type of special respiratory surface. These may be outward-oriented (*evaginated*) or inward-oriented (*invaginated*) extensions of the body surface.

Most multicellular aquatic animals utilize *gills* which have finely subdivided surfaces that expose an immense exchange surface to the water. Most gills contain a rich supply of blood vessels, special carrier pigments of which transport O_2 to the individual cells.

In fish, the water flows over the gills in the opposite direction to that of the blood flow; this *countercurrent exchange system* maximizes the amount of O_2 the blood can pick up from the water.

Because of oxygen's low solubility and slow diffusion rate in water, most aquatic organisms must constantly move water across the exchange surface.

Most terrestrial animals have evolved invaginated respiratory systems, either *lungs* or *tracheae*. Lungs are invaginated gas-exchange organs supplied with blood vessels. The evolution of the lung in higher vertebrates has tended toward increased surface area and an increased blood supply to the exchange surface.

In human beings, air is drawn through the *external nares* into the *nasal cavities*, then moves into the throat, or *pharynx*, a common passage-way for food and air. After leaving the pharynx through the *glottis*, the air enters the *larynx*. During swallowing, the *epiglottis* closes the glottis. The air then moves into the *trachea*, which divides at its lower end into the two *bronchi*, which lead into the two lungs. Each bronchus branches repeatedly, forming *bronchioles*, which branch into smaller ducts that terminate in the *alveoli*. Each alveolus is surrounded by a dense bed of blood capillaries; the diffusion of O_2 and CO_2 takes place here.

Air is drawn into and pushed out of the lungs during the breathing process. When air is forced into the lungs the process is termed *positive-pressure breathing*; when it is drawn into the lungs as a result of an enlarged body cavity it is termed *negative-pressure breathing*. Birds and mammals utilize negative-pressure breathing. In humans, inhalation occurs when the rib muscles contract, drawing the rib cage up and out, and the *diaphragm* contracts, moving downward, thus reducing the air pressure within the cavity below the atmospheric pressure and drawing air into the lungs. Normal exhalation is a passive process; the muscles relax and the rib cage and diaphragm return to their resting position, thus increasing the air pressure in the lungs, which forces air out.

The pattern of air flow in the respiratory system of birds is fundamentally different from that of mammals. The special arrangement of the lungs and their associated *air sacs* permits a continuous unidirectional flow of air through the lungs during both inhalation and exhalation. Because the blood flow in the capillaries is at an angle to the continuous air flow, the blood can pick up the maximum amount of O_2; thus birds extract oxygen from air more efficiently than mammals.

Modern fish do not have lungs as the ancestral fish did; it is believed that the primitive lung evolved into the *swim bladder* of modern fish. The swim bladder is filled with gases and acts to adjust the density of the fish to that of the water at different depths.

The tracheal system typical of land arthropods consists of many small air ducts called *tracheae* that run from openings (*spiracles*) in the body wall and carry air directly to the individual cells. There is no significant transport of O_2 by the blood.

QUESTIONS

Testing recall

Complete the following statements.

1 To obtain enough energy to carry on life processes, most plants and animals require _____ for respiration. (p. 272)

2 Aquatic organisms such as coelenterates have no special respiratory mechanism since gases can move a distance of approximately _____ mm by simple diffusion. (p. 280)

3 Gas exchange between cells and the environment always takes place by _____ across _____ membranes. (p. 273)

4 Gas exchange in the leaf takes place across the cell membrane of the _____ cells. (p. 274)

5 Stomata are usually _____ when the guard cells are carrying out photosynthesis in bright sunlight. (p. 276)

6 In the leaf, excessive water loss by transpiration is prevented by _____. (p. 277)

7 Gaseous oxygen can reach living cells deep within the plant stem by entering the intercellular air space system via openings called _____. (p. 274)

8 Gills are an example of an _____ respiratory surface whereas lungs are an example of an _____ surface. (pp. 281, 285)

9 In the gills of fish, the countercurrent flow of water and blood greatly _____ the pickup of O_2 by the blood. (pp. 282–84)

10 The rate of diffusion is _____ in air than in water (p. 279)

11 Most land organisms utilize an _____ type of respiratory system of which there are two principal types, _____ and _____. (p. 285)

12 Gas exchange in the human takes place by the process of _____ across the thin, moist membranes of the _____ , which are surrounded by a dense network of _____ . (pp. 286–87)

13 In humans, incoming air is warmed, moistened, and filtered in the _____ . (p. 286)

14 Frogs force air into their lungs in a process called _____-pressure breathing whereas birds and mammals _____ air into their lungs using _____-pressure breathing. (p. 290)

15 When the diaphragm contracts, the air pressure in the lungs _____ . (p. 288)

16 In birds, most of the air drawn in during inhalation goes through the bronchus to the _____ . During exhalation, this air moves through the recurrent bronchi into the _____ where gas exchange takes place. (pp. 288–89)

17 Swim bladders probably evolved from the _____ of primitive fish. (p. 291)

18 _____ are the respiratory openings on the body surface of insects. They lead into many smaller air ducts called _____ , which conduct air to internal cells. (pp. 292–94)

19 List the four basic requirements for a specialized respiratory surface.

_____ _____
_____ _____
_____ _____
(p. 274)

Testing knowledge and understanding

Choose the one best answer.

20 The principal site of gas exchange in the leaf is the

a upper epidermis.   d mesophyll.
b cuticle.   e lower epidermis.
c lenticel.   (p. 274)

21 During the day, the guard cells

a lose water and are flaccid.
b synthesize starch from sugar.
c have a relatively low concentration of K^+.
d have a lower osmotic concentration than at night.
e carry out photosynthesis.   (pp. 276–77)

22 Animals like the earthworm must live in moist soil. How is this related to their method of gas exchange?

a They respire through their moist skin.
b Their lungs function most efficiently in a moist habitat.
c Waterlogged soil is rich in oxygen.
d The soil must be moist to keep their unvaginated parapodia moist.   (p. 280)

23 In which one of the following habitats would aquatic organisms be most likely to show special adaptations to low oxygen levels?

a a pool at the base of a woodland waterfall
b the Antarctic Ocean
c a fast-running mountain stream
d a shallow pool in a sunny meadow in Texas
e the Great Lakes   (p. 284)

24 All of the following organisms use gills as a gas-exchange system *except* the

a squid.
b sea star.
c fish.
d marine segmented worm.
e frog.   (pp. 281–82, 290)

25 Water flowing over the gills of the fish

a flows in the same direction as the blood.
b flows in a direction opposite to that of the blood.
c exchanges O_2 and CO_2 with the blood in the capillaries of the gills.
d a and c.
e b and c.   (p. 282)

26 The passageway common to the respiratory and digestive systems of humans is the

a nasal cavity.   d larynx.
b trachea.   e esophagus.
c pharynx.   (p. 286)

Questions 27–29 refer to the following graph, which shows pressure changes in the thoracic (chest) cavity during breathing in a human.

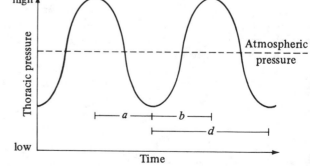

27 Which segment of the curve (*a, b, c,* or *d*) covers the period of inhalation? (p. 288)

28 Which segment of the curve covers the period of exhalation? (p. 288)

29 Which segment of the curve covers the period when the diaphragm muscle is relaxing? (p. 288)

30 Which of the following features do *all* gas-exchange systems have in common?

a The exchange surfaces are moist.
b They are enclosed in a special chamber.
c They are maintained at a constant temperature.
d They are exposed to air.
e They are found only in vertebrates.
(p. 274)

31 Which one of the following statements is *false* for gills operating in water?

a Water can support the delicate gill filaments.
b Most fish must actively pump water across their gills.
c Keeping membranes moist is no problem.
d Water carries more oxygen than air.
(pp. 282−84)

32 The normal air pathway in the human respiratory system is

a trachea → pharynx → alveolus → bronchus.
b bronchus → bronchiole → pharynx → trachea.
c alveolus → bronchus → bronchiole → pharynx.
d trachea → bronchus → alveolus → pharynx.
e pharynx → trachea → bronchus → alveolus. (pp. 286−87)

33 Which one of these statements about human lungs is *false*?

a Gas exchange always takes place across moist membranes.
b The gases move across the exchange membranes by diffusion.
c The total exchange surface area is very large.
d The walls of the alveoli are only one cell thick.
e The concentration of CO_2 is higher in the air in the lungs than in the blood in the alveolar capillaries. (p. 287)

34 All of the following structures are involved in gas exchange in humans *except* the

a nostrils.
b mouth.
c lungs.
d diaphragm.
e skin.
(pp. 286−87)

35 In the frog, which structure is *not* important in gas exchange?

a nostrils
b mouth
c lungs
d diaphragm
e skin
(p. 290)

36 In the respiratory system of birds, oxygen-rich air moves into the lungs

a during inhalation.
b during exhalation.
c during both inhalation and exhalation.
d only when the air sacs contract.
e when air is forced from the mouth into the trachea. (p. 289)

37 Insects carry out respiratory gas exchange

a in specialized external gills.
b in specialized internal gills.
c in the alveoli of their lungs.
d across the membranes of nearly every internal cell.
e across their thin moist skin. (p. 293)

For further thought

1 Explain why plants do not need special gas-transporting systems whereas many animals require some sort of transport mechanism to carry gases from the exchange surface to the rest of the body. (p. 279)

2 Discuss basic requirements for gas-exchange mechanisms, and show how these are fulfilled in a fish, a human, an insect, and a large tree. (pp. 274, 278−79, 286−87, 292−94)

3 Explain why gills are not suitable for life on land. (p. 283)

4 The second largest invertebrate group is the phylum Mollusca. The name means "soft-bodied"; the soft body is wholly or partially surrounded by a thin fleshy mantle and is commonly sheltered in an external limy shell. A familiar example of the mollusc is the marine clam. In the marine clam, gas exchange is performed by two double gills, which hang in the mantle cavity. The gills consist of many interconnected thin plates, or lamellae. Each lamella is supplied with small blood vessels, which bring blood to the gill and return it to the heart. Cilia cover the gills and function to produce a flow of water through the gills.

a For gas exchange, give one advantage and one disadvantage of a marine environment.
b Discuss how the clam meets the basic requirements of a gas-exchange system.
(p. 274)

5 Compare and contrast the respiratory systems of a human being and a terrestrial insect. (pp. 286–87, 292–94)

6 Cigarette smoke has been found to have the following effects on the respiratory system:

a destruction of many of the cilia that line parts of the respiratory tract;

b thickening of the walls of the bronchioles, thus reducing the interior diameter of the tubes;

c rupturing of the walls of some of the alveoli.

For each of these effects, indicate how the *normal* functioning of the respiratory tract is altered by the use of cigarettes. (pp. 286–87)

7 You are probably aware that your exhaled breath contains water vapor. How do you account for this? (pp. 286–87)

ANSWERS

Testing recall

1 oxygen
2 one
3 diffusion, moist
4 mesophyll
5 open
6 the cuticle and closing the stomata
7 lenticels
8 evaginated, invaginated
9 facilitates
10 faster
11 invaginated, lungs, tracheae
12 diffusion, alveoli, capillaries
13 nasal cavities
14 positive, draw, negative
15 decreases
16 posterior air sacs, lungs
17 lungs
18 Spiracles, tracheae
19 large surface area, moist surface, method of transport, protection for fragile surface

Testing knowledge and understanding

20 d	25 e	30 a	34 e
21 e	26 c	31 d	35 d
22 a	27 a	32 e	36 c
23 d	28 b	33 e	37 d
24 e	29 b		

INTERNAL TRANSPORT IN UNICELLULAR ORGANISMS AND PLANTS

A GENERAL GUIDE TO THE READING

This chapter, the first of two on the topic of internal transport, discusses transport in higher plants; in the next we consider transport in animals. As you read Chapter 12 in your text, you will want to give special attention to the following aspects of the topic.

1   Organisms without special transport systems. Though this chapter deals primarily with transport in higher plants, you need to realize that many organisms do not have special transport mechanisms. The section on these organisms (p. 296) makes this point clear.

2   Stems. Focus your attention on the three types of plant stems: monocots, herbaceous dicots, and woody dicots. Notice the similarities and differences. You will find learning about stems easier if you carefully study the photographs and diagrams on pages 298-301 (Figures 12.2−12.7).

3   Structure of xylem and phloem cells. Study carefully the material on the different types of conducting cells, noticing particularly the differences between xylem and phloem cells. Figures 12.11−12.17 (pp. 305−8) are most helpful in learning this material.

4   Hypotheses about the mechanisms of transport. You will want to understand the cohesion theory of xylem transport (pp. 312−14) and the pressure-flow model of phloem transport (pp. 308−17). Figure 12.22 (p. 316) is crucial for understanding the latter.

KEY CONCEPTS

1   Diffusion plays an important role in the movement of materials, but it is a very slow process. Consequently most organisms have

evolved some sort of specialized transport mechanism to distribute substances more rapidly. (pp. 296–97)

2   The successful exploitation of the land environment by plants was dependent on the evolution of specialized conducting tissues that would transport water and nutrients from one part of the plant to another. (p. 297)

3   All tissues produced by the apical meristem are primary tissues. All tissues derived from the vascular cambium are secondary tissues; they contribute to growth in diameter. (p. 300)

4   The two main types of plant vascular tissue are xylem, which conducts water and inorganic ions upward from the roots to the aerial parts of the plant; and phloem, which conducts water and organic solutes from one part of the plant to another, both upwards and downwards. (pp. 305–8)

5   According to the cohesion theory, water lost by transpiration from the aerial parts of plants is replaced by withdrawal of water from the water column in the xylem, and in the process the whole column is pulled upward. (pp. 312–14)

6   According to the pressure-flow, or mass-flow, hypothesis, there is a mass flow of solutes under pressure through the sieve tubes of phloem from an area of high concentration in an area of low concentration. (pp. 316–17)

OBJECTIVES

After studying this chapter and reflecting on it, you should be able to carry out the following objectives.

1   Explain why the unicellular and multicellular algae can get along without a specialized internal-transport system, whereas most higher plants cannot. (p. 296)

2   Give three reasons why the evolution of specialized transport tissue was necessary for the higher plants to exploit the land environment fully. (p. 297)

3   Using cross sections like those in Figures 12.3–12.7 (pp. 299–301), point out the xylem and phloem, and where present, the vascular cambium, pith, epidermis, and cortex. Explain how you can distinguish among herbaceous dicot, woody dicot, and monocot stem. (pp. 297–305)

4   Compare the structure and arrangement of tissues in a typical dicot stem (as shown in Figure 12.5, p. 300) with those of a typical dicot root (see Figure 9.4, p. 228). Do the same for a monocot stem (as shown in Figure 12.3, p. 299) and a monocot root (see Figure 9.7, p. 230). Explain how you can tell whether the cross section you are looking at is a root or a stem.

5   Explain the differences between primary and secondary tissue in a dicot, and describe the process of secondary growth in a dicot stem. Using cross sections like those in Figure 12.7 (p. 301), identify the primary xylem, secondary xylem, vascular cambium, primary phloem, secondary phloem, and periderm. (pp. 300–1)

6   Describe the pattern of growth that leads to the rings seen in cross sections of trees, explaining the difference between heartwood and sapwood. Draw diagrams showing how the structure of a typical twenty-five-year-old tree trunk (as seen in cross section) would differ from that of a stem of the same species during its first season of growth. (pp. 301–5)

7   Contrast the structure of the xylem tissue with that of the phloem with respect to the types of cells present, the structure of the conducting cells, and whether the conducting cells are living or dead. (pp. 305–8)

8   Discuss the mechanisms by which sap moves upward in xylem tissue. In doing so, consider the force necessary to raise water to the tops of the tallest trees, the role of root pressure in the ascent of sap, and the cohesion theory and the role of hydrogen bonding and capillarity in the rise of sap. Explain what is meant by transpiration. (pp. 308–14)

9   Using a diagram such as Figure 12.22 (p. 316), describe the mechanism of phloem transport according to the pressure-flow hypothesis. Define the "source" and the "sink." Indicate whether the movement through sieve tubes would be predominantly upward or downward on a sunny day in summer, and in the same plant at night. (pp. 314–17)

KEY TERMS

The following terms are important in this chapter; you should become familiar with them.

cytoplasmic streaming (p. 296)
herbaceous (p. 297)
woody (p. 297)
monocot (p. 297)

SUMMARY

In general, only very small organisms can rely exclusively on the processes of diffusion and intracellular transport for the movement of substances; most higher organisms, both plant and animal, require a specialized transport system to deliver required materials to the cells and remove the waste products.

The large multicellular land plants, the *vascular plants*, have evolved two types of specialized conducting tissue: *xylem* and *phloem*. These tissues form continuous pathways running through the roots, stems, and leaves, and serve to transport materials from one part of the body to another.

Stems function as organs of transport and support. The arrangement of the vascular tissue varies in stems of different types and ages. In monocot stems the vascular tissue is in discrete bundles scattered throughout the stem, and there is no vascular cambium. In a young dicot stem the vascular tissue may be arranged in a continuous hollow cylinder around the central pith or in a ring of discrete bundles. The phloem always lies

outside the xylem, with a layer of lateral meristematic tissue, the *vascular cambium*, between them. In herbaceous and young woody dicot stems the tissues are composed of cells produced by the *apical meristem* as the stem grows in length; these tissues are *primary tissues*.

In some species of dicots the cambium becomes active and produces new cells; those formed on the outside differentiate as *secondary phloem*, those formed on the inside as *secondary xylem* or wood. All tissues derived from the cambium are *secondary tissues*; they contribute to growth in diameter. Since xylem cells produced early in the growing season are larger than cells produced later, a series of concentric annual rings is formed. Usually only the newer, outer rings, or *sapwood*, are functional in transport; the cells of the older rings, the *heartwood*, have become plugged.

As woody stems grow in diameter, a layer of cells outside the phloem, the *cork cambium*, begins to produce cork cells, which form the *periderm*.

The sequence of tissues (moving from outside toward the center) in an old woody stem is: cork tissue and cork cambium (outer bark), primary phloem and secondary phloem (inner bark), vascular cambium, secondary xylem (wood), primary xylem, and pith.

Xylem contains two types of conductive cells: *tracheids* and *vessel elements*. Both are dead at functional maturity; the cellular contents disintegrate, leaving thick, lignified cell walls that form hollow tubes within which vertical movement of materials can take place. Numerous *pits* in the wall allow the lateral movement of materials from cell to cell. Xylem also contains *fiber* and *parenchyma* cells. The thick-walled fiber cells are supportive elements; the parenchyma cells are often arranged to form *rays*, which function as pathways for the lateral movement of materials.

Phloem also contains both fiber and parenchyma cells in addition to those unique to it: *sieve elements* and *companion cells*. The elongate sieve elements with their perforated end walls—the *sieve plates*—are arranged end to end, forming a *sieve tube* through which vertical transport takes place. At maturity the nucleus of the sieve element disintegrates, but the cytoplasm remains. Companion cells, which retain both their nuclei and their cytoplasm, are intimately associated with the sieve elements in most advanced plants.

Water and inorganic ions absorbed by the roots (sap) move upward through the plant body in the tracheids and vessels of the xylem. According to the *cohesion theory*, water lost by transpiration from the aerial parts of the plant is replaced by withdrawal of water from the water column in the xylem. The water in the xylem forms a continuous chain from the aerial parts of the plant to the roots; these water molecules exhibit great cohesion because of hydrogen bonding. Thus the withdrawal of water molecules from the top of a water column will pull the whole chain of water molecules upward.

In some plants, especially in early spring, *root pressure* may contribute to the upward movement of sap in the xylem. Root pressure is created when ions are actively transported into the stele, and water follows passively in such quantity as to build up pressure in the xylem, pushing the sap slowly upward.

Inorganic ions are transported, or *translocated*,

upward from the roots to the leaves primarily through the xylem; however, the downward transport of these ions is through the phloem. Most organic solutes (carbohydrates and organic nitrogen compounds) are transported, both up and down, through the phloem.

Unlike the xylem transport, the movement of substances in the phloem is through living cells that retain their cytoplasm. The end walls are penetrated only by the tiny pores of the sieve plate. The most widely supported hypothesis of phloem function is the *pressure flow* hypothesis, according to which there is a mass flow of water and solutes under pressure through the sieve tubes from an area of high turgor pressure (the "source") to an area of low turgor pressure (the "sink").

QUESTIONS

Testing recall

1 Complete the following chart; it will be a useful study aid. (pp. 305–8)

	Xylem	Phloem
Conducting cells		
Are cells living or dead at functional maturity?		
Materials transported		
Direction of movement		
Probable mechanism of transport		

Match each item below with the associated part of the plant stem shown in cross section in the diagram. Then refer to the diagram for questions 10 and 11. Answers may be used once, more than once, or not at all.

2 primary phloem (p. 300)

3 secondary phloem (p. 300)

4 vascular cambium (p. 300)

5 primary xylem (p. 300)

6 secondary xylem (p. 300)

7 tissues produced by the vascular cambium (p. 300)

8 cork (p. 304)

9 "wood" (p. 304)

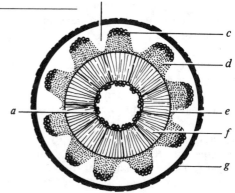

10 As the stem grows, tissue *c*

    a will be pushed outward.

    b will be pushed inward.

    c will remain in the same position. (p. 300)

KEY DIAGRAM

In the following diagrams of monocot and woody dicot plants, color in the following using a different color for each structure. First color the title, then the appropriate structure. (Reference: Figures 12.2, p. 298 and 12.8, p. 302)

Which plant is the monocot and which the dicot? How can you tell the difference between the two?

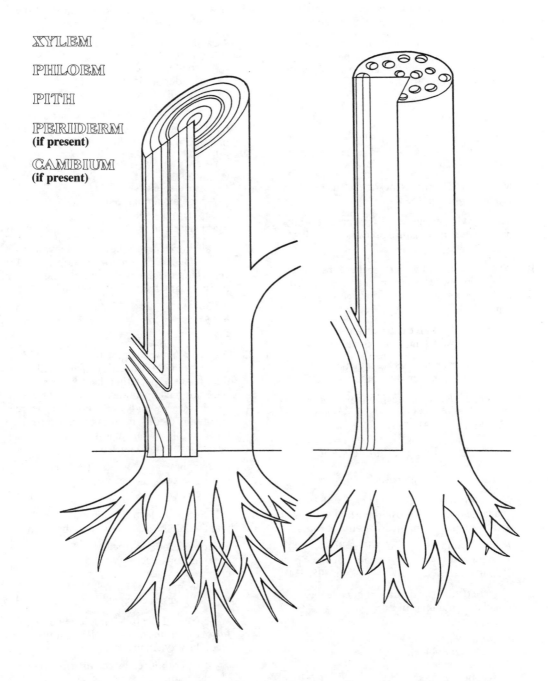

XYLEM

PHLOEM

PITH

PERIDERM
(if present)

CAMBIUM
(if present)

11 The stem shown is

   a a monocot.
   b a herbaceous dicot.
   c a woody dicot.   (p. 301)

Decide whether the following statements are true or false, and correct the false statements.

12 In a stem, the secondary phloem is located outside the layer of primary phloem. (p. 306)

13 As the stem of a woody plant grows through successive years, the cortex and pith become less and less important and are often virtually obliterated. (pp. 301–3)

14 Phloem is the tissue of which wood is principally composed. (p. 304)

15 Mature functioning vessel cells are dead, but mature functioning tracheids are alive. (pp. 305–7)

16 Mature sieve elements lack a nucleus but retain cytoplasm. (p. 308)

17 Mature xylem tracheid cells are metabolically more active than mature phloem sieve cells. (pp. 305, 308)

18 At present, most biologists believe transpiration pull is more important than root pressure in the upward movement of water in xylem. (pp. 310–14)

19 The phloem of a tree can transport organic materials upward only. (pp. 308, 316)

20 Some mineral nutrients are much more mobile in the phloem than others. (p. 315)

21 Current evidence on the mechanism of transport in the phloem favors the mass-flow hypothesis. (pp. 316–17)

Testing knowledge and understanding

Choose the one best answer.

22 Which one of the following parts of the trunk of a large tree is the most important in transport of sucrose from the leaves to the roots?

   a heartwood   *d* outer bark
   b sapwood   *e* cambium
   c inner bark   (pp. 305, 314)

23 In a tree trunk, the tissue just outside the vascular cambium is the

   a cork.
   b primary phloem.
   c secondary phloem.
   d primary xylem.
   e secondary xylem.   (p. 305)

24 The plant tissue responsible for the growth in height of a vascular plant is the

   a vascular cambium.
   b apical meristem.
   c xylem.
   d cork cambium.
   e phloem.   (p. 300)

25 In a twenty-five-year-old tree, most of the "wood" is

   a pith.
   b primary xylem.
   c secondary xylem.
   d secondary phloem.
   e cork cells.   (p. 304)

26 One cell type in a plant stem is particularly hard to study because its contents are under pressure and the cell bursts when cut. This cell is a

   a tracheid.
   b sieve element.
   c vascular cambial cell.
   d epidermal cell.
   e vessel element.   (p. 308)

27 A nail is driven into the trunk of a young 6-foot-tall tree, 3 feet above the ground. Ten years later the tree is 30 feet tall. The nail will be how many feet above the ground?

   a 3 feet   *d* 15 feet
   b 6 feet   *e* 27 feet
   c 10 feet   (p. 300)

28 A beaver gnawed completely around a birch tree but did not proceed further to fell the tree. It was noticed that the leaves retained their normal appearance for several weeks, which were hot and sunny, but that the tree eventually died. It can be concluded that the most peripheral region it is *certain* that the beaver left functional was

   a the phloem.   *d* the bark.
   b the cortex.   *e* the pith.
   c the xylem.   (p. 305)

29 Which one of the following mature cells in a plant would be least active metabolically?

   a root endodermal cell
   b stem cortex cell
   c mesophyll cell
   d phloem companion cell
   e xylem vessel element   (p. 305)

30 Which one of the following statements is *false* concerning a vessel cell in mature secondary xylem in a large tree?

   a It was produced by cell division in the cambium.

b It once contained a nucleus, but no longer has one.

c Its cytoplasm is under the control of companion cells.

d Its thick cellulose walls are impregnated with lignin.

e Its end walls are extensively perforated. (pp. 306–8)

31 Which of the following statements is *false* concerning the xylem?

a Xylem tracheids and vessels fulfill their vital function only after their death.

b The cell walls of the tracheids are greatly strengthened with cellulose fibrils forming thickened rings or spirals.

c Water molecules are transpired from the cells of the leaf and replaced by water molecules in the xylem pulled up from the roots by the cohesion of water molecules.

d Movement of materials is by mass flow; materials move owing to a turgor-pressure gradient from "source" to "sink."

e In the morning sap in the xylem begins to move first in the twigs of the upper portion of the tree and later in the lower trunk. (pp. 305, 312–14)

32 Which of the following mechanisms could raise water to the greatest height in the xylem vessels of a tall tree?

a transpiration from the leaves combined with cohesion of water molecules

b root pressure

c cytoplasmic streaming

d active transport of water

e mass flow along a turgor-pressure gradient (pp. 296, 310–14)

33 Root pressure

a is not sufficient to raise water above ground level in any plant.

b is negative in all but the tallest trees.

c can push water to the top of a 10-foot corn plant but not to the top of a 200-foot tree.

d is the best explanation known for the transport of water to the tops of the tallest trees.

e is the driving force for the mass flow of sugar. (pp. 310–11)

34 What is the principal pathway by which sugar travels in a plant?

a xylem vessels   *d* sieve elements
b companion cells *e* capillaries
c parenchyma cells   (p. 314)

35 According to the pressure-flow hypothesis, materials are moved through the conducting cells as a result of

a a turgor-pressure gradient along the sieve tubes.

b a pressure gradient along the sieve tubes produced by cytoplasmic streaming.

c active transport between adjacent sieve elements.

d a flow along membranous interfaces within the sieve elements.

e a pull exerted by transpiration from the aerial parts of the plant. (pp. 316–17)

36 Which one of the following statements concerning the flow of sap in the xylem of trees is *correct*?

a In the morning sap begins to flow first in the twigs and later in the trunk.

b Flow makes the trunk increase in diameter during the day and decrease at night.

c Flow is driven by the high concentration of sugar in vessel elements.

d Rapid flow of water puts the xylem under pressure much greater than atmospheric pressure.

e Flow from the roots to the twigs would be accelerated if the leaves were removed. (pp. 312–14)

37 Arrange the following five events in an order that explains the mass flow of materials in the phloem.

1 Water diffuses into the sieve elements.
2 Leaf cells produce sugar by photosynthesis.
3 Solutes are actively transported into sieve elements.
4 Sugar is transported from cell to cell in the leaf.
5 Sugar moves down the stem.

a 2 – 1 – 4 – 3 – 5   *d* 1 – 2 – 3 – 4 – 5
b 2 – 4 – 3 – 1 – 5   *e* 4 – 2 – 1 – 3 – 5
c 2 – 4 – 1 – 3 – 5   (pp. 316–17)

38 The "pressure" in the pressure-flow hypothesis comes from

a root pressure.

b the osmotic uptake of water by sieve elements in the leaf.

c the accumulation of minerals and water by the stele in the root.

d transpiration in the leaves. (pp. 316–17)

For further thought

1 As plants evolved to land forms, they were confronted with the problem of support, since they no longer had water to support their weight. How was this problem solved? (p. 304)

2 Trace the route a molecule of water would follow from the place where it entered a vascular plant until it was lost from a leaf by transpiration. (pp. 229–30, 310–14)

3 Explain how the carbohydrates synthesized in the leaves get from the leaves to the root and other parts of the growing plant. (pp. 314–17)

ANSWERS

Testing recall

1

	Xylem	Phloem
Conducting cells	tracheids vessels	sieve elements
Are cells living or dead at functional maturity?	dead	living
Materials transported	water, inorganic nutrients	water, organic materials—sucrose, amino acids, etc.
Direction of movement	up only	up and down
Probable mechanism of transport	transpiration, cohesion	pressure-flow

2	c	6	f	9	f
3	d	7	d,f	10	a
4	e	8	g	11	c
5	a				

12 false—inside
13 true
14 false—xylem
15 false—both are dead
16 true
17 false—less active (they are dead!)
18 true
19 false—upward and downward
20 true
21 true

Testing knowledge and understanding

22	c	27	a	31	d	35	a
23	c	28	c	32	a	36	a
24	b	29	e	33	c	37	b
25	c	30	c	34	d	38	b
26	b						

INTERNAL TRANSPORT IN ANIMALS

A GENERAL GUIDE TO THE READING

This chapter briefly describes the circulatory system in a variety of animals and then focuses on the human circulatory system. Most students find the subject fascinating. As you read Chapter 13 in your text, you will want to concentrate on the following material.

1   The introductory discussion. This part of the chapter (pp. 318–19) covers a number of important topics, including the concept of open and closed circulatory systems.

2   The parts of the human heart and the main blood vessels. You will want to learn these, as shown in Figures 13.3 and 13.4 (p. 321).

3   The circuit in other vertebrates. This section describes the evolution of the vertebrate heart. Learn this material well now as it will be discussed later in Chapter 43. Figure 13.7 (p. 326) nicely summarizes the evolutionary trends.

4   The exchange of materials in the capillaries. This process, one of the more difficult subjects presented in the chapter, is discussed on pages 331–38. Figure 13.15 (p. 335) is the key to understanding it.

5   Thromboembolic and hypertensive disease in human beings. The box on this subject (pp. 334–35) is particularly interesting and timely.

6   The lymphatic system. Students often fail to understand the importance of this system; read pages 338–39 with care.

7   The function of the red blood cells and hemoglobin. The discussion of this topic (pp. 345–51) refers back to the material in Chapter 3 on the structure of the hemoglobin molecule. Make sure you understand the role of hemoglobin in the transport of O_2 and CO_2. Understanding Figures 13.25–13.28 is crucial (pp. 348–49).

KEY CONCEPTS

1 The active way of life and rapid metabolism of animals require that each cell be provided with a continuous, abundant supply of oxygen and nutrients to metabolize for energy. Therefore many complex animals have evolved a true circulatory system to transport materials to and from the body cells more efficiently. (pp. 318–19)

2 The high metabolic rate and active way of life of the warm-blooded birds and mammals are made possible by the four-chambered heart and the evolution of the separate pulmonary and systemic circulations. (p. 325)

3 Nearly all exchange of materials between the blood and the tissue fluid occurs in the capillaries; this exchange is governed by a delicate balance between the blood pressure and osmotic pressure. (pp. 331–38)

4 The cells of higher animals require a fairly uniform, or stable, environment in order to maintain normal function. The circulatory system plays an important role in maintaining this stability within the body. (p. 340)

OBJECTIVES

After studying this chapter and reflecting on it, you should be able to carry out the following objectives.

1 Explain why many very small animals, such as hydra and planaria, can get along without a specialized circulatory system, whereas most higher animals cannot. Give one major way in which the circulatory system of plants differs from that of animals. (pp. 318–19)

2 Differentiate between an open and a closed circulatory system, and compare them with respect to speed and efficiency. Specify the advantage of a closed circulatory system over an open one. Explain why insects can function so successfully with an open circulatory system, whereas animals such as earthworms utilize a closed system. (pp. 319–20)

3 Describe the principal parts of your heart, and the main blood vessels. Using a diagram such as Figure 13.4 (p. 321), and naming the parts of the heart as appropriate, trace the flow of blood from the posterior vena cava, through the heart, to the lungs, back to the heart, and out the aorta. Indicate which portion of this pathway comprises the pulmonary circuit.

4 Describe the evolutionary adaptations that take place in the vertebrate heart from fish, to amphibians, to reptiles, to mammals. (pp. 325–26)

5 Discuss the events that take place during one cycle of contraction and relaxation of the heart. In doing so, trace the path of an electrical impulse from its point of origin to other parts of the heart; indicate whether the heart requires stimulation from the nervous system to contract; and explain the terms diastole and systole. (pp. 327–28)

6 Discuss the structure and function of arteries, veins, and capillaries. In doing so, explain why the blood pressure is considerably lower in capillaries and veins than in arteries, and why the blood pressure is constant rather than fluctuating; specify the mechanisms other than blood pressure that function in moving blood through the veins; indicate whether blood flows more slowly or more quickly through capillaries than through arteries, and explain how this difference in speed is correlated with capillary function; and give two mechanisms by which water and solutes move out of the capillaries. (pp. 320, 328–38)

7 Using a diagram such as Figure 13.15 (p. 335), explain how water and dissolved materials are forced out of the capillaries at the arteriole end of the capillary bed and into the capillaries at the venule end. (pp. 332–36)

8 Describe the structure of the lymphatic system and discuss three important ways in which this system helps maintain the normal functions of the body. (pp. 328–29)

9 Name the substances that make up the two constituents of blood—the plasma and the formed elements. In doing so, list the three major groups of formed elements and give one function of each; then list the solutes found in the plasma and give a major function of each. (pp. 339–42)

10 State two ways in which the blood acts as a protective agent against infection and disease. (pp. 345–50)

11 Review Figure 3.25 (p. 67), which shows the quaternary structure of hemoglobin. Point out where the O_2 molecules bind, and specify how many O_2 molecules can bind to each hemoglobin molecule. (p. 67)

12 Using diagrams such as those in Figures 13.25–13.28 (pp. 348–49), be able to explain the dissociation curves for hemoglobin and describe the effect of pH, temperature

changes, and different types of hemoglobin on the dissociation curve. (pp. 347–49)

13 Explain how CO_2 is transported in the blood and the role hemoglobin plays in this process. (p. 349)

KEY TERMS

The following terms are important in this chapter; you should become familiar with them.

heart (p. 319)
open circulatory system (p. 319)
closed circulatory system (p. 319)
artery (p. 320)
vein (p. 320)
capillary (p. 320)
right atrium ⎫
right ventricle ⎬ parts of the
left atrium ⎭ heart (pp. 321–22)
left ventricle
pulmonary artery (p. 322)
pulmonary vein (p. 322)
aorta (p. 322)
anterior vena cava (p. 322)
posterior vena cava (p. 322)
pulmonary circulation (p. 325
foramen ovale (p. 325)
ductus arteriosus (p. 325)
systemic circulation (p. 327)
sino-atrial, or S-A, node (p. 327)
nodal tissue (p. 327)
atrio-ventricular, or A-V, node (p. 327)
Purkinje fibers (p. 327)
bundle of His (p. 327)
systole (p. 327)
diastole (p. 327)
blood pressure (p. 328)
hydrostatic pressure (p. 330)
lymph (p. 338)
spleen (p. 338)
plasma (p. 339)
erythrocyte (p. 339)
leukocyte (p. 339)
platelet (p. 339)
homeostasis (p. 340)
fibrinogen*
fibrin
prothrombin ⎬ clotting
thromboplastin ⎬ proteins (pp. 342–43)
thrombin

*Some prefixes and suffixes are used repeatedly in biology. For example, the suffix *-ogen* usually means "giving rise to," and the prefix *pro-* (as in the clotting protein *prothrombin*) usually means "coming before." Similarly, the suffix *-ase* almost invariably signifies that the substance is an enzyme.

lymphocytes (p. 344)
antigen (p. 344)
antibody (p. 344)
histamine (p. 344)
hemoglobin (p. 346)

SUMMARY

Most higher animals have a true circulatory system in which blood is moved around the body along a fairly definite path. Animal circulatory systems usually have at least one pumping device, or *heart*. The system is a *closed circulatory system* if the blood is always contained within well-defined vessels; it is an *open circulatory system* if blood leaves the vessels and flows through large open spaces within the body. Closed circulatory systems are characteristic of earthworms and all vertebrates. Open circulatory systems are characteristic of most molluscs and all arthropods.

The open circulatory system of insects consists of a dorsal longitudinal vessel which contracts, forcing the blood out at the anterior end. The blood then flows backward through large sinuses and finally reenters the dorsal vessel.

The closed circulatory system characteristic of vertebrates consists of a heart and numerous vessels: *arteries*, which carry blood away from the heart; *veins*, which carry blood to the heart; and *capillaries*, tiny vessels that connect the arteries to the veins. The actual exchange of materials between the blood and tissues takes place in the capillaries.

The human heart is a double pump; each side of the heart is divided into two chambers, an upper *atrium* that receives the blood and pumps it into the lower chamber, or *ventricle*, which then pumps the blood away from the heart.

The right heart receives deoxygenated blood from all over the body and pumps it via the *pulmonary arteries* to the lungs, where the blood picks up O_2 and gives up CO_2. The oxygenated blood then returns to the left atrium by the *pulmonary veins*. This portion of the circulatory system is called the *pulmonary circulation*.

The left ventricle pumps the blood into the *aorta* and its numerous branches, from which it moves into capillaries, where the exchange of materials takes place, then into veins, and finally back via the *anterior* or *posterior vena cava* to the right side of the heart. This portion of the circulatory system is called the *systemic circulation*. In the mammalian fetus, oxygenated blood is transferred from the pulmonary circuit to the systemic arteries through the *foramen ovale* (located between the atria) and the *ductus arteriosus* (which connects the pulmonary artery and aorta). Both close at birth.

The four-chambered heart, with complete separation of the right and left sides, is characteristic of the "warm-blooded" birds and mammals. The hearts of primitive vertebrates and modern fish have only one atrium and ventricle; the blood is pumped directly to the gills for oxygenation and on into the systemic circulation. From amphibian to mammal there is an increasing separation between the two sides of the heart.

The heartbeat is initiated when a wave of contraction spreads out from the *S-A node* across the atria to the *A-V node*, which sends excitatory impulses down the fibers of the bundle of His, stimulating both ventricles to contract. When the ventricles contract (systole), the blood is forced out of the heart and into the arteries under high pressure. During relaxation (diastole), the blood pressure in the arteries falls. This fluctuation (pulse) between the systolic and diastolic pressure diminishes in the arteries and no longer occurs in the capillaries and veins. In addition, friction causes the blood pressure to decrease as the blood moves farther away from the heart. One-way valves, skeletal muscle action, and the motions of the chest during breathing aid in moving blood in the veins.

The enormous numbers, extensive branching, and small diameters of the individual capillaries ensure that all tissues are supplied with a large capillary surface area for the exchange of materials. Exchange may be accomplished by diffusion through the capillary cell walls, by transport in endocytotic vesicles, and by filtration between capillary cells.

The movement of materials into and out of the capillaries is governed by the balance between the hydrostatic blood pressure and osmotic pressure. At the arteriole end of a capillary the hydrostatic pressure usually exceeds the osmotic pressure, and water is forced out of the capillary; at the venule end the reverse is true, and water moves into the capillary.

The *lymphatic system* helps maintain the osmotic balance of the body fluids by returning excess tissue fluid and proteins to the blood. Lymph nodes located along the major veins act to filter out particles and also as sites of formation of some white blood cells.

Blood consists of the *plasma*, the liquid portion of the blood; and the *formed elements*: red blood cells, white blood cells, and *platelets*. The water in the plasma is the solvent for the inorganic ions, plasma proteins, organic nutrients, nitrogenous waste products, hormones, and dissolved gases that are transported by the blood. The concentration of these solutes, especially the inorganic ions, remains relatively stable, being regulated by a variety of homeostatic mechanisms. *Homeostasis* (stability or equilibrium) is essential to the normal function of the organism.

The five types of white blood cells, or *leukocytes*, defend the body against disease and infection. Some kinds of leukocytes act by carrying on phagocytosis, others by producing enzymes that detoxify dangerous substances, and still others by producing highly specific *antibodies* that destroy or inactivate certain kinds of foreign substances called *antigens*.

The red blood cells, or *erythrocytes*, contain the oxygen-carrying pigment *hemoglobin*. Each hemoglobin molecule can combine loosely with four molecules of O_2 to form oxyhemoglobin. When the partial pressure of O_2 is high (as in the lungs) the hemoglobin picks up oxygen; when the partial pressure of O_2 is low (as in the tissues) the oxyhemoglobin releases O_2. Both the pH of the blood and the temperature affect the affinity of hemoglobin for O_2. Among mammals, the oxyhemoglobin of smaller animals tends to release O_2 more readily than that of larger animals, which is consistent with the fact that smaller animals have a higher metabolic rate and that their tissues therefore require more O_2 per unit time. Because human fetal hemoglobin has a higher affinity for O_2 than maternal hemoglobin, it can obtain O_2 from the maternal hemoglobin.

The blood also transports carbon dioxide from the tissues to the lungs. Most of the CO_2 is carried in the form of the bicarbonate ion (HCO_3^-). The excess H^+ ions are bound to hemoglobin.

QUESTIONS

Testing recall

Decide whether the following statements are true or false, and correct the false statements.

1 Insects have blood capillaries but no red blood cells. (pp. 319–20)

2 A weak aorta is more likely to rupture during systole than during diastole of the cardiac cycle. (p. 322)

3 At the arteriole end of a capillary, the osmotic pressure is greater than the blood pressure. (p. 334)

4 The lymphatic system returns excess tissue fluid to the blood. (p. 338)

5 Blood clotting begins as soon as the platelets are exposed to air. (p. 342)

6 Carbon monoxide is a dangerous poison because it has a strong tendency to bind to hemoglobin. (p. 350)

7 Iron is an essential component of the hemoglobin molecule. (p. 346)

8 Carbon dioxide is transported in the blood plasma principally as dissolved gas (CO_2). (p. 349)

9 The pacemaker (S-A node) of the heart is located in the brain. (p. 327)

10 Earthworms, like insects, have an open circulatory system. (p. 319)

Questions 11–20 refer to the following diagram of the adult heart.

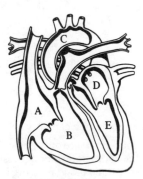

11 A correct sequence for the flow of blood through the heart would be

a A − B − C − E − D.
b A − B − D − E − C.
c D − E − C − B − A.
d B − A − D − E − C.
e C − E − D − B − A. (p. 321)

12 The hearbeat originates in the wall of what part of the heart? (p. 327)

13 Blood going to the lungs leaves from what part of the heart? (p. 322)

14 Blood from all over the body is returned to what part of the heart? (p. 321)

15 Which chamber pumps blood that will be distributed to all parts of the body? (p. 322)

16 The mitral or bicuspid valve is between which two chambers of the heart? (p. 322)

17 Which two chambers contract as a unit? (p. 327)

18 Which chambers pump oxygenated blood? (p. 322)

19 The pulmonary circulation connects which two chambers of the heart? (pp. 322–24)

20 The pacemaker (S-A node) is located in which part of the heart? (p. 327)

Testing knowledge and understanding

Choose the one best answer.

21 Both open and closed circulatory systems

a usually have one-way valves.
b have capillary beds.
c carry respiratory gases.
d are low-pressure systems. (pp. 319–20)

22 Which one of the following blood vessels carries deoxygenated blood in a mammal?

a aorta
b artery to the kidney
c pulmonary artery
d pulmonary vein
e coronary artery (p. 322)

23 In which one of the following blood vessels of a person would you expect to find the highest systolic blood pressure?

a pulmonary artery
b pulmonary vein
c arteriole in the leg
d capillary in the brain
e anterior vena cava (p. 329)

24 In which one of the above vessels would there be the greatest difference between systolic and diastolic pressure? (p. 329)

25 A labeled red blood cell is released into the arterial circulation in the left leg. It is recaptured 30 seconds later in the left lung. What is the minimum number of chambers of the heart it must have passed through?

a 0   *d* 3
b 1   *e* 4
c 2   (p. 321)

26 A red blood cell leaves the left ventricle of a human and travels into an artery leading to the right arm. How many capillary beds will it pass through before it returns to the left ventricle?

a 0   *d* 3
b 1   *e* 4
c 2   (pp. 321–22)

27 A correct continuous sequence for blood flow in the adult human heart is

a right atrium, right ventricle, aorta.
b posterior vena cava, left atrium, left ventricle.
c pulmonary vein, left atrium, left ventricle.
d left atrium, right atrium, aorta.
e left atrium, left ventricle, pulmonary artery. (p. 321)

28 Which one of the animals listed below would *not* show the pattern of circulation: heart → lung or gill capillaries → heart → systemic capillaries → heart?

a dog   d turtle
b frog   e bird
c fish   (p. 326)

29 The driving force that moves blood back to the heart in the veins is

a active transport.
b beating of the heart.
c skeletal-muscle contraction.
d closing of one-way valves.
e stretching of the walls of the veins. (pp. 330–31)

Questions 30–35 refer to the following diagram of two mammalian blood vessels (A and B) connected by a capillary bed. Blood pressure is higher in B than in A. The arrows indicate the direction of net diffusion for O_2 and CO_2.

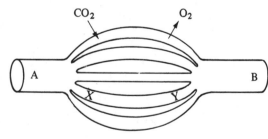

Choose the one best answer.

30 This capillary bed is probably part of

a the systemic circuit.
b the pulmonary circuit.
c either the systemic or the pulmonary circuit. (p. 325)

31 The hydrostatic blood pressure exceeds the osmotic pressure at

a X.
b Y.   (p. 335)

32 Which blood vessel is a vein?

a A
b B   (pp. 334–35)

33 Blood flows from

a A to B.
b B to A.   (pp. 334–35)

34 The concentration of O_2 in the blood will be highest at

a X.
b Y.   (p. 334)

35 Which chemical reaction is occurring in these capillaries? (Hb = hemoglobin)

a $Hb + 4O_2 \rightarrow Hb(O_2)_4$
b $Hb(O_2)_4 \rightarrow Hb + 4O_2$ (p. 347)

36 The following diagram of a capillary shows the hydrostatic blood pressure measured at three points along this particular capillary. Net outward movement of water at the arteriole end and net inward movement at the venule end would be facilitated if the osmotic pressure of the blood (relative to the tissue fluid) were

a 15 mm Hg.
b 20 mm Hg.
c 30 mm Hg.
d 40 mm Hg.
e 50 mm Hg.

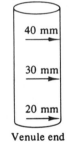

Arteriole end

40 mm

30 mm

20 mm

Venule end
(p. 334)

37 At the venule end of a capillary bed,

a the osmotic pressure of the blood is greater than the hydrostatic blood pressure.
b the hydrostatic blood pressure is greater than the osmotic pressure of the blood.
c the hydrostatic blood pressure and the osmotic pressure are equal.
d water and dissolved materials leave the capillary.
e the hydrostatic blood pressure is higher than it is at the arteriole end. (p. 334)

38 Which one of the following statements concerning the regulation of the human heartbeat is *correct*?

a The heart cannot beat on its own when nervous connections are severed.
b The rate of heartbeat is not influenced by nerves or chemical factors from outside the heart.
c Heartbeat is initiated and coordinated by nodal tissue.
d The right half of the heart beats before the left half.
e The left half of the heart beats before the right half. (p. 327)

39 Assume that the graph describes the pulse pressure and cross-sectional area of a circulatory system whose dynamic interrelationships are similar to those of the corresponding human system. Considering only these two parameters, what would you expect a graph of velocity to look like in such an organism?

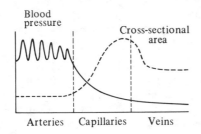

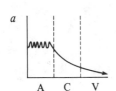

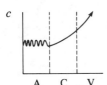

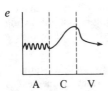

(pp. 329, 331)

40 All *but one* of the following are parts of or contained in the lymphatic system. The exception is the

a capillary networks. d leukocytes.
b arterioles. e nodes.
c valves. (pp. 338–39)

41 Which one of the following would you *least* expect to find in lymph?

a salt d protein
b fat e hemoglobin
c water (pp. 338–39)

42 Which one of the following statements concerning blood clotting is *false*?

a For clotting to occur, the blood must be exposed to molecular oxygen.
b When platelets are mechanically damaged, they release a substance that initiates clotting.
c For clotting to occur, calcium ions must be present.
d The plasma protein fibrinogen is necessary for blood clotting.
e The plasma protein prothrombin is necessary for blood clotting. (p. 342)

43 Disintegrating blood platelets release

a thromboplastin. d calcium ions.
b thrombin. e prothrombin.
c fibrinogen. (p. 342)

44 Which one of the following statements concerning mammalian red blood cells is *false*?

a They are spherical in shape.
b They lack a nucleus when mature.
c They contain hemoglobin.
d They transport O_2.
e They are destroyed by the liver and the spleen. (p. 345)

45 CO_2 released by a mammalian cell would probably be found moving in the blood primarily as

a carboxyhemoglobin.
b bicarbonate ions.
c carbonic acid.
d acid hemoglobin.
e oxyhemoglobin. (p. 349)

46 The typical S-shaped dissociation curve for hemoglobin owes its shape to the fact that

a oxyhemoglobin releases O_2 when the partial pressure of O_2 is high and picks up O_2 when the partial pressure is low.
b an increase in the partial pressure of O_2 favors the release of O_2.
c the binding of the first molecule of O_2 facilitates the binding of subsequent molecules.
d oxyhemoglobin is formed only when the partial pressure of O_2 is very high. (pp. 348–49)

47 The graph shows dissociation curves for the hemoglobin of five different animal species. On the basis of the characteristics of its hemoglobin, which animal would be best adapted for life at high altitudes?

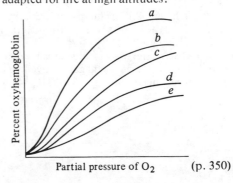

(p. 350)

One of the best ways to learn the human circulatory system is to trace the route a red blood cell follows as it moves through the circulatory system, naming, in order, all the vessels and chambers of the heart that the red cell would pass through. For example:

Question

Trace the path of a red cell moving by the most direct route possible from the capillaries in the lung to a vein in the brain.

Answer

capillaries in lung, pulmonary vein, left atrium, left ventricle, aorta, artery leading to brain, arterioles, capillaries in brain, venules, vein in brain

Now try the problems in questions 48 and 49.

48 Trace the path of a red cell moving by the most direct route possible from an artery in the leg to an artery in the arm. (pp. 321–22)

49 Trace the path of a red cell moving by the most direct route possible from the anterior vena cava to an artery leading to the kidney. (pp. 321–22)

For further thought

1 Distinguish between open and closed circulatory systems. What are the advantages of a closed system? (p. 319)

2 If a person touched a high-voltage electric power line, what would be the effect on his or her heart? Explain. (p. 327)

3 A man goes to his doctor and finds that his blood pressure is 190/110. What do these numbers refer to? What is the effect of high blood pressure on the exchange of materials in the capillaries? What symptoms may result over a long period of time if this condition remains untreated? (pp. 328–31)

4 What role does hemoglobin play in addition to its role in oxygen transport? (pp. 349–51)

5 John Jones was admitted to the hospital with a swollen and painful infected finger. There was a red streak up his arm, and several large, tender lumps in his armpit.

a What caused the lumps under his arm?
b State the factors that contributed to the swelling.
c Discuss the ways in which the human body protects itself against invading bacteria and viruses. (pp. 339, 341, 344–45)

ANSWERS

Testing recall

1 false—no blood capillaries and no red blood cells
2 true
3 false—less
4 true
5 false—clotting initiated only by damaged tissue and disintegrating platelets
6 true
7 true
8 false—as HCO_3
9 false—in the right atrium
10 false—earthworms have a closed circulatory system

11	b	15	E	18	D, E
12	A	16	D, E	19	B, D
13	B	17	B, E	20	A
14	A				

Testing knowledge and understanding

21	a	28	c	35	b	42	a
22	c	29	c	36	c	43	a
23	a	30	a	37	a	44	a
24	a	31	b	38	c	45	b
25	c	32	a	39	b	46	c
26	c	33	b	40	b	47	a
27	c	34	b	41	e		

48 artery in leg, arterioles, capillaries in leg, venules, veins in leg, posterior vena cava, right atrium, right ventricle, pulmonary artery, arterioles, capillaries in lung, venules, pulmonary vein, left atrium, left ventricle, aorta, artery in arm.

49 anterior vena cava, right atrium, right ventricle, pulmonary artery, arterioles, capillaries in lung, venules, pulmonary vein, left atrium, left ventricle, aorta, artery leading to kidney.

REGULATION OF BODY FLUIDS

A GENERAL GUIDE TO THE READING

This chapter discusses the mechanisms by which animals keep their body fluids constant. You may find it helpful to begin by reviewing the material on osmosis (see Chapter 4, pp. 90–94), since understanding osmosis is prerequisite to understanding much of the material in this chapter (pp. 353–55 for example). Then, as you read Chapter 14 in your text, you will want to concentrate on the following topics.

1   Homeostasis. The concept of homeostasis was introduced in the last chapter (see p. 340) but is elaborated on here. It is an important concept for you to know.

2   The differences between plants and animals with respect to the problem of maintaining a constant environment. These differences are presented on pages 354–55.

3   The vertebrate liver. The role of the liver in maintaining homeostasis (pp. 355–58) is often unappreciated; you will want to learn the importance of this vital organ. Pay particular attention to the hepatic portal circulation (Figure 14.3, p. 355), a most unusual arrangement of blood vessels.

4   Excretion and salt and water balance. The discussion of this problem (pp. 358–62) is interesting and presents some difficult concepts, particularly in the section on osmoregulation in fishes. You will find Figure 14.8 (p. 361) a great help in understanding this material.

5   Types of excretory systems. You will want to know the different excretory mechanisms found in various animals (pp. 362–67), and you should concentrate on learning the vertebrate kidney thoroughly (pp. 367–75). Notice too that the insect excretory system (pp. 366–67) is built on an entirely different plan from that of both other invertebrates and vertebrates.

6   The countercurrent-multiplier model. This is probably the most difficult topic in the chap-

ter. Figures 14.20 (p. 372) and 14.21 (p. 373) are most helpful in explaining this model.

7 Sodium-potassium pump. The concept of pumps was introduced in Chapter 4 (see pp. 103–5). You may find it helpful to review Figure 4.18 (p. 101). You should understand how pumps such as the sodium-potassium pump work as you will encounter this pump again in reference to nerve function in Chapter 18.

KEY CONCEPTS

1 The cells of multicellular animals require a relatively nonfluctuating fluid environment in order to carry out their life functions. (p. 353)

2 Living organisms have various mechanisms for maintaining a relatively constant internal environment despite changes in the external environment. This constant internal environment is a prerequisite to their ability to live in a variety of environments. (pp. 353–77)

3 The extracellular fluid of plants differs greatly from that of animals in that the extracellular fluid of plants is not fully distinct from environmental water and cannot be as well regulated as the tissue fluid and blood of animals. (pp. 353–55)

4 Plant cells can withstand much greater fluctuations in the makeup of the fluids bathing them than animal cells. (pp. 353–55)

5 Multicellular animals have various mechanisms for ridding their bodies of metabolic wastes and for regulating their salt and water balance to maintain the constancy of their body fluids. (pp. 355–77)

6 No cells have a mechanism for the active transport of water; when water must be moved, the usual mechanism is the active pumping of ions, with the water following passively. (pp. 371–72)

7 The processes of excretion and osmoregulation depend on the use of active transport pumps in the cell membrane that move substances against osmotic and/or electrical gradients. (pp. 375–77)

OBJECTIVES

After studying this chapter and reflecting on it, you should be able to carry out the following objectives.

1 State the principle of homeostasis and explain why the body's health and survival depend upon the maintenance of homeostasis. (pp. 340, 353)

2 Contrast the extracellular fluids of multicellular plants with those of animals. (pp. 352–55)

3 Discuss why plant cells seem able to withstand greater fluctuations in the makeup of their extracellular fluids than animal cells, explaining what happens to typical plant and animal cells in a hypotonic environment and in a hypertonic environment. (pp. 353–55)

4 Explain the functional significance of the hepatic portal system. In doing so, describe the circulatory connections between the stomach and intestines and the liver. Trace the flow of blood from the aorta through the intestines and liver and into the vena cava, using a diagram such as that in Figure 14.3 (p. 355) or Figure 13.4 (p. 321).

5 Discuss the role of your liver in maintaining homeostasis. Include in your answer the liver's function in carbohydrate, protein, and fat metabolism. (pp. 356–58)

6 List the three main nitrogenous waste products; compare them with respect to toxicity and to the amount of water that must be expelled in order to excrete each one; and indicate where these compounds are synthesized. Specify the main nitrogenous waste product of each of the following animals: freshwater hydra, marine jellyfish, freshwater planaria, freshwater fish, marine fish, mammal, bird, lizard, insect. Explain the correlation between the nitrogenous waste product and the habitat of an organism. (p. 357)

7 Explain the difference between osmoregulation and excretion, and the difference between excretion and elimination (defecation). (pp. 358–59)

8 Contrast the osmoregulatory problems faced by freshwater bony fish with those of marine bony fish and with those of sharks. Discuss the adaptations that enable these animals to cope with their respective problems. (pp. 360–61)

9 For each of the following organisms, specify the type of excretory mechanism possessed, indicate whether the mechanism functions primarily in water regulation or in both water regulation and excretion, and describe the role of the circulatory system in excretion: *Paramecium*, planaria, earthworm, mammal, insect. (pp. 362–67)

10 Using diagrams such as Figures 14.15 (p. 367) and 14.16 (p. 368), point out the following: renal artery, renal vein, kidney, ureter, urethra, bladder.

11 Draw a diagram of the nephron, labeling the following: glomerulus, Bowman's capsule, proximal convoluted tubule, descending limb of the loop of Henle, ascending limb of the loop of Henle, distal convoluted tubule, collecting tubule. (p. 369)

12 Discuss in some detail the process of urine formation in man, explaining

   a the source of urea, and when the urea content of the blood will be highest.

   b what happens in the glomeruli, naming the force that powers this filtration process, and indicating what substances remain in the blood and are not filtered into the glomerulus.

   c the process by which glucose, amino acids, and some salts are reabsorbed.

   d how water is reabsorbed.

   e the roles of filtration, reabsorption, and tubular excretion; and where and how each of these processes occurs.

   (pp. 369–75)

13 Explain how the countercurrent-multiplier system of the loop of Henle acts to concentrate the urine. (pp. 372–73)

14 Summarize the role played by your kidney in maintaining homeostasis. (pp. 369–77)

15 Describe a model for the sodium-potassium pump and explain how such a pump might work in the cells of the gills of a freshwater fish. (pp. 375–77)

KEY TERMS

The following terms are important in this chapter; you should become familiar with them.

extracellular fluid (p. 353)
intracellular fluid (p. 353)
plasmolysis (p. 354)
portal system (p. 355)

portal vein (p. 355)
hepatic vein (p. 356)
deamination (p. 357)
ammonia ⎱
urea ⎰ nitrogenous-waste compounds
uric acid ⎰ (p. 357)
excretion (p. 358)
elimination (p. 358)
osmoregulation (p. 359)
contractile vacuole (p. 362)
flame cell system (p. 364)
nephridia (p. 365)
Malpighian tubules (p. 366)
Bowman's capsule ⎫
glomerulus ⎪
proximal convoluted tubule ⎬ parts of
distal convoluted tubule ⎪ the nephron
loop of Henle ⎪ (pp. 367–68, 371)
collecting tubule ⎭
ureter (p. 367)
cloaca (p. 367)
urethra (p. 367)
urinary bladder (p. 367)
renal artery (p. 368)
glomerulus (p. 368)
renal vein (p. 369)
kidney threshold (p. 374)
tubular excretion (p. 374)

SUMMARY

Living cells require a relatively constant environment to carry out their life functions. As complex multicellular animals arose, body fluids developed that could provide a stable environment for the internal cells. A variety of mechanisms evolved for maintaining the *homeostasis* of the body fluids despite changes in the external environment.

The extracellular fluids of plants are not separate from the environmental water. Consequently plants regulate only the composition of their intracellular fluids whereas animals must regulate the composition of both intracellular and extracellular fluids. Plant cells can withstand much greater fluctuations in their extracellular fluids than animals. Because the rigid cell wall resists expansion, the plant cell can withstand pronounced changes in the osmotic concentration of the surrounding fluids as long as the fluids remain hypotonic to the cell. If the external fluids become decidedly hypertonic to the cell, the cell may lose water and shrink away from the wall (*plasmolysis*).

The vertebrate liver The liver is important in maintaining a relatively constant environment for the body cells. All the blood from the intestine and stomach is collected in the *portal vein* and conducted to the liver, where it flows into a second set of capillaries before it is collected into the *hepatic veins* and emptied into the posterior vena cava. Because of this portal system, the products of digestion are brought directly to the liver cells, where the levels of these digestive products can be regulated.

When the incoming blood is high in glucose, the liver removes the excess and stores it as glycogen or fat; when the incoming blood is low in glucose, the liver reconverts glycogen into glucose to maintain the blood-sugar level. The liver is the center for fat metabolism; fatty acids and other lipid materials are processed here and some plasma lipids are synthesized.

The liver also removes many of the amino acids from the blood, temporarily storing small quantities. Excess amino acids are *deaminated* and the remainder of the molecule is converted into carbohydrate or fat. In the deamination process the amino group ($-NH_2$) is converted into *ammonia* (NH_3). The ammonia may be released directly as a waste product or it may be converted into the less toxic compounds *urea* and *uric acid*. These are released into the blood and must be removed from the body.

In addition to its metabolic functions, the liver also detoxifies many injurious chemicals, manufactures many plasma proteins, stores certain substances, destroys red blood cells, synthesizes bile salts, and excretes bile pigments.

Animals need mechanisms to rid their bodies of metabolic wastes (*excretion*) and to regulate their salt and water balance.

The nitrogenous excretory product characteristic of most aquatic animals is ammonia. Ammonia is poisonous but is easily released from the body if the water supply is plentiful. Many marine invertebrates simply release their wastes across their membrane surfaces. Such organisms are isotonic with their environment and thus have no problems with water balance as long as they remain in the sea.

Freshwater animals have an osmotic concentration that is less than that of seawater but greater than that of freshwater. A freshwater organism tends to gain water and lose salts to the surrounding water. To survive, the organism must have an osmoregulatory structure to pump out the excess water and some mechanism to absorb salts. Freshwater bony fishes compensate by drinking very little, by actively absorbing salts through specialized gill cells, and by excreting copious dilute urine.

Marine bony fishes are hypotonic relative to seawater. They tend to lose water and take in salt. They compensate by drinking constantly, by actively excreting salts across their gills, and by producing little urine.

The marine elasmobranchs (sharks, etc.) compensate in a different way—by maintaining a high concentration of urea in their blood so that their total osmotic concentration is slightly greater than that of seawater.

The greatest problem for a land animal is desiccation. The water lost by evaporation, elimination, and excretion must be replaced. Amphibians and mammals produce urea as their nitrogenous waste product; it must be released from the body in solution, thus draining away needed water. Most reptiles, birds, insects, and land snails excrete uric acid. Little water is lost in the excretion of this highly insoluble compound. The uric acid produced by a developing embryo is precipitated and stored in the eggs of these animals; in this form it is not harmful. Uric acid excretion is correlated with giving birth to living young.

Excretory mechanisms in animals Many unicellular and simple multicellular animals have no special excretory structures; the nitrogenous wastes are excreted across membrane surfaces. Some Protozoa do have a *contractile vacuole*, which collects fluid and expels it from the cell. Its primary function is the elimination of excess water. Flatworms have a simple, tubular *flame-cell system*, to eliminate excess water.

In animals that have evolved closed circulatory systems, the blood vessels have become intimately associated with the excretory organs, making possible direct exchange of materials between the blood and the excretory system. In earthworms, each segment of the body has a pair of excretory organs called *nephridia*, which open to the outside. Tissue fluids move into the open end of this tubular structure and pass through the coiled tubule to the exterior. Blood vessels associated with the tubules also secrete substances into the tubules for excretion.

The excretory organs of insects are called *Malpighian tubules*. They are diverticula of the digestive tract located at the junction between the midgut and hindgut. The tubules are bathed directly by the blood; fluid from the blood moves into the tubules, where the uric acid is precipitated. The concentrated urine moves into the hindgut and rectum, where it is exposed to a powerful water-absorptive action; the urine and feces leave the rectum as dry material.

The vertebrate kidney The vertebrate excretory organ is the kidney, which is composed of *nephrons*. Three processes are important in urine formation: filtration, reabsorption, and tubular excretion. High blood pressure within the *glomerulus* forces small molecules from the blood into the *Bowman's capsule* (filtration). As the filtrate moves through the tubules of the nephron most of the water and substances needed by the body are reabsorbed into a second set of capillaries associated with the tubules. In addition, some chemicals are actively removed from the blood by the tubules and deposited in the filtrate (tubular excretion). The nephrons empty into *collecting tubules*; from here the urine is conducted into the kidney *pelvis*, then down the *ureter* to the urinary bladder for storage, and finally out the *urethra*.

The portion of the nephron called the *loop of Henle* makes it possible to produce concentrated urine; it establishes a salt ion gradient in the tissue fluid of the kidney medulla that favors the osmosis of water out of the collecting tubules into the surrounding tissue fluid.

Most substances reabsorbed by the kidney have a *threshold level*; if the concentration of such a substance exceeds the threshold, the excess is not reabsorbed by the tubules but instead appears in the urine. Thus, the kidneys play a vital role in maintaining homeostasis; they help to regulate the blood-sugar level and the concentration of various inorganic ions in the blood.

The kidneys of vertebrates vary in their ability to produce concentrated urine. Marine fishes, turtles, and birds cannot produce concentrated urine; they excrete excess salt by special cells or glands. Seals and fish-eating whales have kidneys capable of excreting a urine with high urea concentration. Kangaroo rats have extraordinarily efficient kidneys capable of producing extremely concentrated urine. The human kidney is incapable of producing urine with a very high concentration of either salt or urea.

The processes of excretion and osmoregulation depend on active transport. Specific pumps in cell membranes use energy to move substances across the membranes against osmotic and/or electrical gradients. The sodium-potassium pump is a transmembrane protein that changes conformation when phosphorylated by ATP; it moves K^+ from outside the cell into the cell, while Na^+ from inside the cell is moved outside. Many epithelial cells are involved in active transport in osmoregulatory and excretory organs; these cells probably utilize the same basic type of pump.

KEY DIAGRAM

In the following diagram of a human kidney, color in the following, using a different color for each structure. First color the title, then the appropriate structure. (Reference: Figure 14.16, p. 368)

CORTEX

MEDULLA

RENAL PELVIS

URETER

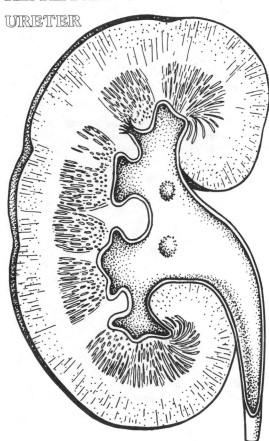

Next, draw in a single nephron in its proper position in the kidney.

QUESTIONS

Testing recall

Complete the following statements with the words greater than, less than, or equal to.

1 If you were to measure the extracellular fluid of a marine alga living in seawater you would find that its osmotic concentration is _____ that of seawater. The osmotic concentration of intracellular fluid of the alga would be _____ that of the extracellular fluid. (pp. 352–53)

2 The osmotic concentration of the extracellular fluid of a marine jellyfish living in the ocean is _____ that of seawater. The osmotic concentration of its intracellular fluid is _____ that of its extracellular fluid. (pp. 352, 359)

3 If you were to measure the osmotic concentration of the extracellular fluids of a marine bony fish you would find it to be _____ that of seawater, and the osmotic concentration of the interior of the fish's body cells would be _____ that of the blood. (p. 361)

4 The osmotic concentration of the extracellular fluid in the cortex of a root of a land plant would be _____ that of soil water and the osmotic concentration of the cortex cells would be _____ that of soil water. (p. 354)

For questions 5–11 use the following table, which lists the appropriate osmotic concentrations of seawater and freshwater, and of the body fluids of various animal groups living in these environments. (pp. 358–62)

	percent
Seawater	3.5
a marine sharks	3.6
b marine invertebrates	3.5
c marine bony fishes	1.5
Freshwater	0.01–0.5
d freshwater invertebrates	0.4 –0.6
e freshwater bony fishes	0.85
f amphibians	0.85
g mammals	0.9
h reptiles	0.9

5 Which organisms will tend to take in excessive water from their environment?

6 Which organisms are essentially isotonic with their environment?

7 Which organisms will tend to lose too much body water to their environment?

8 Which organisms tend to take in too much salt from the environment?

9 Which organisms would have special mechanisms for actively taking up salts from the environment and releasing them into their body fluids?

10 Which organisms probably use ammonia as their main nitrogenous waste product?

11 Which organisms use mainly urea as their nitrogenous waste product?

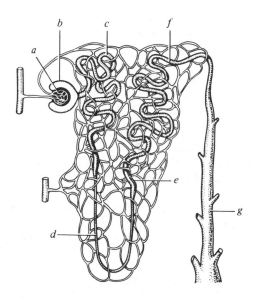

12 Label the parts (*a–g*) in the sketch of the nephron. (p. 369)

13 Referring to the sketch, indicate by letter in which part of the nephron each process occurs.

   a filtration of the blood (p. 369)
   b passive transport of Na^+ and Cl^- (p. 372)
   c reabsorption of most water (p. 372)
   d pumping of Na^+ out of the tubule to establish a concentration gradient by countercurrent multiplication (p. 372)
   e ·conducting of urine to pelvis of the kidney (p. 373)

14 Which parts of the nephron are found in the cortex of the kidney? (p. 368)

Testing knowledge and understanding

Choose the one best answer.

15 The term "homeostasis" is best defined as

a the maintenance of a constant internal environment.
b the maintenance of constant salt concentration in the body.
c adaptation to a harsh environment.
d maintenance of constant body temperature. (pp. 340, 353)

16 A plant cell placed in a hypotonic environment will

a lose water and become flaccid.
b actively transport salts out of the cell.
c take up water and burst.
d take up water and become turgid.
e become impermeable to prevent water loss. (p. 354)

17 An erythrocyte is in an artery leading to the duodenum of a human being. How many capillary beds will it probably pass through before it reaches the right ventricle of the heart?

a one   d four
b two   e five
c three (p. 355)

18 In a person who has recently eaten a large meal, which one of the following blood vessels would you expect to contain blood with the highest concentration of sugar?

a aorta
b posterior vena cava
c renal artery
d hepatic portal vein
e jugular vein (p. 356)

19 In early 1973, Senator John Stennis was shot and severely wounded as he was getting out of his automobile on a Washington street. A later account of the incident published in the *Washington Post* included the following passage:

The most serious wound, which at the outset many thought would cost him his life, was just at the belt-line on the left side. It affected his pancreas, colon, and portal vein, which supplies blood to the stomach. The vein was almost cut in two.

There is a biological error in this passage, which is that

a the portal vein does not supply blood to the stomach.
b the pancreas is not on the left side of the body.

c the colon is nowhere near the indicated wound site.
d a single bullet could not have hit both the pancreas and the colon.
e the listed wounds would not have been sufficiently serious to endanger his life. (p. 355)

20 In the human being, blood leaving the liver in the hepatic vein ordinarily has a higher concentration of _____ than other blood.

a bile   d erythrocytes
b urea   e leukocytes
c oxygen (p. 357)

21 All of the following are functions of the human liver *except*

a removing the excess glucose from the blood and storing it as glycogen.
b inactivating certain drugs and injurious chemicals.
c deaminating amino acids and converting the ammonia thus formed into urea.
d manufacturing many plasma proteins and some plasma lipids such as cholesterol.
e secreting digestive enzymes into the small intestine. (p. 358)

22 The vertebrate liver carries out all of the following functions *except*

a regulation of blood-sugar levels.
b secretion of bile.
c metabolism of amino acids.
d production of cholesterol.
e removal of urea from the blood. (p. 358)

23 Which of the following nitrogen-containing compounds is most toxic to animals, and can be tolerated only in dilute solutions?

a nitrate ion   d ammonia
b urea   e uric acid
c amino acid (p. 357)

24 The nitrogen present in proteins can be eliminated in the form of ammonia, which is highly toxic and must be removed from the body rapidly. In which of the following habitats would you be most likely to encounter organisms using ammonia as their nitrogenous waste product?

a underground   d in the air
b in water   e in deserts
c on land (p. 359)

25 A biologist experimenting with a protozoan noticed that the contractile vacuole stopped contracting although the other parts of the organism seemed healthy and active. Which of the following experiments was the one most likely to have produced this result?

a   transferring the protozoan from a lighted environment to a dark environment

b   transferring the protozoan from fresh-water to seawater

c   transferring the protozoan from seawater to freshwater

d   changing the pH of the medium from 7.0 to 6.5

e   cooling the medium from 20°C to 15°C
(p. 363)

26   A trout in a freshwater stream

a   loses water and salts by diffusion, and compensates by drinking large quantities of water and actively absorbing salts via the gills.

b   takes in much water and loses salts by diffusion, and compensates by excreting copious dilute urine and actively absorbing salts via the gills.

c   loses water and takes in large quantities of salts by diffusion, and compensates by actively absorbing water via the gills and by excreting salty urine.

d   takes in much water and salt by diffusion, and compensates by excreting copious urine and actively excreting salts via the gills.

e   takes in much water and loses salt by diffusion, and compensates by nearly constantly drinking and by actively excreting both water and salts via the gills.   (pp. 360–61)

27   A freshwater fish is placed in a marine environment. What would you expect to happen to this fish eventually?

a   It will dehydrate.

b   It will take up water.

c   It will neither gain nor lose water.

d   It will be in osmotic balance with the seawater.

e   It will actively excrete salt.   (pp. 360–61)

28   A marine bony fish living in the ocean must

a   excrete copious amounts of dilute urine.

b   actively transport salts into the body.

c   drink all the time.

d   produce a hypotonic urine.

e   actively transport water out through the gills.   (p. 361)

29   Dehydration is *not* a problem for which one of the following animals?

a   saltwater fish   d   human
b   freshwater fish   e   desert iguana
c   camel   (p. 360)

30   Which one of the following excretory mechanisms do many of the flatworms, such as planaria, employ?

a   flame cells

b   nephridia

c   loop of Henle

d   bladder

e   contractile vacuole   (p. 364)

31   You are given an unknown animal to study. It is bilaterally symmetrical and has a complete digestive system. It has a circulatory system but the blood carries very little oxygen. The animal's principal nitrogenous excretory waste product is uric acid. Which one of the following animals best fits this description?

a   salmon   d   planarian
b   earthworm   e   grasshopper
c   sparrow   (pp. 319, 357, 366–67)

32   The excretory organs of insects are called

a   Malpighian tubules.

b   nephridia.

c   flame cells.

d   Bowman's capsules.

e   glomeruli.   (p. 366)

33   All of the following are functions of the mammalian kidney *except*

a   removal of excess sodium ions from the blood.

b   reabsorption of salts and sugar from blood filtrate.

c   filtration of the blood.

d   synthesis of urea.

e   the concentration of urine.   (pp. 369–75)

34   Hydrostatic pressure powers the process of

a   filtration across the glomerulus.

b   reabsorption of water and dissolved substances at the venule end of the capillary.

c   sodium gradient maintenance in the kidney tubule.

d   salt and glucose reabsorption in the kidney tubules.

e Na^+ recycling in the loop of Henle.
(p. 370)

35   The liquid collected by Bowman's capsules in a human kidney may best be characterized as

a   concentrated urine.

b   dilute blood.

c   blood plasma minus blood proteins.

d   a solution of urea.

e   blood minus formed elements.   (p. 370)

36   The *second* of the two capillary beds in the kidney

a   reabsorbs nutrients.

b   is called the glomerulus.

c   filters the blood.

d   goes to the liver.

e   carries less concentrated blood than does the first bed.   (p. 373)

37 A person with kidney failure who is undergoing blood dialysis would probably be restricted to a diet low in

    a protein.   *d* calories.
    b carbohydrate.   *e* bulk.
    c fat.   (p. 374)

38 The kidneys of vertebrates function in all of the following ways but one. This exception is

    a excretion of metabolic wastes.
    b regulation of water concentration.
    c regulation of ion concentration.
    d elimination of undigested wastes.
    e elimination of materials in the blood that are in oversupply.   (p. 374)

39 Large molecules such as uric acid, creatinine, and penicillin require ATP energy to be excreted by the kidney. These substances enter the nephron by the process of

    a filtration.   *d* diffusion.
    b osmosis.   *e* tubular secretion.
    c reabsorption.   (p. 374)

40 The production of urine by the cells of the mammalian kidney requires energy, which is used mainly to

    a actively reabsorb solutes from the urine.
    b actively reabsorb water from the loop of Henle.
    c filter the blood into Bowman's capsule.
    d actively "pump" water into the urine.
    e actively "pump" wastes into the urine.   (pp. 373–74)

41 Based on your knowledge of the anatomy and physiology of the kidney, in which of the following blood vessels would you expect to find the highest concentration of plasma proteins?

    a the renal artery
    b the arteriole entering Bowman's capsule
    c the vessel leaving Bowman's capsule
    d the venule leaving a nephron
    e the renal vein   (p. 370)

42 Assume that you have dissected kidneys from the following mammals and measured the lengths of the loops of Henle and the collecting ducts in each. In which animal would you expect to find these structures to be proportionately longest in comparison to the overall size of the animal?

    a a desert rodent
    b a marine bird
    c a marine seal
    d an elephant
    e a forest rabbit   (p. 374)

43 In order for humans to drink saltwater, the hairpin loops of the kidney tubules would have to be

    a longer.
    b shorter.   (p. 374)

44 Which of the following processes that occur in the kidney directly requires active transport?

    a movement of Na^+ into the descending limb of the loop of Henle
    b movement of Na^+ out of the ascending limb of the loop of Henle
    c filtration into Bowman's capsule
    d osmosis of water in the collecting tubule
    e movement of urea into Bowman's capsule   (pp. 370–71)

45 According to the countercurrent-multiplier model for water removal from the kidney tubules,

    a Na^+ is being pumped out of the ascending loop of Henle.
    b Na^+ is being pumped out of the descending loop of Henle.
    c water is actively transported out of the collecting tubules.
    d the filtrate becomes hypertonic to the tissue fluid.
    e water is actively reabsorbed in the convoluted tubules.   (p. 372)

46 According to the current model of the sodium-potassium pump,

    a Na^+ ions are pumped out of the cell and K^+ ions are pumped in.
    b hydrolysis of ATP is required.
    c phosphorylation of the permease induces a conformational change in the permease.
    d all the above are correct.   (pp. 375–77)

For further thought

1 Lake Baikal in the Soviet Union is the only known habitat of freshwater seals; all other seal species inhabit marine environments. What changes are most likely to have occurred in the kidney functions of these freshwater seals? (pp. 360–61, 367–75)

2 A patient was admitted to a hospital with a preliminary diagnosis of viral hepatitis (inflammation of the liver). The following tests were performed:

    a Fasting blood sugar. The blood sugar level was measured six hours after an injection of glucose. Result: blood sugar was 58 mg/100 ml. (Normal: 70–120 mg/100 ml.)

b   Serum albumin test. The level of albumin in the blood was determined. Result: serum albumin was lower than normal.

c   Prothrombin time. The time it takes for blood to clot was measured. Result: clotting time was higher than normal.

d   Bromsulfalein test. This dye was injected into the blood and the amount of dye retained was measured. Result: 50 percent of the dye was retained in 30 minutes. (Normal: 10 percent or less.) (This dye is not excreted by the kidneys.)

On the basis of what you know about liver function, explain the above results. What symptoms do you think the patient might present? (pp. 356–58)

3   A molecule of water is absorbed into a capillary in the small intestine of a human being. Trace the route this molecule will follow as it goes by the shortest possible route to the urinary opening. (Name all the blood vessels and parts of the kidney, etc.) (pp. 321, 369)

4   Would the urine composition of a vegetarian differ from that of a heavy meat eater? (p. 359)

5   Approximately 180 liters of fluid are filtered through the kidney every 24 hours, but only about 1.5 liters of urine are produced. What is the average rate of filtration in ml/minute? (1 liter = 1000 ml.) What will happen to the rate of filtration if the blood pressure is increased? What is the rate of urine production in ml/minute? If the two kidneys together have two million glomeruli, how many ml of fluid are filtered by each glomerulus in 24 hours?

a   9.0 ml   d   8.9 ml
b   0.9 ml   e   0.89 ml
c   0.09 ml   f   0.089 ml
                          (pp. 369–73)

6   What do kidneys and nephridia have in common? Why must insects have an excretory system that operates on an entirely different principle? (pp. 366–74)

ANSWERS

Testing recall

1   equal to, equal to
2   equal to, equal to
3   less than, equal to
4   equal to, greater than
5   freshwater invertebrates (d), freshwater bony fishes (e), amphibians (f)
6   marine sharks (a), marine invertebrates (b)
7   marine bony fishes (c), mammals (g), reptiles (h)
8   marine sharks (a), marine bony fishes (c)
9   freshwater invertebrates (d), freshwater bony fishes (e), amphibians (f)
10 sharks, fishes, invertebrates (a-e)
11 amphibians (f), mammals (g)
12 A glomerulus
    B Bowman's capsule
    C proximal convoluted tubules
    D loop of Henle
    E loop of Henle
    F distal convoluted tubules
    G collecting tubule
13 a A   d E
    b D   e G
    c C
14 A, B, C, F

Testing knowledge and understanding

15	a	23	d	31	e	39	e
16	d	24	b	32	a	40	a
17	b	25	b	33	d	41	c
18	d	26	b	34	a	42	a
19	a	27	a	35	c	43	a
20	b	28	c	36	a	44	b
21	e	29	b	37	a	45	a
22	e	30	a	38	d	46	d

CHEMICAL CONTROL IN PLANTS

A GENERAL GUIDE TO THE READING

This chapter is the first of three on chemical control mechanisms in multicellular organisms. The focus of this chapter is on plants; that of the next two will be on animals. You will learn that hormonal control in plants differs from hormonal control in animals in a variety of ways; the fundamental one—which you should try to keep in mind—is that plant hormones are involved primarily in nearly all aspects of growth and development, whereas animal hormones are primarily involved in maintaining homeostasis. As you read Chapter 15 in your text, you will want to concentrate on the following topics.

1   Auxins. Considerable emphasis is placed on auxins; the other plant hormones are not discussed in as much detail. Auxins have been emphasized for two reasons. First, they were the earliest investigated of the groups of plant hormones, and are the best known. Second, they can serve as a model for hormonal activity in plants; by studying the activity of

one group of hormones you can learn the fundamental principles of hormonal control in plants. For these reasons you will want to concentrate your study on auxins (pp. 380–89).

2   Chemical control. A summary of chemical control in plants is on page 394; you will find this summary extremely helpful.

3   Photoperiodism. This concept is introduced in the section "Control of Flowering" (pp. 395–402). Notice that there is much about the flowering process yet to be learned.

KEY CONCEPTS

1   Chemical control mechanisms play an important role in the coordination of the myriad functions of living organisms. Hormonal control is common to plants and animals. (pp. 378–79)

2   Plant hormones are not highly specific in their action but rather participate in some fashion in nearly all aspects of growth and development. (pp. 379–80)

3   A plant's response to a given hormone depends on the tissue, the concentration of the hormone, and the concentration of other plant hormones that may be present. (pp. 380–94)

4   Cell division is stimulated by cytokinins and auxins, and by other factors that enhance their activity. Cell division is inhibited by a variety of substances. The balance between the cytokinins and the inhibitors determines whether a cell will divide or not. (pp. 388, 391–93)

5   Control of cell enlargement involves substances such as auxins and gibberellins, which promote elongation. (pp. 384, 390–91)

6   The various plant hormones, by their mutual interactions and other differential effects on various parts of the plant body, help integrate and coordinate the development of form and function. (p. 394)

7   The aging process, leading to death, is brought on by various senescence-inducing substances, of which ethylene is one of the most important. (pp. 393–94)

8   In many plants, flowering is affected by photoperiodism, the response to the duration and timing of light and darkness. (pp. 395–97)

OBJECTIVES

After studying this chapter and reflecting on it, you should be able to carry out the following objectives.

1   Discuss the role of auxin in phototropism, describing the effect of light on the distribution of auxin and how auxin distribution is related to phototropism. (pp. 380–84)

2   Describe how each of the following scientists contributed to our understanding of phototropism: Charles and Francis Darwin, P. Boysen-Jensen, A. Paal, and F. Went. (pp. 380–84)

3   Discuss the role of auxin in geotropism in stems, explaining the mechanism by which plants sense gravity, the effect of gravity on auxin distribution, and how auxin distribution is related to geotropism. (pp. 384–85)

4   Discuss the role of auxin in apical dominance, using a drawing such as Figure 15.11 (p. 386) to explain this phenomenon. Indicate why pinching off the growing tip of some plants causes the plants to become more bushy. (pp. 385–86)

5   Name a major effect that each of the following has on plant growth and development: gibberellins, cytokinins, ethylene, abscisic acid. (pp. 390–94)

6   Contrast the effect of auxin and gibberellins on cell elongation in the plant stem. (pp. 384, 390–91)

7   Explain how auxin and cytokinins influence cell division and growth. (pp. 388, 391–92)

8   Discuss the roles of the hormones auxin and ethylene in the development and ripening of fruit. (pp. 386–87, 393–94)

9   Discuss the role of the plant growth inhibitors in maintaining dormancy and preparing the plant for overwintering. Indicate how dormancy is broken. (pp. 392–93)

10 Describe the role that abscisic acid plays in geotropism in roots, indicating how the root cells sense gravity, the effect of gravity on abscisic acid distribution, and how this is related to geotropism. (p. 393)

11 Describe the process of leaf abscission and how auxin, abscisic acid, and ethylene influence this process. (p. 388)

12 Differentiate among short-day plants, long-day plants, and day-neutral plants, specifying which will flower when nights are shorter than a certain critical length and which will flower when nights are longer. Describe the effect upon the flowering of a short-day plant of a flash of light in the middle of the night. (pp. 395–97)

13 Discuss the relationship among florigen, phytochrome, and photoperiodism in the control of flowering. (pp. 397–402)

KEY TERMS

The following terms are important in this chapter; you should become familiar with them.

hormone (p. 379)
tropism (p. 380)
phototropism (p. 380)
coleoptile (p. 381)
auxin (p. 382)
indoleacetic acid (p. 383)

SUMMARY

Three steps are involved in the flow of information in organisms; reception of the stimulus, communication of the information to the site of response, and the response itself. Plants pass information by means of chemicals; control chemicals called *hormones* move from one part of the plant to the target cells where a response is elicited. Plant hormones are produced most abundantly in the actively growing parts of the plant body: the apical meristems, young growing leaves, and developing seeds. The known plant hormones are primarily involved in regulating growth and development.

Auxins are a class of plant hormones that produce a variety of effects, the most important of which is control of cell elongation. Experiments show that auxin is produced by the apical meristem and moves downward, promoting cell elongation in the stem.

Light influences the distribution of auxin within the stem by causing migration of the auxin away from the lighted side. Consequently the illuminated side of the plant grows more slowly than the shaded side and the plant turns towards the light—a phenomenon called *phototropism*. Gravity also causes an unequal distribution of auxin within the stem. The plant shoot turns away from the pull of gravity, showing *negative geotropism*.

Auxin produced in the terminal bud moves downward in the shoot and inhibits the development of the lateral buds (*apical dominance*). Auxin produced by the pollen grain and developing seed stimulates the development of the floral ovary into the fruit. Auxin also acts to prevent the abscission (dropping) of flowers, fruits, and leaves by inhibiting formation of the *abscission layer*. In early spring it stimulates renewed cell division in the cambium. It also initiates the formation of lateral roots from the root pericycle and promotes the development of adventitious roots from cuttings. Auxin-like chemicals such as 2,4-D and 2,4,5-T are used as broad-leaved weed killers.

The most dramatic effect of another class of plant hormones, the *gibberellins*, is stimulation of rapid stem elongation in dwarf plants. Because gibberellins can move freely throughout the plant, most of its effects on the pattern of growth are different from those of auxin. Gibberellins are also active in breaking dormancy, inducing the formation of a starch-hydrolyzing enzyme in germinating seeds, stimulating flowering in some biennials, inducing flowering in some long-day plants, and stimulating fruit set.

Cytokinins are a class of hormones that promote cell division. In the normal growing plant, cytokinins and auxins may act synergistically in some situations and antagonistically in others. Cytokinins also play a role in chloroplast development, breaking dormancy in some seeds, enhancing flowering, promoting fruit development, and delaying the aging of leaves.

Auxins, gibberellins, and cytokinins are growth-promoting hormones, but plants also produce some growth inhibitors, the most important of which is the hormone *abscisic acid*. Inhibitors are important in maintaining dormancy of buds and seeds when conditions for growth are unfavorable, as in winter or very dry periods. Dormancy is broken when the inhibitors have been degraded, destroyed, or leached away by water. Abscisic acid also is involved in promoting leaf abscission and flowering in certain species, inducing stomatal closing, and preparing the plant for overwintering. Another important function is controlling the positive geotropism of roots. Gravity causes the amyloplasts to sink to the bottom of cells, where they interact with the membrane and cause an increase of abscisic acid on the lower side of the root, inhibiting growth on that side.

The hormone *ethylene* is a gas that induces the onset of the climacteric in fruit, causing ripening. It also contributes to leaf abscission and various other changes involved in the aging process.

Control of flowering The flowering response in many plants is initiated by *photoperiodism*, the length of light and dark periods. Most plants can be classified as: (1) short-day plants, which flower only when the night is *longer* than a critical value; (2) long-day plants, which flower only when the night is *shorter* than a critical value; and (3) day-neutral plants, which are independent of day and night length. The critical night length is a *maximum* value for a long-day plant and a *minimum* value for flowering in a short-day plant.

It is believed that an inducing photoperiod causes the leaves to produce the hormone *florigen*, which moves through the phloem to the buds and stimulates flowering. A noninducing photoperiod causes the leaves of many plants to inhibit florigen in some way. All attempts to isolate florigen have failed and many biochemists suggest that flowering is controlled by a balance of different hormones—that there is no florigen.

Light is detected by a receptor pigment in the plasma membrane called *phytochrome*, which exists in two forms, one that absorbs red light (P_r) and one that absorbs far-red light (P_{fr}). The red and far-red light interconverts the phytochrome between its two forms. P_{fr} is the less stable form in the dark; it reverts to P_r over time, and some of it is enzymatically destroyed.

When phytochrome is exposed to both red and far-red light simultaneously (as in sunlight) the red light dominates and most of the pigment is converted into P_{fr}. During the night the P_{fr} supply dwindles as a result of reversion and destruction. Thus the plant has a way to detect whether it is day or night.

The mechanism that enables the plant to measure the length of the dark period is apparently tied to an "internal clock." The phytochrome mechanism determines whether it is day or night while the internal clock measures the length of the dark period. According to the florigen hypothesis, once these have indicated to the plant that the photoperiod is appropriate, florigen is produced and flowering is initiated.

Phytochrome participates in many other light-induced functions such as the germination of seeds, gibberellin-controlled stem elongation, expansion of new leaves, breaking of dormancy, the formation of plastids, and detecting the extent of shading.

QUESTIONS

Testing recall

Match each function below with the associated plant hormone. Answers may be used once, more than once, or not at all.

a	abscisic acid	*d*	ethylene
b	auxin	*e*	florigen
c	cytokinin	*f*	gibberellin

1 induces ripening of fruit and aging of the plant (pp. 393–94)

2 stimulates cell division (p. 392)

3 can cause rapid elongation when applied to intact stems of dwarf plants (p. 390)

4 inhibits growth of lateral buds (pp. 385–86)

5 inhibits seed germination (p. 393)

6 stimulates flowering when the photoperiod is appropriate (p. 397)

7 controls geotropism in roots (p. 393)

8 induces bending of coleoptiles (p. 381)

9 participates in phototropism and geotropism in stems (pp. 380–85)

10 induces changes that prepare the plant for overwintering (p. 393)

Testing knowledge and understanding

Choose the one best answer.

11 Plant hormones, in contrast to most animal hormones,

 a always stimulate and never inhibit various processes.
 b are required in large amounts.
 c are not produced in specialized glands.
 d do not coordinate activities of the organism.
 e are pigmented green. (pp. 378–79)

12 A response to changes in the length of daylight is known as

 a photoperiodism. *d* phototaxis.
 b phototropism. *e* photolysis.
 c photosynthesis. (p. 395)

13 In the experiment shown in the following drawing, a transparent cap is placed on a grass shoot from which the apical meristem has been removed; also, an opaque cylinder is placed around the base. If the sun is positioned as shown, what result would you expect?

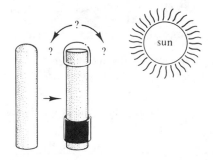

a slight movement of shoot toward sun
b pronounced movement of shoot toward sun
c slight movement of shoot away from sun
d pronounced movement of shoot away from sun
e no movement of shoot toward or away from sun (p. 381)

14 A horizontal root turning downward under natural conditions is demonstrating

a ethylene stimulation.
b negative geotropism.
c positive phototaxis.
d positive chemotropism.
e abscisic acid inhibition. (p. 393)

15 The hormone that stimulates the ripening of fruit is

a ethylene.   d abscisic acid.
b auxin.   e gibberellin.
c cytokinin.   (p. 393)

16 Phototropism

a cannot occur unless tissues are capable of differential growth.
b occurs because the concentration of auxin (indoleacetic acid) is higher on the side of the apex that is illuminated, thus inhibiting cell growth.
c occurs because the tolerance limits of the shoot apex to auxin concentrations are very low.
d results in the elongation of internodes due to an increase in cell division and elongation.
e none of the above (p. 395)

17 The pigment responsible for detecting the presence or absence of light is

a auxin.   d gibberellin.
b phytochrome.   e florigen.
c chlorophyll.   (p. 400)

18 An agar block, divided into X and Y halves by an impermeable partition, is inserted underneath the growing tip of an oat coleoptile. The coleoptile is then exposed for a day to light from the right, as shown in the drawing.
The agar block is next removed, and the X and Y halves are placed asymmetrically atop two decapitated coleoptiles and left for two days in the dark. Which one of the following diagrams best represents the expected condition of the decapitated coleoptiles after two days?

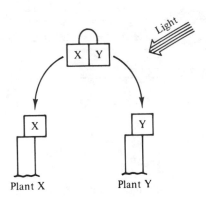

Plant X   Plant Y

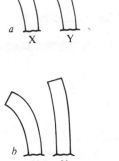

(pp. 381–83)

19 Suppose you took two plants, a long-day plant (LDP) with a critical night length of 12 hours and a short-day plant (SDP) also with a critical night length of 12 hours, and exposed the two to several "days," each consisting of a 14-hour dark period and a 10-hour light period. You would predict that

a the LDP would flower; the SDP would not.
b the SDP would flower; the LDP would not.
c neither plant would flower.
d both plants would flower.   (p. 395)

20 A certain short-day plant flowers only when the night is at least 14 hours long. Under which one of the following light regimes will this plant flower? (pp. 396–97)

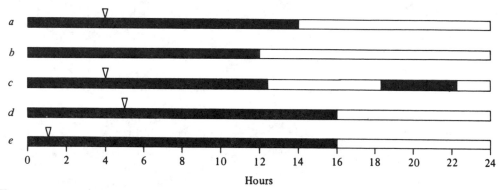

Key
   White bars: daylight
   Black bars: night
   White triangles: intense flashes of light

21 A certain short-day plant has a critical night length of 14 hours. Under which one of the following light regimes will this plant flower? (pp. 396–97)

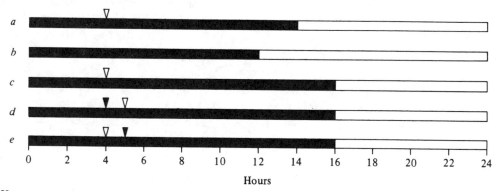

Key
   White bars: daylight
   Black bars: night
   White triangles: intense flashes of red light
   Black triangles: intense flashes of far-red light

For further thought

1 Explain the role of the hormones auxin, gibberellin, abscisic acid, and cytokinin in the germination of the seed and the growth of the young plant.

2 The poinsettia is a short-day plant with a critical night length of 12 hours. How could you get the plant to flower in June?

ANSWERS

Testing recall

1	d	5	a	8	b
2	c	6	e	9	b
3	f	7	a	10	a
4	b				

Testing knowledge and understanding

11	c	15	a	19	b
12	a	16	a	20	e
13	e	17	b	21	e
14	e	18	b		

CHEMICAL CONTROL IN ANIMALS

A GENERAL GUIDE TO THE READING

This chapter describes the major hormonal control mechanisms in animals, with particular emphasis on those in vertebrates. The chapter begins with a discussion on the differences between plant and animal hormones and goes on to consider hormones in invertebrate animals. The remainder of the chapter is devoted to a consideration of hormones in vertebrates, particularly humans. As you read Chapter 16 in your text you will find the presentations of the following material particularly helpful.

1 The major differences between plant and animal hormones. The paragraph summarizing these differences (p. 403) reviews and amplifies the corresponding discussion at the beginning of Chapter 15.

2 Hormonal interaction in insects. The process of molting in insects is discussed as an example of hormonal action in invertebrates. Note particularly the close relationship between the nervous system and the endocrine system.

3 The location of the major human endocrine organs. Figure 16.2 (p. 406) will help you to learn this material.

4 Hormonal control of digestion. You should recognize that digestion is a complicated process, precisely controlled through a series of neural and hormonal mechanisms.

5 The major functions of important mammalian hormones. Table 16.1 (pp. 408–9) is a wonderful summary of these functions; you will want to refer to it frequently.

6 The known actions of insulin. The summary of these actions (p. 411) is most useful. You will find the various effects of this hormone easier to learn if you keep in mind that the overall effect of insulin is to lower blood sugar.

7 Double glands. Notice that both the adrenals (pp. 412–16) and the pituitary (pp. 419–22) are double, each being composed of two different glands, anatomically and functionally distinct.

8 The control function of the hypothalamus. This topic is crucially important; you will want to study the section on it (pp. 420–22) very carefully, with special attention to Figure 16.14 (p. 422), which is key to understanding the material.

9 The mechanism of hormonal action. The section on this topic (pp. 424–31) is difficult but rewarding. Figure 16.16 (p. 425) will help your understanding of how hormones transmit messages. Pay particular attention to the material on the second-messengers.

10 Local chemical mediators. Notice that our definition of hormones is broad and includes a discussion of the effect of local chemical mediators such as histamine, prostaglandins, and endorphins.

11 The evolution of hormones. The last section in this chapter provides an interesting insight into the evolution of hormonal control.

KEY CONCEPTS

1 Chemical control mechanisms play an important role in the coordination of the myriad functions of living organisms. Hormonal control is common to plants and animals. (pp. 378–79, 403)

2 Animal hormones, notably those found in mammals, not only guide growth and development, but also play a large role in regulating metabolism and in maintaining general homeostasis. (pp. 408–9)

3 The endocrine tissues that produce animal hormones are often intimately involved with the nervous system. (pp. 406, 413, 420, 424)

4 The secretion of many animal hormones is regulated by a negative-feedback control mechanism; low concentrations of a particular hormone result in increased secretion of that hormone, and when the concentration reaches a certain level in the blood, the secretion is reduced. Feedback loops are an extremely important mechanism in living organisms for maintaining homeostasis. (pp. 422–23)

5 Many hormones, the peptide hormones, act on target cells indirectly, by interacting with specific receptors on the cell membrane, thus opening an ion channel or activating cytoplasmic enzyme systems in the cell. Other hormones, the steroid hormones, act more directly, by moving into the target cell and interacting with the genetic material or with the process of protein synthesis. (pp. 426–30)

6 Local chemical mediators are secreted by cells that are not part of specialized chemical-control organs, and in some cases are so rapidly destroyed that they affect only cells in their immediate neighborhood. (pp. 431–33)

7 The specialized hormones and receptor systems in higher plants and animals may have evolved by modifying or building on systems and structures found in more primitive, unspecialized organisms. (p. 433)

OBJECTIVES

After studying this chapter and reflecting on it, you should be able to carry out the following objectives.

1 Define the terms hormone and endocrine. Explain how hormones in higher animals are transported from the site of synthesis to the target tissue. (pp. 378–79, 403)

2 Explain how juvenile hormone, brain hormone, and molting hormone interact to control the molting process in some insects. (pp. 404–5)

3 List the major endocrine organs in the human body and indicate their location. (p. 406)

4 Explain how hormones act to coordinate the secretion of enzymes in the digestive process. (pp. 406–12)

5 Give the standard procedure for determining whether an organ has an endocrine function. (p. 410)

6 Give the location of the endocrine gland that secretes insulin; using a drawing such as Figure 16.3 (p. 410), point out and name the cells where insulin is produced; describe the function of the α islet cells; and indicate what other cells are present in the pancreas and what their function is. (pp. 407–12)

7 List four actions of insulin in reducing the concentration of glucose in the blood. Contrast its action with that of a second hormone from the pancreas, glucagon. Name one other hormone that acts as an insulin antagonist. (pp. 411–12, 414)

8 Give the location of the adrenal glands, and in a cross section like that in Figure 16.5 (p. 412), point out the cortex and the medulla. Indicate which portion secretes adrenalin,

and which portion secretes steroid hormones. (pp. 412–16)

9 Describe the effect of adrenalin on the blood pressure, heartbeat, blood-sugar level, oxygen consumption by cells, blood supply to skeletal muscle and heart muscle, and blood supply to the skin and the digestive tract. (p. 413)

10 Name the three categories of steroid hormones produced by the adrenal cortex, and give a major function of each. (p. 414)

11 Discuss the role of the thyroid hormones (T_3 and T_4, together known as TH) in the human body. In doing so, specify the location of the thyroid gland, the effect of TH on oxidative metabolism of the cells of the body, and the symptoms of excessive TH secretion (hyperthyroidism) and of insufficient TH secretion (hypothyroidism). (pp. 416–18)

12 Discuss the roles of calcitonin and parathyroid hormone (PTH) in the control of calcium and phosphate levels in the blood. In doing so, specify where each hormone is synthesized; describe the effect of PTH on the blood-calcium level, the effect of calcitonin on the blood-calcium level, and the effect of PTH on the blood-phosphate level; and name the three target organs of PTH. (pp. 418–19)

13 Describe the relationship between the hypothalamus and the posterior pituitary. Indicate where the hormones oxytocin and vasopressin are produced; from what organ they are released, and how they get from one place to another. Give one function of vasopressin and oxytocin. (pp. 419–20)

14 Give the location of the pineal gland, and contrast the function of the pineal in lower vertebrates with its function in mammals. (pp. 406, 423–24)

15 For each of the following hormones produced by the anterior pituitary, describe a major function and specify the target organ: prolactin (PRL), growth hormone (STH), thyrotropic hormone (TSH), adrenocorticotropic hormone (ACTH), gonadotropic hormone (FSH, LH). (pp. 420–22)

16 Explain the structural and functional relationships between the hypothalamus and the anterior pituitary. In doing so, be sure to describe the portal system between the hypothalamus and the anterior pituitary; specify the way releasing hormones produced in the hypothalamus affect the activity of the anterior pituitary; and using a diagram such as Figure 16.15 (p. 423), describe the basic features of a feedback loop involving the thyroid gland. (pp. 420–23)

17 Describe the second-messenger model of hormonal control, indicate the intracellular effects of the extracellular binding of the hormone, and show how this model explains why only small quantities of hormone are required for normal functioning and why only certain cells are influenced by a given hormone. (pp. 426–30)

18 Explain how steroid hormones enter their target cell and discuss their mode of action. (pp. 424–26)

19 Explain how local chemical mediators differ from hormones, and for each of the following, indicate where it is produced and its target cells, and give one function: histamine, prostaglandins, and endorphins. (pp. 431–33)

KEY TERMS

The following terms are important in this chapter; you should become familiar with them.

brain hormone (p. 404)
ecdysone (p. 404)
juvenile hormone (p. 404)
endocrine (p. 406)
gastrin (p. 407)
secretin (p. 407)
cholecystokinin (p. 407)
α islet cell ⎫
β islet cell ⎭ islets of Langerhans (p. 409)
insulin (p. 410)
glucagon (p. 412)
adrenal cortex (p. 412)
adrenal medulla (p. 412)
adrenalin (p. 412)
noradrenalin (p. 413)
glucocorticoids (p. 413)
mineralocorticoids (p. 414)
cortical sex hormones (p. 414)
thyroxin, or T_4 ⎫ thyroid hormone,
triiodothyronine, or T_3 ⎭ or TH (p. 417)
calcitonin (p. 418)
parathyroid hormone, or PTH (p. 418)
hypothalamus (p. 420)
posterior pituitary (p. 420)
oxytocin (p. 420)
vasopressin (p. 420)
anterior pituitary (p. 420)
prolactin, or PRL (p. 420)
growth hormone, or STH (p. 420)
thyrotropic hormone, or TSH (p. 421)

SUMMARY

The tissues that produce and release hormones in animals are termed *endocrine* tissues. The hormones are secreted directly into the blood, which then transports them to other parts of the body.

Hormonal control mechanisms have been found in a variety of invertebrates; those in insects have been most extensively studied. In invertebrates the nervous system and endocrine function are usually intimately associated. In insects the brain produces a *brain hormone* that stimulates glands in the prothorax to secrete *ecdysone*, which induces molting. Levels of *juvenile hormone*, which is secreted by glands closely associated with the brain, determine whether the larva molts into another larval stage or to a pupa or adult.

In mammals, hormones produced by the mucosal lining of the stomach and small intestine act to coordinate the digestive process. The Russian physiologist Pavlov showed that gastric secretion is partly controlled by hormones; partially digested food in the stomach stimulates the gastric mucosa to release the hormone *gastrin* into the blood, where it is carried to the gastric glands and stimulates the secretion of gastric juice. Likewise, the presence of food in the small intestine stimulates the intestinal lining to release the hormones *secretin* and *cholecystokinin*. Secretin stimulates the secretion of pancreatic juice by the pancreas; cholecystokinin the release of bile from the gall bladder.

The *islet cells* of the pancreas secrete *insulin*, which acts to reduce the blood-glucose concentration by stimulating glucose absorption in muscle and adipose cells, by promoting glucose oxidation and glycogen synthesis in liver and muscle cells, and by inhibiting glycogen hydrolysis. It also promotes the synthesis of fat and proteins while inhibiting their breakdown.

The pancreas secretes another hormone, *glucagon*, which has effects opposite to those of insulin; it causes an increase in the blood-glucose concentration.

The two *adrenal glands*, located above the kidneys, consist of an inner *medulla* and outer *cortex*, which remain functionally distinct.

The adrenal medulla secretes two hormones, *adrenalin* and *noradrenalin*, whose effects are similar. Both help to prepare the body for emergencies by stimulating reactions that increase the supply of glucose and oxygen to the skeletal and heart muscles ("fight-or-flight" response).

The adrenal cortices produce many different steroid hormones. The cortical hormones may be grouped into three functional categories: (1) those regulating carbohydrate and protein metabolism, the *glucocorticoids*; (2) those regulating salt and water balance, the *mineralocorticoids*; and (3) those that function as sex hormones. The cortical hormones may be involved in the body's reaction to stress.

The *thyroid gland* is located just below the larynx. It has a great affinity for iodine, which is used to synthesize two thyroid hormones (TH), *thyroxin* and *triiodothyronine*. These hormones stimulate the oxidative metabolism of most tissues in the body. Hyperthyroidism—excessive TH secretion—produces an increase in the metabolic rate with high body temperature, high blood pressure, profuse perspiration, irritability, and weight loss.

Hypothyroidism—decreased TH secretion—leads to the opposite symptoms. It can be caused by dietary iodine insufficiency or by malfunction of the thyroid itself. Continued iodine deficiency causes an enlargement of the gland known as goiter. Untreated hypothyroidism in newborn children is called cretinism; such children show retarded physical, sexual, and mental development. Cretins also show abnormal protein distribution; apparently TH also plays a role in regulating protein synthesis and distribution.

The thyroid also secretes the hormone *calcitonin*, which prevents the excessive rise of calcium ions in the blood.

The *parathyroids* are four small pealike organs located on the surface of the thyroid. The parathyroid hormone (PTH) regulates the calcium-phosphate balance between the blood and other tissues; it acts primarily on the kidneys, the intestine, and the bones.

The *posterior pituitary* is connected to a part of the brain, the hypothalamus, by a stalk. It stores and releases two hormones, *oxytocin* and *vasopressin*, which are produced in the *hypothalamus* and flow along nerves in the stalk to the posterior pituitary. The hormones are released upon nervous stimulation from the hypothalamus. Oxytocin stimulates the contraction of uterine muscles. Vasopressin has two effects: it causes constriction of the arterioles with a consequent rise in blood pressure, and it stimulates the kidney tubules to reabsorb more water.

The *anterior pituitary* produces many hormones with far-reaching effects. The hormone *prolactin* stimulates milk production by the mammary glands and also participates in reproduction, osmoregulation, growth, and metabolism of carbohydrates and fats. *Growth hormone* (STH or somatotropic hormone) promotes normal growth. A serious deficiency in a child results in stunted growth; an oversupply results in a giant. Growth hormone is a powerful inducer of protein anabolism and acts as an insulin antagonist.

The anterior pituitary also secretes a number of hormones that help control other endocrine organs. *Thyrotropic hormone* stimulates the thyroid, *adrenocorticotropic hormone* (ACTH) stimulates the adrenal cortex, and the two *gonadotropic hormones* (FSH and LH) act on the gonads. The interaction between the anterior pituitary and these glands is an example of *negative feedback*. For example, when the thyroxin level in the blood is low, the anterior pituitary releases thyrotropic hormone, which stimulates the thyroid to increase production of TH. The increased TH level then inhibits the secretion of more thyrotropic hormone by the pituitary.

The activity of the anterior pituitary is, in turn, regulated by the hypothalamus, which produces special peptide *releasing hormones*. These hormones are carried by a special blood portal system to the anterior pituitary, where they stimulate its secretory activity. The hypothalamus is the point at which information from the nervous system influences the endocrine system and is also one of the major sites of feedback from the endocrine system.

The *pineal*, a lobe in the forebrain, secretes a hormone called *melatonin*. In some lower vertebrates melatonin lightens the skin and is involved in the control of circadian rhythms. In mammals, the amount of light influences the pineal's secretion of melatonin in an inverse relationship: the more light, the less melatonin. Melatonin, in turn, influences the secretion of gonadotropic hormones.

Hormones enter target cells in various ways. Steroid hormones, being lipid-soluble, can move through the plasma membrane and enter the cytoplasm of the target cell. There the hormone binds to a receptor molecule, and the complex then enters the nucleus, where it influences synthesis of RNA by the genes. Other hormones can pass into the cell through special channels or by active transport. However, most hormones do not enter the cell at all, but instead bind to extracellular receptors. This binding causes an intracellular effect. The *second-messenger model* explains this mechanism more fully.

According to this model, the binding of hormone to a receptor on the membrane results in the intracellular activation of the enzyme *adenylate cyclase*, which is attached to the membrane receptor. The adenylate cyclase catalyzes conversion of ATP into cAMP, which in turn activates a protein *kinase*. Kinase, in its turn, activates a second enzyme, which activates a third, etc. This cascade of enzyme-catalyzed reactions makes possible a great *amplification* of the effects of the original hormone-binding effect. The presence or absence of receptors on the membrane determines whether a cell will respond to a given hormone. Moreover, a cell may respond in the same way to two or more different hormones if it has cyclase-activating receptors for each of those hormones. A variety of other chemicals also influence cells by their effects on cAMP levels. Some work by activating adenylate cyclase; others act by influencing the enzyme *phosphodiesterase*, which breaks down cAMP. Many hormones do not use the cAMP system. Insulin induces a rise in cGMP, a related substance with effects opposite to those of cAMP.

Ions can also act as second messengers inside the cell. The binding of certain hormones to specific receptors on the cell membrane may open an ion (e.g. Ca^{++} ion) channel. The ions then rush in and bind to and activate one or more specific enzymes in the cell.

Local chemical mediators such as histamine, prostaglandins, and endorphins are secreted by cells that are not part of specialized chemical-control organs. In some cases these are so rapidly destroyed that they affect only cells in their immediate neighborhood; in other cases they can be transported by the bloodstream to distant targets. *Histamine* signals nearby capillaries to dilate and leak, thus allowing more blood to reach the site and cells of the immune system to leave the capillaries and help. *Prostaglandins* are synthesized from phospholipids in the cell membrane and are released into the bloodstream or surrounding fluids and exert a wide variety of effects on their target sites. *Endorphins* are chemical mediators that bind to specific receptors on nerve cells and affect the electrical activity of these cells.

The specialized hormones and receptor systems in higher plants and animals may have evolved from modifying or building on systems and structures found in more primitive, unspecialized organisms.

KEY DIAGRAM

The following coloring exercise is designed to familiarize you with the anatomical relationships between the hypothalamus and anterior and posterior pituitary. First color the titles and then the appropriate structure, using a different color for each structure. (Reference: Figure 16.4, p. 422)

HYPOTHALAMUS
ANTERIOR PITUITARY
POSTERIOR PITUITARY
NERVE CELLS
CAPILLARIES

Neurosecretory cells producing:

RELEASING HORMONES
OXYTOCIN, VASOPRESSIN

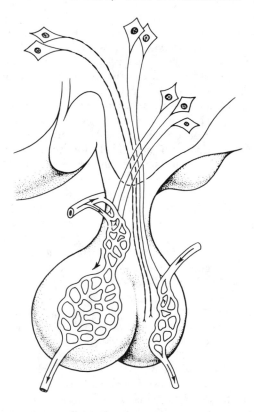

QUESTIONS

Testing recall

Match each function below with the associated endocrine gland or glands shown in the figure. Several glands may match the same statement. Answers may be used once, more than once, or not at all. (Fig. 10.2, p. 406)

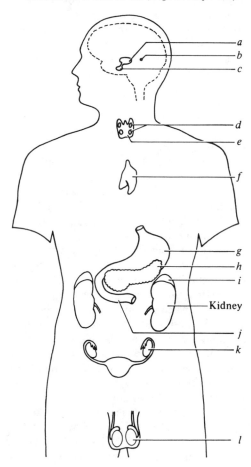

1 secretes a variety of steroid hormones (p. 414)

2 secretes a hormone that prevents an excessive rise in blood calcium (p. 418)

3 secretes the hormones that stimulate the "fight-or-flight" reaction (p. 413)

4 regulated by hormones from the anterior pituitary (p. 421)

5 produces vasopressin and oxytocin (p. 420)

6 produces growth hormone (p. 420)

7 secretes a hormone that lowers blood-glucose concentration (p. 411)

8 secretes a hormone that raises blood-glucose concentration (p. 414)

9 secretes a hormone that stimulates oxidative metabolism (p. 411)

10 secretes a hormone that regulates calcium-phosphate balance (p. 418)

11 secretes melatonin (p. 423)

12 produces releasing hormones that stimulate the anterior pituitary (p. 420)

Testing knowledge and understanding

Choose the one best answer.

13 All of the following are characteristic of hormones *except*

a acts within the gland that synthesizes it
b present in small quantities
c carried in the bloodstream
d sometimes synthesized by nervous tissue
e secreted by glands without ducts
(pp. 403, 406)

14 Many insects go through a series of developmental stages before molting into an adult. Which one of the following manipulations will delay, or prevent, the onset of pupation and keep the insect in a perpetual larval stage?

a injections of juvenile hormone
b injections of ecdysone
c injections of brain hormone
d removal of the corpora allata
e removal of the prothoracic glands
(pp. 404–5)

15 The proper metamorphosis of the moth *Cecropia* (and presumably many other insects as well) involves the antagonistic interaction of two hormones. These are

a brain cell hormone and ecdysone.
b brain cell hormone and the hormone produced by the corpora allata.
c thyroxin and a pituitary hormone.
d brain cell hormone and the hormone produced by the prothoracic gland.
e juvenile hormone and ecdysone. (p. 404)

16 The hormone released by the mucosa of the pyloric region of the stomach, which is carried by the blood to the gastric glands and stimulates them to secrete gastric juice, is called

a secretin.
b insulin.
c glucagon.
d gastrin.
e cholecystokinin.
(p. 407)

17 The pancreas
a has a duct leading to the intestine and therefore is not an endocrine gland.
b is a major gland in mammals in which digestive enzymes are made.
c produces a hormone that lowers the blood-sugar level.
d is located next to the pineal gland.
e two of the above (pp. 407–12)

18 Which of the following endocrine organs has no known function except hormone secretion?
a pancreas
b adrenals
c gonads (sex organs)
d hypothalamus (p. 412)

19 In a normal person, insulin is secreted by the pancreas immediately after ingestion of carbohydrates. Which *one* of the following is *not* an effect of insulin?
a It promotes the transport of glucose from the blood into muscle cells.
b It stimulates glycogen synthesis from glucose by the liver.
c It promotes the synthesis of fats from glucose in adipose cells.
d It stimulates the convoluted tubules of the kidney to reabsorb more sugar.
e It acts to lower blood-glucose levels.
(p. 411)

20 In persons who have an abnormally high level of sugar in the urine, the production of insulin by the pancreas is usually
a abnormally high.
b abnormally low. (p. 411)

21 Which one of the following organs does not play an important role in the regulation of blood-glucose levels?
a parathyroid   d anterior pituitary
b pancreas   e hypothalamus
c adrenal cortex   (pp. 418–19)

22 Which one of the following is *not* an effect of insulin?
a stimulates absorption of glucose from the blood by muscle and adipose cells
b promotes oxidation of glucose in liver and muscle cells
c inhibits breakdown of stored glycogen in liver and muscle cells
d promotes synthesis of glucose from fats in adipose cells
e alleviates symptoms of diabetes mellitus
(p. 411)

23 Glucocorticoids are produced by the

  a   thyroid.   d   adrenal medulla.
  b   pituitary.   e   pancreas.
  c   adrenal cortex.   (p. 414)

24 An iodine deficiency in the human diet is most likely to lead to

  a   an elevated blood-calcium level.
  b   an elevated blood-glucose level.
  c   excessive inflammation reactions.
  d   a lowered metabolic rate.
  e   an increase in insulin secretion.   (p. 417)

25 An increase of thyrotropic hormone in the blood of a mammal causes

  a   an increased blood supply to the thyroid.
  b   increased secretion of thyroxin.
  c   regression of the thyroid gland.
  d   increased nervous stimulation of the thyroid.
  e   reduced secretion of thyroxin.
                          (pp. 417–18, 421)

26 A physician examines an obese patient with low blood pressure. The patient complains of feeling very lethargic and mentally dull much of the time. Of the following endocrinological malfunctions, the patient's condition is most likely to be

  a   diabetes.
  b   hypoparathyroidism.
  c   hyperparathyroidism.
  d   hypothyroidism.
  e   hyperthyroidism.   (p. 418)

27 When a person takes thyroid tablets to increase his blood level of thyroid hormone,

  a   the secretion of TRH by the hypothalamus is inhibited.
  b   the secretion of thyrotropic hormone increases.
  c   the thyroid gland enlarges.
  d   the anterior pituitary is stimulated to produce more hormones.
  e   the thyroid gland produces more TH.
                                    (p. 418)

28 A patient has an endocrine malfunction that results in excessive irritability of the muscles and nerves, which respond even to minor stimuli with tremors, cramps, and convulsions. Blood tests show an abnormally high concentration of phosphate and an abnormally low concentration of calcium. This condition is probably

  a   hyperthyroidism.
  b   hypothyroidism.
  c   hyperparathyroidism.
  d   hypoparathyroidism.
  e   hyperadrenocorticalism.   (p. 419)

29 Which one of the following hormones regulates the rate of cellular respiration?

  a   insulin
  b   thyroid hormone (TH)
  c   parathyroid hormone
  d   adrenocortical hormones
  e   vasopressin   (p. 411)

30 Parathyroid hormone

  a   is under the control of the adrenal gland.
  b   is a steroid.
  c   serves to increase the concentration of calcium in the blood.
  d   mobilizes release of calcium and phosphate from bone.
  e   two of the above   (pp. 418–19)

31 Which one of the following hormones is mismatched with the stated function?

  a   melatonin—inhibits gonadotropin secretion
  b   parathyroid hormone—regulates calcium-phosphate balance
  c   ACTH—stimulates the adrenal cortex
  d   oxytocin—stimulates water reabsorption by the kidneys
  e   prolactin—stimulates milk production by the mammary glands   (pp. 423–24)

32 The primary connection between the nervous system and the endocrine system is

  a   adrenalin.
  b   the pancreas.
  c   the brain.
  d   the hypothalamus.
  e   cyclic AMP.   (p. 420)

33 Which one of the following hormones is *not* secreted by the anterior pituitary?

  a   FSH
  b   thyrotropic hormone
  c   cortisone
  d   LH
  e   growth hormone   (pp. 420–21)

34 Vasopressin is synthesized in the hypothalamus. Where is it stored?

  a   pineal gland
  b   anterior pituitary
  c   posterior pituitary
  d   parathyroids
  e   adrenal cortex   (p. 420)

35 The injection of a small quantity of posterior pituitary extract into the blood is likely to cause

  a   an increase in the volume of urine.
  b   a decrease in the volume of urine. (p. 420)

36 A person with a hypofunctioning anterior pituitary would probably show all of the following symptoms *except*

a decreased metabolic rate.
b decreased activity of the adrenal cortex.
c sexual immaturity.
d increased urine output.
e decreased growth-hormone secretion.
(pp. 420–21)

37 According to the proposed second-messenger model for the mode of action of animal hormones,

a the hormone enters the target cell by combining with a carrier molecule and moves to the nucleus, where it reacts with genetic material.
b .the hormone enters the target cell, where it causes production of prostaglandins which then interact with the genetic material in the nucleus.
c the hormone interacts with a specific receptor site on the outer surface of the cell so as to influence the concentration of cyclic AMP within the cell.
d the first hormone is produced by the pituitary and it in turn causes the secretion of a second hormone from another endocrine gland.
(p. 426)

38 Which one of the following hormones does *not* work by the second-messenger system?

a TSH   d glucagon
b aldosterone   e ACTH
c adrenalin   (p. 426)

39 Which one of the following hormones requires the involvement of cAMP in order to bring about a change in the target cells?

a mineralocorticoids
b sex hormones
c all plant hormones
d thyroxin
e adrenalin   (p. 426)

40 A polypeptide hormone affects muscle but not liver cells. The most likely reason is that

a liver cells lack a receptor for that hormone.
b liver cells do not have adenylate cyclase in their plasma membrane.
c liver cells do not make cAMP.
d liver cells do not contain enzymes.
(p. 428)

41 Which one of the following statements concerning steroid hormones is *true*?

a They bind to a receptor that is located on the plasma membrane.
b They bind to a cytoplasmic receptor and migrate into the cell nucleus.
c They activate the enzyme adenylate cyclase.
d They travel within the body in the lymphatic rather than the circulatory system.   (p. 426)

42 Which one of the following may function as a second messenger in a chemical control system?

a Ca^{++} ions
b adenylate cyclase
c ATP
d adrenalin
e thyrotropic hormone (TSH)   (p. 429)

43 By the technique of autoradiography it is possible to determine the location of hormones in a target tissue. Where do steroid hormones act to alter the physiology of target cells?

a plasma membrane
b mitochondria
c nucleus
d cytosol
e endoplasmic reticulum   (p. 426)

For further thought

1 List the hormones that help control (*a*) carbohydrate metabolism and (*b*) protein metabolism.

2 Discuss the relationship between the hypothalamus and the anterior and posterior pituitary. Describe the role of the anterior pituitary and hypothalamus in coordinating the activity of the thyroid gland.

3 At the age of nine years a girl was taken to a doctor who suspected that she had some sort of endocrine deficiency. At that time she was only 36 inches tall, though correctly proportioned. By 17, she was only as tall as a nine-year-old. There was no sexual development during her adolescent years.

a What type of endocrine deficiency might this child have?
b What hormones is she not producing?
c What other symptoms might she have?
d What hormonal therapy might be prescribed?

ANSWERS

Testing recall

1	*i*	5	*a*	9	*e*
2	*e*	6	*c*	10	*d, e*
3	*i*	7	*h*	11	*b*
4	*e, i, k, l*	8	*c, h, i*	12	*a*

Testing knowledge and understanding

13	*a*	21	*a*	29	*b*	37	*c*
14	*a*	22	*d*	30	*e*	38	*b*
15	*e*	23	*c*	31	*d*	39	*e*
16	*d*	24	*d*	32	*d*	40	*a*
17	*e*	25	*b*	33	*c*	41	*b*
18	*b*	26	*d*	34	*c*	42	*a*
19	*d*	27	*a*	35	*b*	43	*c*
20	*b*	28	*d*	36	*d*		

Chapter 17

HORMONES AND VERTEBRATE REPRODUCTION

A GENERAL GUIDE TO THE READING

Most students find this chapter fascinating, since it focuses on the human reproductive system. As you read Chapter 17 in your text, concentrate on the following material.

1   The introductory section, "The Process of Sexual Reproduction." This section (pp. 434–37) presents many important terms and concepts that you will need to know.

2   The eggs of land vertebrates. Carefully study Figure 17.4 (p. 437), "The embryonic membranes in a bird's egg," which shows a so-called amniotic egg, or land egg—an important evolutionary advance because it provides a fluid-filled chamber in which the embryo can develop even when the egg is laid in a dry place. The evolution of the amniotic egg enabled land vertebrates (first reptiles and later birds and mammals) to be truly terres-

trial, since reproduction no longer required laying eggs in a very moist place.

3   The human male and female reproductive tracts. These are thoroughly discussed in this chapter. Figures 17.5 and 17.6 (p. 438) will help you learn the anatomy of the male reproductive tract, and Figures 17.11 and 17.12 (p. 442) will help you learn the anatomy of the female reproductive tract.

4   The role of hormones in mammalian reproduction. The section "Hormonal Control of Sexual Development in the Male" (pp. 440–41) is straightforward, but "Hormonal Control of the Female Reproductive Cycle" (pp. 443–48) is more complicated. You will want to study the latter section carefully and refer frequently to the summary diagram, Figure 17.14 (p. 444).

KEY CONCEPTS

1 Sexual reproduction in higher animals always involves the union of two gametes, an egg and a sperm; once united in fertilization, these form the first cell of the new individual. (p. 434)

2 Both the ovaries and testes are controlled by a negative feedback loop involving the hypothalamus and anterior pituitary gland. (pp. 440, 443–45)

OBJECTIVES

After studying this chapter and reflecting on it, you should be able to carry out the following objectives.

1 Differentiate between self-fertilization and cross-fertilization, and specify which is more common in animals. (p. 434)

2 Explain the differences between external fertilization and internal fertilization, and show how the number of gametes produced by a particular organism, necessity for water, and the type of environment inhabited are related to this difference. Indicate whether each of the following organisms uses external or internal fertilization: fish, bird, snake, frog, butterfly, earthworm, mammal, lizard. (pp. 435–37)

3 Using a diagram of an amniotic egg such as Figure 17.4 (p. 437), identify the amnion, chorion, allantois, and yolk sac, and give a function of each.

4 Using diagrams such as Figures 17.6 (p. 438) and 17.11 (p. 442), trace the path of a sperm from the testis of a male to an egg in the oviduct of a female. In doing so, identify the vagina, ovary, vas deferens, cervix, oviduct, bladder, rectum, uterus, urethra, penis, seminal vesicle, and Cowper's gland. Give two functions of the seminal fluid produced by the seminal vesicles, prostate, and Cowper's glands. (pp. 437–43)

5 Explain the role of the hypothalamus, GnRH, LH, and FSH in the human male at the time of puberty. Give some of the effects of testosterone. Using structural formulas as given in Figure 17.8 (p. 440), point out one difference between the male sex hormone testosterone and the female sex hormone progesterone. (pp. 440–41)

6 Using a diagram such as Figure 17.14 (p. 444), describe the sequence of events in the menstrual cycle of a human female. Be sure to specify the successive levels of the hor-

mones involved, and the effect that each hormone has on the follicle, the uterine lining, or the production of other hormones. (pp. 443–48)

7 For each of the birth-control methods listed below, indicate the way in which it prevents conception, its relative effectiveness, and whether it is normally reversible.

diaphragm with spermicidal jelly or cream
rhythm method
vasectomy
tubal ligation
intrauterine device (IUD)
the pill   (pp. 440–45)

8 Outline the events that occur between the time an egg is fertilized in the oviduct and the birth of the baby. In doing so, use the following terms: corpus luteum, oxytocin, placenta, oviduct, progesterone, luteinizing hormone (LH), human chorionic gonadotropin (HCG), uterine lining, follicle-stimulating hormone (FSH), lactation, estrogen, and implantation. (pp. 441–50)

9 Contrast the embryonic membranes in a human, as shown in Figure 17.15 (p. 449), with those of a reptile or a bird, as shown in Figure 17.4 (p. 437). Specify which of the following are present in both: amnion, chorion, shell, allantois, placenta, yolk sac. (pp. 437, 449)

KEY TERMS

The following terms are important in this chapter; you should become familiar with them.

gamete (p. 434)
cross-fertilization (p. 434)
self-fertilization (p. 434)
external fertilization (p. 435)
internal fertilization (p. 435)
amnion (p. 437)
chorion (p. 437)
allantois (p. 437)
yolk sac (p. 437)
inguinal canal (p. 437)
scrotal sac (p. 437)
cavernous body (p. 439)
seminal vesicle
prostate
Cowper's gland
vas deferens
epididymis   parts of the male
seminiferous tubule reproductive tract
interstitial cells   (pp. 438–39)
testis
urethra
penis

SUMMARY

Sexual reproduction in higher animals involves bringing together two *gametes*, an egg and a sperm, which then unite in the process of fertilization to form the first cell of the new individual.

Most aquatic organisms use external fertilization. The gametes are shed directly into the water and the sperm must swim to the egg. Generally, these animals release large numbers of gametes and often go through elaborate behavioral sequences to ensure that both sexes shed gametes simultaneously.

Most land animals use internal fertilization, in which the egg cells remain in the female reproductive tract until they have been fertilized by sperm inserted by the male. The sperm swim through the fluid in the female reproductive tract. Once fertilized, the egg in its fluid-filled chamber is enclosed by membranes. In reptiles, birds, and mammals the membranes are: the *amnion*, which surrounds a fluid-filled chamber housing the embryo; the *allantois*, which functions in waste storage and gas exchange; the *yolk sac*, which

stores food; and the *chorion*, which encloses the embryo and its membranes. The embryo and its membranes are then either surrounded by a protective shell and released, or held within the female's body until embryonic development is completed.

The male gonads are the *testes*, which lie in the *scrotal sac*. Each testis has two functional components: the *seminiferous tubules*, in which the sperm are produced, and the *interstitial cells*, which secrete male sex hormone. Mature sperm move into the much-coiled *epididymis*, where they are stored and activated. During copulation the sperm move into the *vas deferens*, which conducts them to the *urethra*, which passes through the *penis* and empties to the outside. Seminal fluid from the *seminal vesicles*, the *prostate*, and the *Cowper's glands* is added to the sperm to form the *semen*. Semen provides a fluid medium for transport, for lubrication, and for protection of the sperm from acids in the female genital tract. Its sugar provides energy for the active sperm.

During embryonic development the testes begin secreting small amounts of the male sex hormone, *testosterone*, which is crucial to the differentiation of male structures. At puberty, the hypothalamus sends releasing hormone (GnRH) to the anterior pituitary, stimulating it to release *FSH* and *LH*. The LH induces the interstitial cells to produce more testosterone; this plus the FSH induces the maturation of the seminiferous tubules and causes sperm production to begin. Once testosterone reaches appreciable levels, it stimulates the development of the secondary sexual characteristics.

The female gonads are the *ovaries*, located in the lower abdominal cavity. The ovaries produce the egg cells (*oocytes*) and secrete sex hormones. Each oocyte is enclosed within a small *follicle*. During maturation, the follicle fills with fluid. When ovulation occurs, the outer wall ruptures and the oocyte and fluid are expelled into the abdominal cavity. Cilia lining the adjacent *oviduct* create a current drawing the egg into it. If sperm are present, fertilization occurs in the oviduct.

Each oviduct empties into the muscular *uterus*. If the egg is fertilized it becomes implanted in the uterine wall, where the embryo develops. At its lower end the uterus connects with the tubular *vagina*, which leads to the outside. The vagina is the receptacle for the penis during copulation.

Puberty in the female begins when the hypothalamus sends more GnRH to the anterior pituitary, stimulating it to release FSH and LH. These hormones cause maturation of the ovaries, which begin secreting the female sex hormones, *estrogen* and *progesterone*. Estrogen stimulates maturation of the reproductive structures and development of the secondary sexual characteristics. The changing hormonal balance triggers the onset of the menstrual cycles.

Rhythmic variations in the secretions of gonadotrophic hormones in most mammals lead to *estrous cycles*—rhythmic variations in the reproductive tract and sex urge. The reproductive cycle differs somewhat in humans: the female is receptive to the male throughout her cycle, and the thickened uterine lining is not completely reabsorbed (as in most other mammals) if no fertilization occurs; instead part of the lining is sloughed off during *menstruation*.

At the beginning of the menstrual cycle the uterine lining is thin and there are no ripe follicles. Under stimulation of GnRH from the hypothalamus, FSH is released and stimulates the maturation of the follicles, which begin to secrete estrogen. Estrogen stimulates the uterine lining to thicken. This follicular phase lasts about nine days. The high level of estrogen apparently stimulates a surge of LH from the pituitary, triggering ovulation. The LH converts the follicle into the *corpus luteum*, which continues to secrete estrogen and begins to secrete *progesterone*. Progesterone prepares the uterus to receive the embryo by activating its many glands and by inducing other chemical changes. The high level of progesterone also suppresses the growth of new follicles. If no fertilization occurs, the corpus luteum atrophies about eleven days after ovulation and progesterone secretion falls. When this happens, the thickened uterine lining can no longer be maintained, and menstruation occurs, lasting about five days. The low level of progesterone frees the hypothalamus from inhibition and allows it to stimulate the pituitary to increase FSH secretion; thus another cycle begins.

The egg cell must be fertilized within 12 hours after ovulation. The fertilized egg, or *zygote*, moves down the oviduct and, in humans, becomes implanted in the uterine wall 8−10 days after fertilization. The embryonic membranes then develop and the *placenta* is formed. Exchange of materials between the blood of the mother and that of the embryo takes place by diffusion through the placenta. The placenta soon begins to secrete *human chorionic gonadotrophin*, which maintains the corpus luteum and its secretion of progesterone, thus sustaining the pregnancy. Later the placenta secretes estrogen and progesterone directly.

In late pregnancy, estrogen secretion by the placenta increases. This causes a rise in the placenta's prostaglandin production; this increase apparently initiates the birth process. *Oxytocin* probably contributes to the induction of labor. The hormone *relaxin* aids by enlarging the birth canal.

Initiation and maintenance of lactation by mature mammary glands after birth seems to be controlled primarily by prolactin and glucocorticoids. Oxytocin from the posterior pituitary causes the milk to be ejected into the ducts of the nipple.

QUESTIONS

Testing recall

Questions 1–8 refer to the following diagram of the human female reproductive system.

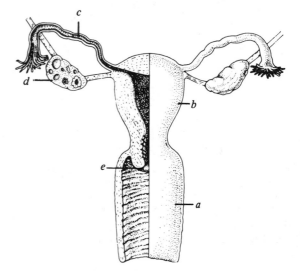

1 Where is the egg produced? (p. 441)

2 Where does fertilization occur? (p. 442)

3 Where would implantation of a fertilized egg take place? (p. 442)

4 Where is a tubal ligation performed? (p. 440)

5 Where does a diaphragm act to prevent conception? (p. 443)

6 Where are estrogen and progesterone produced? (p. 443)

7 Where does an IUD act to prevent conception? (p. 443)

8 What part receives the male penis during copulation? (p. 442)

Match each function below with the associated part or parts of the human male reproductive system shown in the figure.

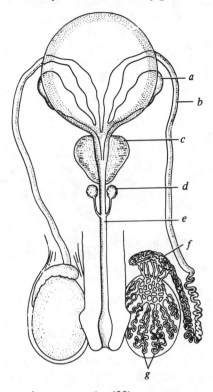

9 produces sperm (p. 438)

10 conducts the sperm through the penis to the outside of the body (p. 438)

11 produces seminal fluid (p. 439)

12 connects the epididymis with the urethra (p. 438)

13 stores sperm (p. 438)

Fill in the blanks.

Most mammals show rhythmic variations in the reproductive tract and sex urges called the

(14) _____ cycle (p. 443).
At the beginning of the menstrual cycle in humans the hormone

(15) _____ from the anterior pituitary stimulates the maturation of the follicle in the ovary (p. 444). The growing follicle begins to secrete the hormone

(16) _____, which causes the lining of the uterus to thicken (p. 444). High levels

of this hormone stimulate an abrupt production of high levels of (17) _____ from the pituitary; this triggers the process of

(18) _____ (p. 445). The old follicle is converted into the

(19) _____, which secretes the hormones (20) _____ and

_____ (p. 445). These hormones cause the uterine lining to become thicker and more glandular. High levels of the

hormone (21) _____ suppress

the release of (22) _____ from the hypothalamus, thereby limiting

(23) _____ and

_____ secretion by the pituitary (p. 445).
If no fertilization occurs the corpus luteum begins to atrophy and its secretion of

(24) _____ falls (p. 446). Now the thickened lining of the uterus can no longer be maintained; part of it is reabsorbed and part of it is sloughed off during

(25) _____ (p. 446). The low levels of the sex hormones free the hypothalamus and immature follicles from inhibition, and the pituitary begins to secrete more

(26) _____, beginning a new cycle (p. 446).
The follicular phase of the cycle lasts about

(27) _____ days, the luteal

phase about (28) _____ days, and the flow phase about

(29) _____ days (p. 444).
If fertilization occurs, the embryo becomes implanted in the wall of the uterus about

(30) _____ days after fertilization (p. 448). The (31) _____ forms from the embryonic membranes and the uterine lining; here the exchange of materials between mother and embryo takes place (p. 449). Part of this organ begins to secrete the hormone

(32) _____; this preserves the corpus luteum, which continues to secrete

(33) _____ (p. 449). Later this organ secretes the hormones

(34) _____ and

_____, which sustain the pregnancy. (p. 449)

Near the time of birth the secretion of the hormone (35) _____ increases (p. 449). The hormone

(36) _____ secreted by the posterior pituitary causes powerful uterine contractions, which aid in the birth process (p. 450).

37 List, in order, all the structures through which a sperm passes from the place where it is formed to the place where fertilization occurs. (pp. 437–42)

Testing knowledge and understanding

Choose the one best answer.

38 In comparison with organisms that use external fertilization, organisms that use internal fertilization usually

  a have fewer progeny that reach adulthood.
  b produce fewer gametes.
  c produce fewer zygotes.
  d live on land.
  e all of the above   (p. 436)

39 The hormones testosterone and progesterone act by

  a combining with receptor molecules and moving to the nucleus, where they react with the genetic material.
  b combining with a receptor molecule on the membrane surface, which activates adenylate cyclase.
  c directly activating enzyme systems within the cell.
  d stimulating the hypothalamus and anterior pituitary to release hormones.
  e inducing a rise in cAMP within the cell.
      (p. 440, Ch. 16, p. 426)

40 Which one of the following animals produces eggs that lack an amnion?

  a human   *d* black snake
  b cow   *e* codfish
  c robin   (p. 437)

41 The prostate gland plays a major role in the production of

  a seminal fluid.   *d* urine.
  b sperm.   *e* amniotic fluid.
  c testosterone.   (p. 439)

42 The role of FSH in humans is to

  a stimulate the development of the follicle in the ovary and the production of sperm in the testes.
  b stimulate the growth of the corpus luteum in the ovary and the production of sperm in the male.

  c stimulate the interstitial cells of the male testes.
  d stimulate the menstrual flow.
  e stimulate uterine contractions at birth.
      (pp. 440, 444)

43 In the menstrual cycle, blood progesterone levels are highest during

  a the follicular (growth) phase.
  b the time of ovulation.
  c the luteal (secretory) phase.
  d the first part of the flow phase.
  e the last part of the flow phase.   (p. 445)

44 Which of the following is the correct sequence of organs involved in producing progesterone?

  a hypothalamus–anterior pituitary–ovary
  b hypothalamus–posterior pituitary–ovary
  c anterior pituitary–adrenal cortex–uterus
  d posterior pituitary–adrenal medulla–ovary
  e hypothalamus–ovary–uterus   (p. 443)

45 Which of the following hormones stimulates the maturation of reproductive structures in *both* male and female humans?

  a estrogen
  b follicle stimulating hormone (FSH)
  c progesterone
  d testosterone
  e oxytocin   (pp. 440, 443)

46 Which of the following statements is *false* regarding progesterone in the human female menstrual cycle?

  a It promotes the development of follicles.
  b It maintains the thickened uterine lining which forms in preparation for embryo implantation.
  c It is secreted by the corpus luteum.
  d It inhibits release of hormones by the hypothalamus.
  e It is a component of birth control pills.
      (p. 445)

47 Which of these contraceptives does *not* prevent fertilization?

  a condom
  b diaphragm
  c intrauterine device (IUD)
  d "pill" (oral contraceptive)
  e spermicidal foam   (p. 442)

48 Which one of the following statements is *false*?

  a Most birth-control pills are mixtures of estrogen and progesterone (or close analogs) designed to inhibit the hypothalamus.

b Vasectomy does not alter a man's endocrine system, and hence should not reduce either sexual drive or sexual competency.

c Tubal ligation does not alter a woman's endocrine system, and hence should not reduce either sexual drive or sexual responsiveness.

d The diaphragm is a birth-control device that, when properly fitted, covers the cervix and prevents sperm from entering the uterus.

e In the rhythm method of birth control, sexual abstinence on days 14–16 in each menstrual cycle reliably prevents pregnancy.   (pp. 441–43, 445, 447)

49 The birth-control pills most commonly used in the United States today are mixtures of estrogen-like and progesterone-like compounds. They probably prevent pregnancy mainly by

a stimulating the release of FSH from the anterior pituitary.

b preventing the glands of the uterine lining from becoming secretory.

c decreasing the mobility of the sperm by inhibiting uterine contractions.

d blocking secretion of gonadotropic releasing hormone (GnRH) by the hypothalamus.

e causing premature release of the egg from the follicle.   (pp. 445–46)

50 Which one of the following describes a function the placenta does *not* perform?

a removes waste materials from the fetus

b supplies oxygen to the fetus

c secretes hormones

d provides the fetus with nutrients

e replaces used fetal erythrocytes with new ones from the mother   (p. 449)

51 If gonadotropins (HCG, LH) from human placenta were injected into a mature male, one might expect to see

a an increase in estrogen and progesterone.

b an increase in testosterone.

c a decrease in the activity of the testes.

d a 28-day cycling of male hormones.

e increased sperm production by the testes.   (p. 440)

For further thought

1 Explain how each of the following methods of birth control works, indicating for each the part of the reproductive process that it interferes with: vasectomy, tubal ligation, IUD, diaphragm and spermicidal jelly or cream, birth-control pills, the rhythm method.

ANSWERS

Testing recall

1	*d*	6	*d*	11	*a, c, d*
2	*c*	7	*b*	12	*b*
3	*b*	8	*a*	13	*f*
4	*c*	9	*g*	14	estrous
5	*e*	10	*e*	15	FSH

16 estrogen
17 LH
18 ovulation
19 corpus luteum
20 estrogen, progesterone
21 progesterone
22 GnRH
23 FSH, LH
24 progesterone
25 menstruation
26 FSH
27 9–10
28 13–15
29 4–6
30 8–10
31 placenta
32 chorionic gonadotropin
33 progesterone
34 progesterone, estrogen
35 estrogen
36 oxytocin
37 seminiferous tubules, ducts in epididymis, vas deferens, urethra, vagina, uterus, oviduct

Testing knowledge and understanding

38	*e*	43	*c*	48	*e*
39	*a*	44	*a*	49	*d*
40	*e*	45	*b*	50	*e*
41	*a*	46	*a*	51	*b*
42	*a*	47	*c*		

NERVOUS CONTROL

A GENERAL GUIDE TO THE READING

In Chapter 15 we learned that there are two principal types of control mechanisms in animals—chemical and nervous. This chapter and the one that follows are concerned with the second of these mechanisms, nervous control. As you read Chapter 18 in your text, you will want to concentrate on the following topics.

1 The contrast between hormonal control and nervous control. The first paragraph of the chapter presents this contrast clearly, and explains why hormonal control is inadequate for active multicellular animals.

2 The evolution of nervous systems. The discussion of this topic is interesting. You will want to concentrate on the evolutionary trends (pp. 456–63) and on the differences between the central nervous system of annelids and arthropods and that of vertebrates (pp. 462–63). Pay particular attention to the information on *Aplysia* since this organism will be discussed further later in the chapter.

3 The structure of the neuron. To understand the discussions of reflex arcs, nerve-impulse transmission, and synaptic transmission later in the chapter, you need to know the basic structure of neurons and the names of their various parts (pp. 452–55).

4 Nerve-impulse transmission. This topic is difficult but indispensable and interesting. Figures 18.16 (p. 465) and 18.19 (p. 467) are particularly helpful. Pay particular attention to the material on voltage-gated channels (see Figure 18.20, p. 468).

5 Restoration of the initial ionic balance in the neuron. Study Figures 18.21 (p. 469), 18.22 (p. 470), and 18.23 (p. 471) with care as these will help you understand the roles of diffusion, electrostatic attraction, and the sodium-potassium pump.

6 Transmission across synapses. It is important that you understand the process of synaptic transmission and the processes of summation and integration since this is what is involved in processing information in the nervous system.

7 The reflex arc. You need to understand the components of the reflex arc. Figures 18.32–18.34 (pp. 480–82) are helpful.

8 The autonomic nervous system. The section on the autonomic nervous system (pp. 483–87) is conceptually difficult for most students; you may have to read through it more than once and to study carefully Figure 18.35 (p. 484).

9 Neural control of more complex behavior. Study pages 482–90 with care, especially page 488, since these describe motor programs, which are responsible for much of an organism's behavior.

KEY CONCEPTS

1 Responses to stimuli from the environment generally involve four components: detection of the stimulus, conduction of a signal, processing of the signals, and response by an effector. (p. 451)

2 The nervous system is composed of cells that are specialized to detect changes in the environment and to conduct information to the effectors, which produce a rapid response. (pp. 452–54)

3 The brains of animals consist almost entirely of interneurons arranged in complex, highly specialized networks. Interneurons typically collect excitatory or inhibitory input from many cells and pass on the resulting information to its target cells. (p. 458)

4 Inhibition is essential to information processing since the nervous system usually operates using an antagonist strategy; both excitatory and inhibitory signals are sent, and the ratio between them determines the target cell's response. (p. 458)

5 Increasing the number of conducting cells in the nervous pathway, by making possible more alternative routes, increases the flexibility of the response to a stimulus. (p. 460)

6 Animals with more advanced nervous systems have evolved more complex pathways, a greater degree of centralization, and better developed sense organs than animals with simple nervous systems. (p. 460)

7 Nerve impulses are conducted along the neuron by electrochemical changes; they are transmitted across the synapse by chemicals. (pp. 464, 472)

8 The propagation of an action potential is a membrane phenomenon; it depends on an initial electrostatic gradient across the membrane followed by a coordinated series of ion-specific changes in permeability. (pp. 466–67)

9 The membrane proteins responsible for creating the action potential are voltage-gated channels; that is, they open and close in response to changes in the electrostatic gradient across the membrane. (p. 468) When an impulse passes and the membrane is depolarized, diffusion and electrostatic attraction restore the electrochemical balance between Na^+ outside and K^+ inside almost instantaneously. In addition, the sodium-potassium pump actively extrudes Na^+ ions and takes up K^+ ions. (pp. 468–71)

10 The development of sodium-potassium pumps, combined with the evolution of ion-specific voltage-gated channels, has been the basis for the evolution of neural transmission. (pp. 467–71)

11 Synapses are points of resistance in neural circuits; the receiving cell integrates all the excitatory and inhibitory signals it receives, and either fires or remains silent. (pp. 477–79)

12 Reflex arcs are simple neural pathways linking a receptor and effector. They also interconnect with other neural pathways; interneurons synapse with pathways leading to other parts of the spinal cord and to the brain, and the brain can send impulses that modify the reflexes. (pp. 480–83)

13 The central nervous system of vertebrates is a coordinating system for two kinds of pathways—somatic and autonomic. Somatic pathways control the voluntary activities whereas autonomic pathways innervate involuntary activities. (p. 483)

14 The basis of the more complex behavior patterns in most animals is motor programs. Motor programs are self-contained neural circuits that coordinate muscle movement; they are automatically fine-tuned by sensory feedback and under direct control of the brain. (p. 488)

OBJECTIVES

After studying this chapter and reflecting on it, you should be able to carry out the following objectives.

1 List the four stages involved in a response to a stimulus and explain how each of these takes place in unicellular organisms such as the bacterium, *E. coli*, and *Paramecium*. (p. 452)

2   Describe the structure of a typical neuron and, using a diagram such as Figure 18.2 (p. 453), point out the axon, dendrite, cell body, and myelin sheath in the different types of neurons. Indicate the path of information flow and point out a synapse and neuromuscular junction. (pp. 452–54)

3   Using diagrams such as Figures 18.4 (p. 454) and 18.5 (p. 455), explain the role that the neuroglia and Schwann cells play in the nervous system and how the myelin sheath is formed. Indicate the function of the myelin sheath. (p. 454)

4   Using diagrams as given in Figures 18.10 (p. 459), 18.11 (p. 460), 18.12 (p. 461), 18.13 (p. 461), 18.14 (p. 462), and 18.17 (p. 465), compare the nervous systems of coelenterates (such as hydra), flatworms, annelids, and arthropods. In doing so, indicate for each type of organism whether it is radially symmetrical or bilaterally symmetrical; what degree of centralization its nervous system has reached, as shown by the absence or presence of major longitudinal nerve cords, and by their location—dorsal or ventral; what degree of cephalization it has reached; and whether ganglia are present. (pp. 459–65)

5   Contrast the nervous system of annelids and arthropods with that of vertebrates with respect to the location, form, and structure of the longitudinal nerve cord or cords; the degree of centralization; and the extent of dominance of the brain. (pp. 460–63)

6   Explain how a nerve impulse is conducted along the neuron, using the terms stimulus, threshold, action potential, and all-or-none response. (pp. 463–68)

7   Discuss the basis for the polarization of the nerve cell membrane, considering the relative amounts of sodium, potassium, and negatively charged ions inside and outside the neuron, and state whether the outside of the neuron is charged positively or negatively with respect to the inside (see Figure 18.16, p. 465). (pp. 464–67)

8   Explain in some detail how an impulse is transmitted along a neuron fiber; specify which ions move and in what order when the fiber is stimulated, and explain what is meant by voltage-gated channels. Using a diagram such as Figure 18.19 (p. 467), explain how the nerve impulse is propagated along the neuron. (pp. 466–67)

9   Explain how diffusion, electrostatic attraction, and the sodium-potassium pump act to reestablish the original ionic balance and keep the neuron functioning. (Figures 18.21 (p. 469), 18.22 (p. 470), and 18.23 (p. 471) may be helpful.) (pp. 467–71)

10 Using a diagram such as Figure 18.26 (p. 433), identify the synaptic terminal, the presynaptic membrane, postsynaptic membrane, synaptic cleft, and synaptic terminal. Describe the events occurring at a synapse when an action potential arrives, and explain how the impulse is transmitted across the synapse and what must happen for an action potential to be induced in the postsynaptic neuron. Name three transmitter substances. (pp. 472–77)

11 Compare events at excitatory and inhibitory synapses and indicate the effect of presynaptic inhibition and facilitation. Explain what is meant by summation and integration, and describe how neurological drugs can effect nerve-impulse transmission. (pp. 476–79)

12 Contrast impulse transmission across the synapses with impulse transmission across the neuromuscular junction. (pp. 472–80)

13 Using a diagram such as Figure 18.33 (p. 481), trace the flow of information through a reflex arc, beginning with a sensory receptor cell and ending with an effector cell. In doing so, be sure to identify the sensory neuron, indicating where the cell body for this neuron is located; specify the part of the nervous system in which information is relayed to interneurons; describe the role of interneurons in the reflex arc, and explain how information gets to the brain; and identify a motor neuron, indicating where the cell body for this neuron is located. (pp. 480–83)

14 Compare the somatic and autonomic pathways, specifying for each whether it is under voluntary or involuntary control, whether the muscle innervated is skeletal or smooth, and how many motor neurons are in the pathway; then indicate which system (somatic or autonomic) would control each of the following: heartbeat, peristalsis of the digestive tract, walking, sweating, lifting a heavy object in your arms, the response to stepping on a tack. (pp. 483–87)

15 Using a diagram such as those in Figure 18.36 (p. 486), compare the four different arrangements for chemical control and explain why the nervous and endocrine systems are so closely related. Give one reason why the sympathetic nervous system and the hormones of the adrenal medulla have similar effects on the body. (pp. 485–87)

16 Illustrate the function of the sympathetic and parasympathetic nervous system by describing the control of the rate of heartbeat. (pp. 486–87)

17 Explain what motor programs are, how they are fine-tuned and controlled, and their importance in the control of an organism's behavior. Illustrate the above using feeding behavior in *Aplysia* as an example. (pp. 488–90)

KEY TERMS

The following terms are important in this chapter; you should become familiar with them.

chemotaxis (p. 452)
neuron (p. 452)
cell body (p. 453)
dendrite (p. 453)
axon (p. 453)
neuroglia (p. 454)
myelin sheath (p. 454)
nodes (p. 454)
Schwann cell (p. 454)
synapse (p. 454)
reflex (p. 456)
adaptation (p. 457)
sensory neuron (p. 457)
motor neuron (p. 458)
interneuron (p. 458)
inhibition (p. 458)
ganglia (p. 458)
nerve (p. 458)
nerve net (p. 459)
central nervous system (p. 460)
brain (p. 460)
cephalization (p. 460)
nerve impulse (p. 464)
threshold (p. 464)
all-or-none response (p. 464)
action potential (p. 467)
voltage-gated channel (p. 468)
sodium-potassium pump (p. 468)
synaptic terminal (p. 472)
synaptic cleft (p. 472)
synaptic vesicle (p. 473)
transmitter (p. 473)
postsynaptic membrane (p. 473)
acetylcholine (p. 473)
acetylcholinesterase (p. 473)
habituation (p. 474)
sensitization (p. 474)
excitatory postsynaptic potential, or EPSP (p. 476)
inhibitory postsynaptic potential, or IPSP (p. 476)

presynaptic inhibition (p. 477)
presynaptic facilitation (p. 477)
summation (p. 477)
integration (p. 477)
neuromuscular junction (p. 479)
reflex arc (p. 480)
dorsal-root ganglion (p. 481)
somatic pathway (p. 483)
autonomic pathway (p. 483)
sympathetic system (p. 484)
parasympathetic system (p. 485)
motor program (p. 488)
reciprocal inhibition (p. 490)

SUMMARY

Because hormonal control is too slow for the rapid integration of sensory information and coordination of movement, electrical communication has evolved to fill this need.

Responses to stimuli from the environment generally involve four components: detection of the stimulus, conduction of a signal, processing of the signals, and response by an effector. All four components may be carried on within unicellular organisms. For example, *E. coli* samples the environment using twenty-four different receptors; the different receptors release different second messengers. Processing involves measuring the various messengers; the response is a change in direction of rotation of the flagella.

The typical nerve cell, or *neuron*, consists of the *cell body*, which contains the nucleus, and one or more long *nerve fibers* that extend from it. The *dendrites* receive impulses and conduct them to the cell body, and the axons conduct impulses away from the cell body. Within the central nervous system the neurons are associated with vast numbers of *neuroglia* cells. The function of many of the neuroglia is uncertain, but some are known to wrap around and around axons, forming the *myelin sheath*. Outside the central nervous system, many vertebrate axons are enveloped by *Schwann cells*, which may also form a myelin sheath. Myelin sheaths speed up the rate of conduction.

All multicellular animals (except sponges) have evolved some form of nervous system. Most nervous pathways have at least three separate cells: receptor, conductor, and effector cells. More complex pathways have additional conductor cells; these increase the flexibility of response. *Sensory neurons* lead from receptor cells, *motor neurons* lead to effector cells, and *interneurons* lie between the sensory and motor neurons. Junctions between neurons are called *synapses*.

Radially symmetrical animals, such as coelenterates, have a simple nerve net with little central control. We can see six major trends in the evolution of the nervous system of bilaterally symmetrical animals:

1 increased centralization by the formation of longitudinal nerve cords;
2 increased complexity of pathways;
3 the formation of distinct functional areas and structures;
4 increased cephalization;
5 one-way conduction;
6 increased number and complexity of sense organs.

These evolutionary trends are most developed in the vertebrates and in the annelids and arthropods. The main difference between the central nervous system (brain and spinal cord) of vertebrates and those of annelids and arthropods is that vertebrates have a single dorsal, hollow nerve cord whereas annelids and arthropods have two ventral, solid nerve cords. The vertebrate brain is also more highly developed and exerts more dominance over the entire nervous system.

A nerve impulse is a wave of electrical activity moving along a nerve fiber. The potential stimulus must be above a critical intensity, or *threshold* value, to initiate an impulse. If the axon fires, it will fire maximally or not at all—an *all-or-none response*.

The inside of a resting nerve fiber is negative relative to the outside because the ratio of negative to positive ions is higher inside the cell than outside. The inside has a high concentration of potassium ions (K^+) and negative organic ions; the outside has a high concentration of sodium ions (Na^+). When a fiber is stimulated and the membrane depolarizes at least the precise amount, specific gates open, exposing Na^+ channels, and Na^+ ions cross the membrane into the cell, making the inside positively charged relative to the outside. Once enough Na^+ ions have flowed through to depolarize the membrane completely, the Na^+ channel gates close and the K^+ gates open, exposing K^+ channels, and K^+ ions then rush out of the cell, restoring the original charge. This cycle of electrical charges is known as the *action potential*. The action potential at the point of stimulation alters the permeability at adjacent points and initiates the same cycles of changes there. Sodium-potassium exchange pumps actively extrude Na^+ ions and take up K^+ ions.

The axon of one neuron usually synapses with the dendrites or cell bodies of other neurons. Each tiny branch of an axon usually terminates in a *synaptic terminal*. A few synapses are electrical; in these the terminal and adjoining cell membranes are connected by a gap junction, permitting direct electrical transmission from the first neuron to the second. Most synapses, however, are chemical. When an impulse traveling along the axon reaches the synaptic terminal, special voltage-gated calcium channels open and Ca^{++} ions diffuse into the terminal, causing the *synaptic vesicles* to discharge their stored *transmitter* chemical into the synaptic cleft. The transmitter molecules diffuse across the cleft and alter the polarization of the postsynaptic membrane of the next neuron. Synaptic transmission is slower than impulse conduction along the neuron. It is the chemical synapses that make transmission along the neural pathways one-way.

The transmitter chemical in the peripheral nervous system is *acetylcholine*. After acetylcholine has exerted its effect on the postsynaptic neuron it must be destroyed by the enzyme *acetylcholinesterase*. Transmitter chemicals inside the central nervous system include *acetylcholine, noradrenalin, serotonin, dopamine,* and *GABA*. Synaptic transmitter substances act by binding to receptor proteins in the membrane, thus opening the gates of a channel, allowing specific ions to cross. This ion movement results in the alteration of the postsynaptic neuron's membrane potential. Transmitter chemicals can be excitatory or inhibitory. An *excitatory* transmitter slightly reduces the polarization of the postsynaptic membrane, creating an excitatory postsynaptic potential (*EPSP*). If the EPSP reaches threshold, it triggers an impulse. An *inhibitory* transmitter increases the polarization of the postsynaptic membrane, a condition called an inhibitory postsynaptic potential (*IPSP*). This makes the neuron harder to fire. Both excitatory and inhibitory synapses are subject to *presynaptic inhibition* and *presynaptic facilitation*, which alter the number of synaptic vesicles that release transmitter.

Synapses are points of resistance in neural circuits; enough excitatory transmitter must be released within a short time to build up sufficient EPSP to initiate an impulse. The cell integrates all the excitatory and inhibitory signals it receives, and either fires or remains silent.

Synapses are responsible for information processing in the nervous system. Their operation depends on a delicate balance between transmitter substance, deactivating enzyme, and membrane sensitivity. Synaptic malfunctions have been implicated in several mental disorders. Neurological drugs can alter synaptic function in a variety of ways.

The gap between the axon and the muscle it innervates is called the *neuromuscular junction*. Transmission across this gap is also via transmitter chemicals. Acetylcholine is the transmitter at neuromuscular junctions of vertebrate skeletal muscle.

A *reflex arc* is a simple neural pathway linking a receptor and effector. Most somatic reflex arcs begin with a *sensory neuron* that conducts the impulse from the receptor to the dorsal portion of the spinal cord, where the sensory neuron synapses with *interneurons*. These in turn synapse with *motor neurons* in the cord, and the impulses are conducted along their axons to the effectors (usually skeletal muscles), which respond to the stimulus. Reflex arcs always interconnect with other neural pathways; the interneurons connect with pathways leading to the brain, and the brain can send impulses that modify the reflexes.

A *nerve* consists of a number of neuron fibers bound together. Some nerves have only sensory fibers and are therefore *sensory nerves*, others are purely *motor*, and still others have both types of fibers and are *mixed nerves*.

The central nervous system of vertebrates is a coordinating system for two kinds of pathways—*somatic* and *autonomic*. Somatic pathways control the voluntary activities whereas autonomic pathways innervate involuntary activities. The *autonomic nervous system* (ANS) consists of nervous pathways that conduct impulses from the *central nervous system* (CNS) to various internal organs. The autonomic nervous system regulates the body's involuntary activities. Autonomic pathways usually have two motor neurons. The first neuron exits from the central nervous system and synapses with a second that innervates the target organ. The autonomic nervous system is separated into two parts, *sympathetic* and *parasympathetic* systems. Most internal organs are innervated by both, with the two systems usually functioning in opposition to each other. The sympathetic system prepares an animal for emergency action whereas the parasympathetic system restores order or passivity after the crisis is over. Noradrenalin is the transmitter in the sympathetic system. This explains the similar effects of the sympathetic nervous system and the hormones of the adrenal medulla; both release noradrenalin.

The rate of heartbeat is under autonomic control. The rate is determined by the relative activity of the cardiac accelerating and decelerating centers in the medulla, which are in turn influenced by the amount of excitation they receive from stretch receptors and chemoreceptors in the arteries. These autonomic reflex circuits fine-tune the heart rate by negative feedback loops.

The control of complex rhythmic or sequential behaviors in vertebrates and invertebrates is based on *motor programs*. Motor programs are self-contained neural circuits that coordinate muscle movement. The program is automatically fine-tuned by sensory feedback and is under direct control of the brain. Much of an organism's behavior is accomplished by interacting groups of neural circuits.

KEY DIAGRAM

*The following coloring exercise is designed to help
you learn the anatomy of vertebrate neurons.
Three types of neurons are shown below. Color
the title and the appropriate structures, using a
different color for each. (Reference: Figure 18.2,
p. 453)*

DENDRITE   CELL BODY   MYELIN SHEATH

AXON   NUCLEUS   MUSCLE

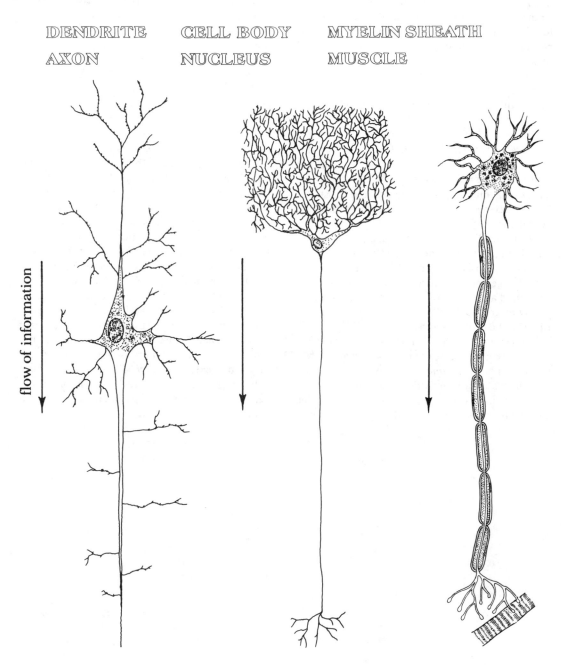

flow of information

QUESTIONS

Testing recall

Select the correct term or terms to complete each statement.

1 Responses to stimuli involve (sensory reception, conduction, response). (p. 451)

2 Increasing the number of conductor cells in a nervous circuit (increases, decreases) the flexibility of response. (p. 460)

3 The central nervous system consists of the (brain, spinal cord, autonomic nerves, spinal nerves). (p. 463)

4 Radially symmetrical animals such as coelenterates have a nervous system that is (highly centralized, a diffuse nerve net, organized into longitudinal nerve cords). (p. 459)

5 The nervous system of the advanced flatworms is characterized by (a nerve net, two ventral nerve cords, a dorsal nerve cord, a highly developed brain, simple sense organs at the anterior end). (p. 460)

6 The nerve cord(s) of an arthropod is (are) (single, double), (solid, hollow), and (dorsal, ventral). (p. 462)

7 The brain of an arthropod exerts (limited, extensive) dominance over the entire nervous system. (pp. 462–63)

8 In the vertebrate neuron, the (axons, dendrites) conduct impulses toward the cell body and the (axons, dendrites) conduct impulses away from the cell body. Each neuron usually has (one, many) dendrite(s) and (one, many) axon(s). (Sensory neurons, Motor neurons, Interneurons) conduct impulses to the central nervous system; (sensory neurons, motor neurons, interneurons) conduct impulses within the central nervous system. (pp. 452–54, 460)

9 Within the central nervous system special satellite cells called (Schwann cells, neuroglia) are associated with the neurons; frequently these wrap around and around the neuron, forming the (myelin sheath, axon, nerve). (p. 454)

10 The nerves of the autonomic nervous system innervate the (skeletal muscles, blood vessels, digestive tract, respiratory system, reproductive system). (p. 483)

11 The autonomic nerve pathways usually have (one, two, three, many) motor neuron(s). The parts of the autonomic nervous system are the (somatic, sympathetic, parasympathetic) divisions. (pp. 483–85)

12 The resting nerve fiber is polarized with the inside (positive, negative) compared to the outside. When a stimulus is above threshold, the membrane first becomes permeable to (Na^+, K^+, organic) ions, which rush (into, out of) the cell; the inside of the cell is now charged (positively, negatively) with respect to the outside. Next the membrane becomes permeable to (Na^+, K^+, organic) ions and these rush (into, out of) the cell, restoring the original polarization. (p. 466)

13 The original ionic balance is restored by (the sodium-potassium pump, active transport, electrostatic attraction, voltage-gated pumps, diffusion). (pp. 468–71)

14 Myelinated fibers conduct impulses (slower, faster) than unmyelinated fibers. (p. 454)

15 Most synapses are (electrical, chemical). When an impulse traveling down the axon reaches the terminal, it makes the membrane of the terminal permeable to (Na^+, K^+, Ca^{++}), which diffuses into the terminal and promotes the release of (Na^+, K^+, transmitter chemicals) into the synaptic cleft. Synaptic transmission is much (slower, faster) than impulse conduction along the axon. (pp. 472–73)

16 The transmitter substance at neuron–neuron synapses outside the central nervous system is generally (noradrenalin, serotonin, acetylcholine). (p. 473)

17 An excitatory transmitter substance slightly (reduces, increases) the polarization of the postsynaptic membrane, making it (easier, more difficult) to trigger an impulse. If the transmitter had been inhibitory, the membrane would have become (depolarized, more polarized), a condition called (IPSP, EPSP). (pp. 476–77)

18 If both excitatory and inhibitory transmitters converge on a cell at the same time, their effects are combined in the process called (addition, integration). (p. 477)

19 The gap between the end of the motor axon and the effector it innervates is called the (synapse, neuromuscular junction). The transmitter substance released by the motor neurons at their junction with the effectors in both somatic and parasympathetic neurons is (acetylcholine, noradrenalin), whereas the motor neurons of the sympathetic pathways release (acetylcholine, noradrenalin). (p. 485)

Questions 20–24 refer to the following diagram showing the effect of transmitter substance on the membrane potential of a neuron.

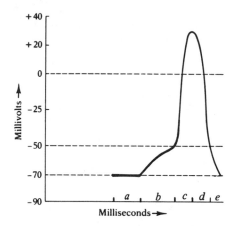

20 In the diagram, what is the resting potential?

a +50 mV c −50 mV
b 0 mV d −70 mV (p. 476)

21 In the diagram, what is the threshold potential?

a +50 mV c −50 mV
b 0 mV d −70 mV (p. 476)

22 During which interval (a, b, c, d, or e in the diagram) would the transmitter substance be active? (p. 476)

23 During which intervals would sodium ions be entering the neuron? (p. 476)

24 During which interval would potassium ions be leaving the neuron? (p. 476)

25 Suppose you pricked your finger with a pin. On the accompanying diagram, draw in and label the components of a reflex arc that originates in the finger. Include the sensory neuron, interneurons, motor neuron, receptor, and effector. Label the parts of each neuron (axon, dendrites, cell body).

Testing knowledge and understanding

Choose the one best answer.

26 Chemotaxis in bacteria involves

a specific receptor molecules in the membrane.
b conduction using second messengers.
c processing through use of second-messenger concentrations.
d response by changing the directional rotation of the cells.
e all of the above are correct (p. 452)

27 Animal control centers tend to be concentrated in the head region; this is known as

a autonomic control.
b corticalization.
c cephalization.
d rationalization. (p. 460)

28 Another evolutionary development that paralleled the above tendency in the vertebrates was

a the increased size and length of the spinal cord.
b the concentration of major sense organs in the head region.
c the increased coordinating ability of the endocrine system.
d the development of efficient locomotor appendages in higher animals.
e the development of the nerve net. (p. 460)

29 One evolutionary advance found in the annelids (such as earthworms) but *not* in the flatworms is

a longitudinal nerve cords.
b prominent ganglia with connecting nerves.
c a nerve net.
d one-way conduction.
e radial symmetry. (p. 462)

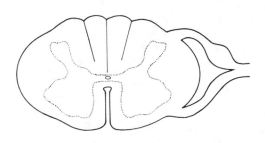
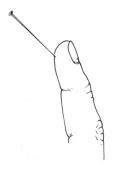

30 Myelin sheaths

   a are found on all axons.
   b are found on all dendrites.
   c speed up conduction.
   d are made of protein.
   e wrap around all sensory neurons. (p. 454)

31 Which of the following structures in nerve cells is specialized for receiving and integrating synaptic inputs from a large number of nerve cells?

   a the axon   *d* neuroglia
   b myelin sheath   *e* dendrites
   c nodes of Ranvier   (p. 453)

32 Which of the following statements most accurately describes what is happening at the point indicated by the X on the diagram of the oscilloscope trace of an action potential shown here?

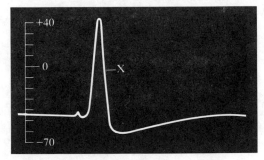

   a K^+ ions are flowing out of the cell.
   b K^+ ions are flowing into the cell.
   c Na^+ ions are flowing into the cell.
   d Na^+ ions are flowing out of the cell.
   e none of the above   (p. 476)

33 Which one of the following statements is *false* concerning the axon in a resting (non-conducting) stage?

   a The membrane is relatively impermeable to sodium ions.
   b The membrane has a positive charge on the outside and a negative charge on the inside.
   c The membrane is highly permeable to large negatively charged organic ions and allows them to leak out.
   d Potassium ions are in higher concentration inside the axon than outside.
   e Sodium ions are in higher concentration outside the axon than inside.   (p. 466)

34 Which one of the following events occurs *first* when a neuron is stimulated?

   a ATP is hydrolyzed.
   b Na^+ channels open and Na^+ rushes inside.
   c K^+ channels open and K^+ rushes inside.

   d The sodium-potassium pump exchanges Na^+ and K^+.
   e Negatively charged organic ions rush outside.   (p. 466)

35 Of the following activities, which is the *second* event to occur in the depolarization of a nerve cell?

   a Na^+ channels open and Na^+ rushes inside.
   b Na^+ channels open and Na^+ rushes outside.
   c K^+ channels open and K^+ rushes inside.
   d K^+ channels open and K^+ rushes outside.
   e Negatively charged ions rush outside.   (p. 466)

36 You have set up two electrical recording instruments that will record any electrical changes that occur. One is placed at each end of a single neuron 10 cm long. If at the first pair of electrodes the magnitude of the action potential is recorded at 100 mV, what will be the magnitude of the action potential when it reaches the second pair of electrodes, at the other end?

   a 10 mV
   b less than 100 mV but cannot be more exact
   c 100 mV
   d more than 100 mV but cannot be more exact
   e 1000 mV   (p. 467)

37 Upon stimulation of a neuron, the influx of sodium ions is the result of

   a active transport.
   b diffusion.   (p. 466)

38 When a nerve cell is stimulated, the magnitude of the potential difference across the plasma membrane

   a increases.
   b decreases.   (p. 466)

39 Which one of the following statements concerning the sodium-potassium pump is *false*?

   a The protein complex has a binding site for sodium and a separate binding site for potassium.
   b Sodium is transported out of the cell and potassium is transported into the cell.
   c The pump transports sodium either into or out of the cell.
   d The energy for the pump is provided by the hydrolysis of ATP.
   e The protein complex has two different possible conformations.   (pp. 470–71)

40 Suppose an intracellular electrode records a resting potential for a given neutron of −70 millivolts. Which one of the recordings below best portrays what will happen if an inhibitory neurotransmitter substance is applied to the neuron? (The arrows indicate time of application.) (p. 476)

a

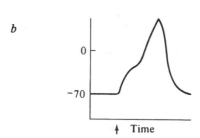

b

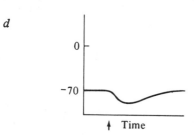

c

d

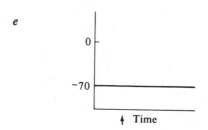

e

41 Suppose you administer a drug that efficiently removes calcium. The effect of this drug on the nervous system would be to

  a prevent action potentials from traveling along axons.
  b prevent transmitter substance in the synaptic cleft from depolarizing the postsynaptic membrane.
  c inhibit the sodium-potassium pump.
  d prevent synaptic vesicles from discharging their contents into the synaptic cleft.
  e prevent transmitter substance from causing the depolarization of the presynaptic membrane. (p. 473)

42 A drug that produces which one of the following effects would probably be *least* effective as a tranquilizer?

  a interferes with uptake of noradrenalin into synaptic vesicles
  b interferes with synthesis of noradrenalin in nerve cells
  c prevents release of noradrenalin from synaptic vesicles
  d blocks receptor sites for noradrenalin on postsynaptic membranes
  e enhances the sensitivity of neurons to noradrenalin (p. 485)

43 The drug lithium carbonate is effective in the treatment of manic-depressive individuals. Its action is to reduce the release of noradrenalin and enhance its reuptake. The effect of ingesting this drug would be to

  a decrease the firing of the postsynaptic neurons at noradrenalin synapses.
  b enhance the sensitivity of the postsynaptic neurons to noradrenalin.
  c block receptor sites for noradrenalin on postsynaptic membranes.
  d stimulate uncontrolled firing of the postsynaptic neurons at noradrenalin synapses.
  e slow down the rate of action-potential movement in the presynaptic neuron. (p. 485)

44 Integration in the nervous system depends upon the addition of _____ from different sources.

  a action potentials
  b specific ions
  c excitatory and inhibitory postsynaptic potentials
  d synaptic vesicles
  e resting potentials (p. 477)

45 Many poisons and drugs act at synapses. One kind of snake venom binds irreversibly with acetylcholine and thus, in an acetylcholine synapse, would be expected to

a prevent the postsynaptic membrane from depolarizing.
b cause the postsynaptic membrane to depolarize only once.
c cause the postsynaptic membrane to depolarize many times in succession.
d produce muscular spasms.
e prevent cholinesterase production.
(p. 473)

46 The diagram below depicts a typical vertebrate reflex arc involving a single receptor (R), a muscle (M), and three neurons. Assume that you apply an electrical stimulus at point X. What is the *most distant* point from X at which you would expect to detect a signal?

(p. 460)

47 What is meant by a reflex arc in the nervous system?

a an inherited behavior pattern that functions through a certain neural pathway
b a functional unit consisting of a receptor, neural pathways, and an effector
c peripheral nerves, spinal cord, and brain
d a homeostatic system of sensory nerves, synapses, and motor nerves
e automatic responses of the central nervous system
(pp. 480–81)

48 A nerve

a is the same as a neuron.
b is a pathway within the somatic nervous system only.
c is a bundle of neurons.
d usually contains only motor neurons.
e usually contains only sensory neurons.
(p. 458)

49 Where A stands for axon, D for dendrite, S for synapse, and CB for cell body, a typical sequence of structures between a receptor and an effector is

a D–CB–A–S–D–CB–A.
b A–D–CB–S–A–D–CB.
c D–CB–A–S–A–CB–D.
d D–A–S–CB–D–A–CB.
e A–CB–D–S–D–CB–A.   (pp. 453–54)

50 A certain neuron is encountered during dissection of the muscles in the leg of a pig. It is part of the somatic (voluntary) nervous system. Which one of the following observations indicates that it is a motor neuron, not a sensory neuron?

a Its cell body is located inside the spinal cord.
b It exhibits an all-or-none response to stimuli.
c The end of its axon secretes a neurotransmitter.
d Its axon is myelinated.   (p. 481)

51 Which one of the following statements most closely fits the knee-jerk reflex arc?

a It involves more than two synaptic junctions, but no brain control.
b It involves one synaptic junction and no brain control.
c It normally involves two synaptic junctions and can be modified by the brain.
d It normally involves one synaptic junction and can be modified by the brain.
e The stretch receptor connects with the brain stem and sends impulses to the dorsal root ganglion motor neurons.
(p. 481)

52 The rate of the human heartbeat is controlled by which of the following? (Pick the most complete and specific answer.)

a sympathetic nervous system
b parasympathetic nervous system
c somatic nervous system
d autonomic nervous system
e peripheral nervous system   (p. 487)

53 The major function of the autonomic nervous system is to

a transmit impulses from the brain to the central nervous system.
b regulate and control the peripheral nervous system.
c control the contraction of skeletal muscles.
d innervate the internal organs.
e coordinate nerve impulses.   (p. 483)

54 A true *difference* between the somatic and the autonomic nervous systems is that

a acetylcholine is found as a neurotransmitter only in the somatic system.

b only the somatic system has motor neurons.

c only the autonomic system connects to the central nervous system.

d only the somatic system can cause muscle contraction.

e the somatic system innervates skeletal muscle whereas the autonomic system innervates smooth muscle. (p. 483)

55 A certain neuron in a cat is located entirely outside the central nervous system. The synaptic vesicles of its axon release a transmitter substance that is not destroyed by acetylcholinesterase. This neuron is

a a sensory neuron of the somatic (voluntary) portion of the nervous system.

b a motor neuron of the somatic system.

c a second motor neuron of the sympathetic system.

d a first motor neuron of the parasympathetic system.

e a second motor neuron of the parasympathetic system. (pp. 483–85)

56 All of the following are functions of the autonomic nervous system *except*

a causing an endocrine gland to secrete hormones.

b contracting certain blood vessels and dilating others.

c causing changes in the heart rate.

d causing the muscles of the intestinal wall to contract.

e causing the muscles of the arm to contract in response to a pinprick. (pp. 483–87)

57 All of the following are functions of the sympathetic nervous system *except*

a dilation of intestinal capillaries.

b acceleration of the rate of heartbeat.

c dilation of capillaries in some skeletal muscle.

d erection of hairs on the skin.

e decrease of peristalsis of the digestive tract. (pp. 484–87)

58 The control of many complex rhythmic behaviors is based on

a chains of reflexes.

b motor programs.

c habituation.

d input from stretch receptors alone.

e autonomic control. (p. 488)

For further thought

1 As you are crossing a street, a car suddenly swerves toward you. Explain how the nervous system enables you to meet this emergency.

2 Certain drugs act as metabolic poisons and prevent the synthesis of ATP. Suppose you treated a neuron with one of these drugs. Would the neuron conduct an impulse? Explain.

3 Cocaine is a drug that inhibits the active reuptake of noradrenalin into the brain's presynaptic neurons. It also binds inside sodium channels and blocks them. Would you classify cocaine as a stimulant or a depressant? What effect might cocaine have on a person's heart rate, blood pressure, and mental awareness?

4 "Voodoomen" in the Caribbean islands often fed pufferfish to their enemies. The effect was dramatic; the individuals appeared dead. They were left in the fields and within a few days, would wander back to the villages as mindless "zombies." The voodoomen then took control of them and put them to work in the fields. Pufferfish is now known to contain a poisonous substance, tetrodotoxin (TTX), that blocks voltage-sensitive sodium channels in both nerve and muscle. Explain the effects of TTX on the individual.

ANSWERS

Testing recall

1 sensory reception, conduction, response
2 increases
3 brain, spinal cord
4 a diffuse nerve net
5 two ventral nerve cords, simple sense organs at the anterior end
6 double, solid, ventral
7 limited
8 dendrites, axons, many, one, sensory neurons, interneurons
9 neuroglia, myelin sheath
10 blood vessels, digestive tract, respiratory system, reproductive system
11 two, sympathetic, parasympathetic
12 negative, Na^+, into, positively, K^+, out of
13 sodium-potassium pump, active transport, electrostatic attraction, diffusion
14 faster
15 chemical, Ca^{++}, transmitter chemicals, slower
16 acetylcholine
17 reduces, easier, more polarized, IPSP
18 integration
19 neuromuscular junction, acetylcholine, noradrenalin

20 *d*   22 *b*   24 *d*
21 *c*   23 *b, c*

25

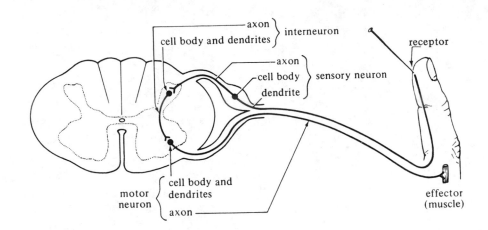

Testing knowledge and understanding

26	e	35	d	43	a	51	d
27	c	36	c	44	c	52	d
28	b	37	b	45	a	53	d
29	b	38	b	46	e	54	e
30	c	39	c	47	b	55	c
31	e	40	d	48	c	56	e
32	a	41	d	49	a	57	a
33	c	42	e	50	a	58	b
34	b						

SENSORY RECEPTION AND PROCESSING

A GENERAL GUIDE TO THE READING

This chapter, the final chapter in the series on control mechanisms, describes the major senses and the structure and function of the vertebrate brain. Though the nervous systems of other animals are discussed, the major emphasis is on the sense organs and brain of the human being. It is after all our brain that sets us apart from other animals. As you read Chapter 19 in your text, you will want to concentrate on the following topics.

1 Receptor function. Try to relate the material on receptor function (pp. 491–510) to the material presented in Chapter 18, particularly in the sections on reflex arcs (pp. 480–83) and on nerve-impulse transmission (especially pp. 464–65 and 472–73). Study carefully Figures 19.4–19.9 (pp. 495–97), 19.20 (p. 503), and 19.26 (p. 508) to gain a good understanding of the various receptor types.

2 Vision. Vision is one of the two senses treated in some detail in this text; the other is hear-

ing. Figures 19.13 (p. 500), 19.17 (p. 501), 19.19 (p. 502), and 19.21 (p. 504) are helpful in understanding the anatomy of the different types of eyes.

3 Hearing. Many students find the mechanism of hearing a little difficult to grasp. You may need to devote extra time to this important topic (pp. 507–511).

4 Sensory specializations. This section (pp. 512–13) describes some special senses: infrared vision, electric, and magnetic senses.

5 Sensory processing. Read carefully the material on lateral inhibition and spot detectors (pp. 514–16) since an understanding of these will help you understand sensory processing in the cerebral cortex (pp. 523–25) at the end of the chapter.

6 Evolution of the vertebrate brain. Pay particular attention to the section on this topic (pp. 517–19), noting the changes at each stage from fish through amphibian and reptile to mammal.

7 The mammalian forebrain. The text (pp. 520–25) presents a fascinating account of the main divisions of this vital organ; here you will learn about the basic structure and functioning of the human brain.

8 The material on sensory processing in the cerebral cortex (pp. 523–25) is complex. You will need to read this section carefully and study Figures 19.43–19.47 (pp. 523–24) in order to fully understand this material.

KEY CONCEPTS

1 Each type of sensory receptor functions as a transducer, converting the energy that constitutes the particular stimulus to which it is attuned into neural stimuli. (p. 491)

2 Regardless of specialization, each type of receptor translates the stimulus into a change in membrane polarization. In most sensory cells, the stimulus opens or closes specific ion channels in the membrane. (p. 492)

3 The receptors of taste and smell are chemoreceptors; they are sensitive to solutions of different kinds of chemicals, which can bind to them by weak bonds. (pp. 495–98)

4 The mechanism of light detection is similar in all animals that respond to light; light energy produces changes in a light-sensitive pigment. (pp. 498–507)

5 Sensory hair cells in the inner ear are responsible for detecting vibrations, acceleration, and position changes that affect equilibrium. (pp. 507–11)

6 Sensation depends on the structure of the brain; the interpretation of incoming action potentials is based on the destination in the brain of the axons carrying them, not on the actual external stimulus or on some special quality of the impulses themselves. (p. 514)

7 The major evolutionary change in the vertebrate brain has been the steady increase in the size and importance of the cerebrum, with a corresponding decrease in the relative size and importance of the midbrain. (pp. 517–22)

8 In the human brain, the cerebral cortex has taken over many of the functions of other parts of the brain and has become dominant. (pp. 521–25)

9 As specific areas of the brain have increased in size, their internal organization has become considerably more complex. The structured organization makes possible complex synaptic organization and information processing. (p. 519)

10 In higher vertebrates, the cortex performs most of the complex processing by which the sensory information relayed by the thalamus is analyzed and used. (pp. 523–25)

OBJECTIVES

After studying this chapter and reflecting on it, you should be able to carry out the following objectives.

1 Explain how environmental changes lead to changes in membrane polarization, and how sensory cells translate the strength of a stimulus and the temporal pattern of the stimuli into neural stimuli. (pp. 491–9)

2 Describe the pain, touch, and deep-pressure receptors of the skin and the stretch receptor of muscle, and explain how a stimulus can lead to a generator potential and to an action potential in the nerve fiber. (pp. 494–95)

3 Contrast the receptors for taste and for smell with respect to location, basic function, type of receptor cell, and possible mode of action. (pp. 495–98)

4 Using diagrams such as Figures 19.13 (p. 500), 19.14 (p. 500), 19.18 (p. 502), and 19.19 (p. 502), compare the structure of the compound eye with that of the camera-type eye. In doing so, list similarities and differences between the two kinds of eyes; give the advantages and disadvantages of each, and briefly describe the functioning of the two kinds of eyes. (pp. 498–502)

5 Compare and contrast the pinhole eye with the lens eye, giving the advantages and disadvantages of each type. (pp. 501–2)

6 Describe, using diagrams such as Figures 19.18 (p. 502), 19.19 (p. 502), 19.20 (p. 503), and 19.21 (p. 504), the structure and function of the human eye. Contrast the structure, location, and function of the two principal types of visual receptor cells. (pp. 502–6)

7 Describe what happens when photons of light strike the light sensitive pigment rhodopsin, and how this leads to a change in the polarity of the membrane of the rod cell. (pp. 504–5)

8 Using diagrams such as Figures 19.21 (p. 504) and 19.35 (p. 515), point out some of the connections between receptor cells and other

types of neurons in the human retina. Explain the significance of this complex wiring arrangement, describing the role of spot detectors and lateral inhibition in visual processing in the retina. (pp. 504, 514–16)

9   Using a diagram such as Figure 19.26 (p. 508), identify each of the following, if shown, and give a function of each: outer ear, middle ear, inner ear, organ of Corti, cochlear canal, vestibular canal, semicircular canals, utriculus, sacculus, tympanic membrane, Eustachian tube, oval window, round window, basilar membrane, tectorial membrane. Indicate which of the listed parts are involved in hearing, which in maintaining equilibrium, and which in detecting acceleration. (pp. 507–14)

10 Describe the processes by which sound vibrations in the air are transmitted to the hair cells of the organ of Corti. Then explain how we distinguish different pitches. (pp. 510–11)

11 Explain how the inner ear determines the position of the head with respect to gravity, and how it detects acceleration. (pp. 511–12)

12 Describe how special senses detect aspects of the environment such as heat, electricity, and magnetism and explain how these sensory specializations are used in the life of the animals that possess them. (pp. 512–13)

13 Using diagrams such as Figures 19.37 (p. 517) and 19.38 (p. 518), discuss the evolution of the vertebrate brain, pointing out the changes at each stage from fish through amphibian and reptile to mammal. In doing so, identify and give a major function of each of the following: cerebrum, cerebellum, medulla, thalamus, hypothalamus, optic holes (midbrain), olfactory bulb. (pp. 517–19)

14 Using a diagram such as Figure 19.39 (p. 519), compare the areas of the cerebral cortex devoted to sensory and motor functions and to association in the cat, the monkey, and the human being. (p. 519)

15 List one major function for each of the following parts of the mammalian forebrain: thalamus, reticular formation, hypothalamus, limbic system. (pp. 520–21)

16 Trace the flow of visual information from the ganglion cells in the retina to the primary visual cortex, naming in order all the structures involved. (p. 522)

17 Briefly explain the role of feature detectors and sensory processing in the visual system. (pp. 523–25)

KEY TERMS

The following terms are important in this chapter; you should become familiar with them.

sensory transduction (p. 491)
generator potential (p. 492)
basal rate (p. 492)
phasic receptors (p. 493)
tonic receptors (p. 493)
chemoreceptors (p. 495)
taste buds (p. 496)
olfaction (p. 496)
eye cup (p. 499)
compound eye (p. 499)
ommatidium (p. 499)
camera-type eye (p. 500)
rhabdomere (p. 500)
pinhole eye (p. 501)
lens eye (p. 502)

sclera
cornea
choroid
ciliary body
iris
pupil
lens
suspensory ligament
rod cells
cone cells
retina
} parts of the human eye (pp. 502–3)

rhodopsin (p. 504)

outer ear
middle ear
inner ear
tympanic membrane
Eustachian tube
oval window
round window
cochlea
organ of Corti
basilar membrane
tectorial membrane
semicircular canals
utriculus
sacculus
} parts of the human ear (pp. 509–11)

infrared vision (p. 512)
electric sense (p. 512)
magnetic sense (p. 513)
lateral inhibition (p. 515)
spot detector (p. 516)

forebrain
midbrain
hindbrain
medulla oblongata
cerebellum
optic lobes
cerebrum
thalamus
hypothalamus
cerebral cortex
neocortex
} parts of the human brain (pp. 517–18)

SUMMARY

Specialized receptor cells are an animal's principal means of gaining information about its environment; they function as transducers, converting the energy of a stimulus into neural stimuli. Each type of receptor is maximally responsive to just one kind of stimulus. Regardless of specialization, each type of receptor translates the stimulus into a change in membrane polarization. In most sensory cells, the stimulus opens or closes specific gated channels in the membrane.

Stimulation of a sensory receptor produces a local depolarization (*generator potential*) of the receptor cell membrane. When the generator potential reaches threshold level, it triggers an action potential in the nerve fiber. The response of most sensory receptor cells to a stimulus usually falls between the two extremes of receptor response—phasic and tonic responses. A *tonic* receptor fires at a continuous rate proportional to the strength of the stimulus and immediately returns to its unstimulated state once the stimulus is removed. By contrast, a *phasic* receptor returns to its basal firing rate almost immediately after the onset of the stimulus, even while the stimulus is still being applied. Removal of the stimulus causes the firing rate to drop below basal level until adaptation takes place. Tonic cells provide information about the magnitude of a stimulus whereas phasic cells are specialized to detect change.

The skin contains sensory receptors for touch, pressure, heat, cold, and pain. These receive information from the outside environment. Receptors inside the body receive information about the condition of the body itself.

The receptors of taste and smell are *chemoreceptors*; they are sensitive to solutions of different kinds of chemicals, which can bind to them by weak bonds. The receptor cells for the four taste senses are located in taste buds on different areas of the tongue. The sensations we experience are produced by a blending of these four basic sensations. The receptor cells for smell are true neurons;

they are located in the upper part of the nasal passages. These are specialized to detect odors from distant sources.

Almost all animals respond to light stimuli. Most multicellular animals have evolved specialized light receptor cells containing a pigment that undergoes a chemical change when exposed to light.

The light receptors of many invertebrates simply detect the presence of light, and perhaps differences in intensity. Receptor organs capable of detecting the direction of a light source frequently contain many sensory cells oriented at different angles. More complex eyes usually include a lens capable of concentrating light on the receptor cells. Lenses made possible the evolution of image-forming eyes.

There are two basic types of image-forming eyes: *camera-type eyes* (in some molluscs and vertebrates) and *compound eyes* (in insects and crustaceans). A compound eye is made up of many closely packed functional units called ommatidia, each of which acts as a separate receptor. Image formation depends on the light pattern falling on the surface of the compound eye. Since each ommatidium points in a slightly different direction, each is stimulated by light coming from different points.

There are two types of camera eyes—the pinhole eye and the lens eye. In *pinhole eyes*, light passes through a tiny opening and projects an image on the retina. A great disadvantage is that very little light can enter. The *lens eye* uses a single-lens system to focus light on the many receptor cells that make up the *retina*.

In humans, the light rays coming into the eye are focused by the *cornea* and *lens* on the light-sensitive *retina*, which contains the specialized receptor cells, the *rods* and *cones*. The sensitive rod cells function in dim light, the cones in bright light. The cones enable us to detect color.

The light-sensitive pigment in the rods (*rhodopsin*) is converted into a different form when struck by light, and is regenerated in the dark. The pigment conversion leads to a change in membrane permeability and an impulse is generated in associated neurons. Cone vision is more complex; there are three types of cones, each containing a different pigment and each sensitive to different wavelengths of light.

Receptors for the sense of hearing are specialized for detection of vibrations. In humans, vibrations in the air pass down the *auditory canal* of the outer ear and strike the *tympanic membrane*, causing it to vibrate. The vibrations are increased in force as they are transmitted by three small bones across the middle ear to the *oval window*. The resultant movement of the oval window produces movement of the fluid in the canals of the *cochlea*, causing the *basilar membrane* to move up and down and rub the hair cells of the *organ of Corti* against the overlying *tectorial membrane*. The stimulus to the hair cells is passed into the associated sensory neurons, which carry impulses to the auditory centers of the brain.

The upper portion of the inner ear consists of three *semicircular canals* and a large *vestibule* that connects them to the cochlea. The sensory hair cells lining the two cavities of the vestibule send information to the brain about the position of the head relative to gravity. The three semicircular canals are concerned with sensing changes in the speed or direction of movement.

Many animals have senses that allow them access to information from other sources, such as heat, electricity, and magnetism. Certain snakes have specialized sensory structures that allow them to locate very faint infrared sources, enabling them to hunt prey in the dark.

Three classes of fishes use electric fields. Strongly electric fishes use modified muscle cells to produce an electric charge which they may use to stun or kill prey or predators; weakly electric fishes produce an electric field around themselves which they monitor for signs of objects that disturb it; and passive electric fish use their electrosensory apparatus to monitor their surroundings.

Various organisms, such as sharks, rays, honey bees, homing pigeons, various migratory birds, and some fishes, can sense the earth's magnetic field. Magnetotactic bacteria have chains of magnetite crystals that rotate them into alignment with the earth's magnetic field.

Sensation depends on the structure of the central nervous system (CNS); the interpretation of incoming action potentials is based on the destination in the brain of the axons carrying them, not on the actual external stimulus or on some special quality of the impulses themselves.

Lateral inhibition, or its equivalent, is involved in the processing of visual information in the CNS in many animals. Certain receptors, when illuminated, fire at a more rapid rate while the adjacent receptors are inhibited from firing by their neighbors' excitation. *Spot detectors* are involved in detecting the stimulus. Similar circuits encode color information by comparing the output of adjacent cone cells.

The brain is the organ of massive sensory processing as well as of motor coordination and control. The brains of primitive vertebrates consist of three irregular swellings at the anterior end of the nerve cord. These three regions, the *forebrain, midbrain,* and *hindbrain*, have been much modified in the evolution of the more advanced vertebrates.

Very early in its evolution the forebrain was divided into the *cerebrum* and the more posterior *thalamus* and *hypothalamus*. The midbrain became specialized as the *optic lobes*, and the hindbrain was modified to form the ventral *medulla oblongata* and the *cerebellum*. The hindbrain has changed little over the course of evolution; the medulla continues to be a control center for some autonomic and nervous pathways, and the cerebellum is still concerned with equilibrium and muscular coordination. The major evolutionary change has been the enormous increase in the size and relative importance of the cerebrum, with a corresponding decrease in the midbrain.

In certain advanced reptiles, a new area of the cerebal cortex, the *neocortex*, evolved. In mammals the neocortex expanded to cover the surface and dominate the other parts of the brain.

As specific areas of the brain have increased in size, their internal organization has become more complex. Primitive brains have unstructured areas, more evolved have neurons grouped into *nuclei*, and highly evolved have the nuclei subdivided to form *laminations*. This structure makes possible complex synaptic organization and information processing. The proportion of the cerebral cortex taken up by purely motor and sensory areas is smaller in humans than in other animals; the association areas occupy the greatest proportion of the cortex.

The mammalian forebrain consists of the thalamus, hypothalamus, and cortex. The thalamus contains part of the important *reticular system*, a network of neurons that runs through the medulla, midbrain, and thalamus. The reticular system receives inputs from "wiretaps" on all incoming and outgoing brain communication channels. It activates the brain upon receipt of stimuli and is an indispensable filter that lets only a few of the major sensory inputs reach the brain's higher centers.

Besides serving as a crucial link between the nervous and endocrine systems, the *hypothalamus* is also the control center for the visceral functions of the body and a major integrating region for emotional responses.

In higher vertebrates, the cortex performs most of the complex processing by which the sensory information relayed by the hypothalamus is analyzed and used. The cortex contains discrete areas activated by different sensory systems. The area of the cortex devoted to each body part is proportional to the importance of that part's sensory or motor activities.

The optic nerves from each eye meet in the *optic chiasm*, where the fibers from the medial half of each eye cross over so that the fibers from the left half of each retina go to the lateral geniculate nucleus on the left side of the thalamus and those from the right half to the right side. From there each nucleus sends axons to the *primary visual cortex* where the two images are integrated and processing occurs. In addition to lateral inhibition of the spot detectors, many vertebrates have other feature-detector circuits such as spot-motion detectors and line detectors. The much edited picture of the world in the left or right visual cortex is passed to several nearby places in the cortex where further transformations are made. Complex organization like that of the visual system appears to be the rule rather than the exception in the brain.

KEY DIAGRAM

The following coloring exercise is designed to help you learn the different parts of the brain and their relative sizes in the brains of different vertebrates. The diagrams below show the brains of the codfish, frog, alligator, and horse. Color the titles and the appropriate structures on each of the diagrams, using a different color for each. (Reference: Figure 19.38, p. 518)

CEREBRUM

OPTIC LOBE **(except on horse)**

CEREBRUM

CEREBELLUM

MEDULLA

OLFACTORY BULB

OPTIC NERVE

Codfish

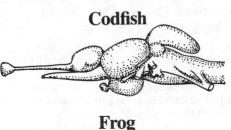

Frog

Alligator

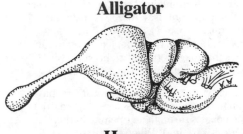

Horse

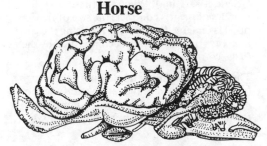

QUESTIONS

Testing recall

Select the correct term or terms to complete each statement.

1 Sensory receptors may be (portions of nerve cells, specialized cells); most are maximally responsive to (one, many) kind(s) of stimulus. (p. 491)

2 The sensation experienced depends on the (type of stimulus, neural message, part of the brain stimulated). (p. 492)

3 When a sensory receptor is stimulated, a small local depolarization called the (action potential, generator potential) is produced. When this reaches threshold level, it triggers a(n) (action potential, generator potential) in the nerve fiber. The (generator potential, action potential) increases in direct relation to the increase in intensity of the stimulus. (p. 492)

4 Sensory receptors that fire at a continuous rate proportional to the strength of the stimulus are (phasic, tonic) receptors, whereas those that adapt almost instantly are (phasic, tonic). (Phasic, tonic) cells provide information about the magnitude of a stimulus and (phasic, tonic) cells are specialized to detect changes. (p. 493)

5 Taste and smell receptors are (touch receptors, chemoreceptors). Taste receptors are (neurons, specialized receptor cells), whereas olfactory receptors are (neurons, specialized receptor cells). (p. 495)

6 (Compound eyes, camera-type eyes) are capable of forming images. Camera-type eyes are found in (molluscs, insects, crustaceans, vertebrates). (p. 501) The eye of the chambered nautilus is a (pinhole, lens) eye; in such an eye the image is (dim, bright) since (little, much) light is admitted. (pp. 499–502)

7 Light entering the human eye must first pass through the transparent (sclera, cornea, iris); it then passes through an opening in the (sclera, iris, retina). When the circular muscles of the iris contract, the size of the pupil is (enlarged, reduced). Focusing the light rays on the retina is accomplished by the (lens, cornea). (pp. 502–3)

8 The rod cells are more numerous in the (center, periphery) of the retina and function in (dim, bright) light. They are responsible for (color, black and white) vision. (p. 503)

9 Within the retina there is (are) (one, several) set(s) of synapses that enable the eye to modify information. (p. 504)

10 Low-frequency sounds stimulate the hair cells near the (base, tip) of the cochlea. (pp. 510–11)

11 The hair cells of the semicircular canals function in sensing (the position of the head, acceleration). (p. 511)

12 Some animals, such as rattlesnakes, have pit organs that are sensitive to (red light, blue light, infrared, magnetism), which enables them to detect prey in the dark. (p. 512)

13 Below are listed nine parts of the human body involved in hearing. Arrange them in the order in which they function in sound reception.

a	auditory centers in the brain	f	oval window
b	auditory canal	g	pinna
c	cochlear fluid	h	sensory neurons
d	middle ear bones	i	tympanic membrane
e	organ of Corti		(pp. 509–11)

Match each function and description below with the associated part or parts of the brain.

a	cerebellum	e	optic lobes
b	cerebrum	f	thalamus
c	hypothalamus	g	reticular system
d	medulla oblongata		

14 part of the forebrain (pp. 520–21)

15 part of the midbrain (p. 517)

16 part of the hindbrain (p. 517)

17 a major sensory-integration center in lower vertebrates (p. 517)

18 filter for incoming sensory information (p. 520)

19 area concerned with balance, equilibrium, and muscular coordination (p. 517)

20 important in regulating the endocrine system (p. 521)

21 site of primary visual cortex (p. 522)

22 vital centers for breathing and heart rate (p. 517)

23 associated with memory and learning (pp. 521–22)

24 centers for thirst and hunger (p. 521)

25 concerned with smell in lower vertebrates (p. 518)

Testing knowledge and understanding

Choose the one best answer.

26 The sensory receptors for smell in the human nasal cavity are

  a closely associated with sensory neurons.
  b similar to light receptors.
  c chemoreceptors.
  d hair cells sensitive to mechanical distortion. (p. 496)

27 When a single ommatidium in a compound eye is stimulated, the receptors in neighboring ommatidia fall silent in a phenomenon called

  a lateral inhibition.
  b spot detection.
  c Mach bands.
  d sensory inhibition.
  e optic chiasm. (p. 515)

28 In a compound eye, incoming light rays will pass through all the following structures *except*

  a rhabdomeres
  b crystalline cone
  c pigment cell
  d lens
  e microvilli of receptor cells (p. 499)

29 The light receptors of the human eye are located in the

  a choroid. *d* cornea.
  b sclera. *e* retina.
  c lens. (p. 503)

30 The rods of the human eye

  a are concentrated in the center of the retina.
  b are sensitive to different colors of light.
  c function only in bright light.
  d contain rhodopsin.
  e two of the above (p. 503)

31 Light entering the normal human eye

  a is bent by the cornea.
  b is bent by the lens.
  c is focused behind the eyeball.
  d is focused on the pupil.
  e *a* and *b* (p. 502)

32 It is easier to see very dim objects (such as certain stars) by using peripheral vision because

  a images are not focused at night.
  b the rods are on the periphery of the retina.
  c small eye movements are necessary for vision.

  d the pupil enlarges at night.
  e the image is formed upside-down on the retina. (p. 503)

33 The photoreceptor molecule rhodopsin is

  a composed of a protein and a non-protein prosthetic group called retinal.
  b a light-sensitive protein without a non-protein prosthetic group.
  c a derivative of vitamin A which is light-sensitive.
  d a component of the cone cell membrane.
  e found only in insects. (p. 504)

34 Rod cells within the retina of a human eye

  a give rise to more sharply defined images than do cone cells.
  b are more sensitive in dim light than are cone cells.
  c are more abundant toward the center of the retina.
  d connect directly to the optic region of the brain with a single intervening interneuron.
  e Two of the above are correct (p. 503)

Below is a diagram of the inner ear. Use the letters on the diagram to answer the following five questions. More than one answer may be correct.

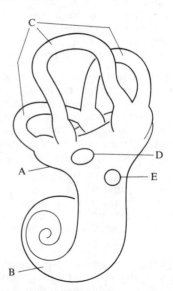

35 senses a change in the direction of motion (p. 511)

36 contains the organ of Corti (p. 508)

37 connects to stapes (p. 508)

38 contains sensory hair cells (pp. 510, 511)

39 sense of hearing located here (p. 509)

40 In humans, the axons from cells in the left
    half of the retina in each eye

    a   travel together to the left side of the
        thalamus.
    b   cross over in the optic chiasm.
    c   are mapped on the right primary visual
        cortex.
    d   lead to the midbrain.   (p. 522)

41 The sensory cells for balance and equilibrium
    are located in the

    a   cochlea.
    b   cerebellum.
    c   semicircular canals.
    d   sacculus and utriculus.
    e   organ of Corti.   (p. 511)

42 Which one of the following groups consists
    of structures all derived from the forebrain?

    a   cerebral cortex, thalamus, cerebellum
    b   hypothalamus, cerebellum, medulla
    c   midbrain, cerebrum, hypothalamus
    d   thalamus, hypothalamus, cerebral cortex
    e   cerebral cortex, medulla, cerebellum
                                    (pp. 517–18)

43 Over the course of vertebrate evolution the
    position of the chief control center has
    shifted from the

    a   hindbrain to the midbrain.
    b   midbrain to the forebrain.
    c   cerebellum to the cerebrum.
    d   cerebrum to the cerebellum.
    e   hindbrain to the cerebrum.   (p. 518)

44 Function maps of the surface of the human
    cerebral cortex reveal a particular relation-
    ship between the sensory and motor capabili-
    ties of various body parts and the size of the
    areas devoted to them. In general, which one
    of the following areas of the cortex surface
    would you expect to be the *smallest*?

    a   sensory area for the lips
    b   sensory area for the hands
    c   sensory area for the tongue
    d   motor area for muscles of the fingers
    e   motor area for muscles of the ankle
                                        (p. 522)

45 A portion of the brain especially concerned
    with filtering incoming sensory information
    and allowing only some of it to reach centers
    of consciousness is the

    a   reticular system.   d   cerebellum.
    b   hypothalamus.   e   cerebrum.
    c   medulla.   (p. 520)

46 Which one of the following portions of the
    brain is incorrectly matched with a function?

    a   reticular system—filtering incoming
        stimuli
    b   medulla—control of breathing
    c   hypothalamus—control of emotions
    d   frontal lobes of the cerebrum—hearing
    e   cerebellum—motor coordination
                                    (pp. 517–18)

47 The area of the brain that acts as the center for
    behavioral drives in higher vertebrates is the

    a   cerebral cortex.   d   midbrain.
    b   hypothalamus.   e   thalamus.
    c   cerebellum.   (p. 521)

48 A person who has a stroke affecting the visual
    area on the left side of the visual cortex may
    be rendered

    a   completely blind.
    b   unable to see on the left side.
    c   unable to see on the right side.
    d   unable to interpret visual information
        from the left side.   (p. 522)

49 Areas of the brain where neurons are grouped
    together with cell bodies and neurons in the
    center and axons on the periphery are called

    a   laminations.   d   nuclei.
    b   spot detectors.   e   reticulations.
    c   feature detectors.   (p. 519)

50 A group of brain areas located in the brain-
    stem that are related functionally in giving
    rise to feeling and emotions are known as the

    a   reticular formation.
    b   limbic system.
    c   neocortex.
    d   hypothalamus.
    e   midbrain.   (p. 521)

For further thought

1   Distinguish between the sense of smell and
    the sense of taste. How does the sense of
    smell contribute to the sense of taste?

2   When you walk into a dark room from bright
    sunlight, you cannot see anything. However,
    after some time your vision returns. Explain
    the initial lack of vision and why the vision
    returns.

3   A person in an automobile accident sustained
    severe head injuries. A brain scan showed
    damage to the central portion of the cere-
    brum on the right side? What symptoms
    would you expect this person to show?

ANSWERS

Testing recall

1 portions of nerve cells, specialized cells, one
2 part of brain stimulated
3 generator potential, action potential, generator potential
4 tonic, phasic, tonic, phasic
5 chemoreceptors, specialized receptor cells, neurons
6 compound eyes, camera-type eyes, molluscs, vertebrates, pinhole, dim, little
7 cornea, iris, reduced, lens, cornea
8 periphery, dim, black and white
9 several
10 tip
11 acceleration
12 infrared
13 *g, b, i, d, f, c, e, h, a*

14	*b, c, f, g*	18	*g*	22	*d*
15	*e, g*	19	*a*	23	*b*
16	*a, d, g*	20	*c*	24	*c*
17	*f*	21	*b*	25	*b*

Testing knowledge and understanding

26	*c*	33	*a*	39	*b*	45	*a*
27	*a*	34	*b*	40	*a*	46	*d*
28	*c*	35	*c*	41	*d*	47	*b*
29	*e*	36	*b*	42	*d*	48	*c*
30	*d*	37	*d*	43	*b*	49	*d*
31	*e*	38	*a, b, c*	44	*e*	50	*b*
32	*b*						

EFFECTORS

A GENERAL GUIDE TO THE READING

This chapter is designed to give you an appreciation of the role of muscles and skeletons in effecting movement. Here, as in other chapters, a variety of animals are discussed, but special attention is given to vertebrates—in this instance to the human skeleton and muscles. As you read Chapter 20 in your text, you will want to concentrate on the following topics.

1   Skeletons. Many students fail to recognize the importance of a skeleton in effecting movement. Muscles can contract only if they have a firm base or skeleton against which they can pull. Try to keep this in mind as you read pages 527–33, which describe the three basic skeletal types.

2   Vertebrate muscle. The section on the three types of vertebrate muscle (pp. 533–35) should be studied carefully. You will want to pay particular attention to the material on skeletal muscle, since this is prerequisite to

understanding the later discussion of muscle contraction, which deals exclusively with skeletal muscle.

3   Muscle contraction. The two sections on this topic, "The Molecular Basis of Contraction" (pp. 538–44) and "The Electrochemical Control of Contraction" (pp. 544–49), are difficult, and you may need to read them over several times. Pay close attention to the figures that accompany the text; these are an essential part of the material, and will help you learn it.

4   Movements produced by microtubules. This section focuses primarily on the movements of cilia and flagella. As you read these pages, contrast the movement of muscle with that of cilia.

KEY CONCEPTS

1 The underlying mechanism of effectors of motion depends on either microfilaments or microtubules. Muscular movement is produced by the action of microfilaments. (p. 527)

2 The principal effectors of movement in higher animals are the muscle cells—elongate cells specialized for contraction. (p. 527)

3 A skeleton plays an important role in effecting movement; it provides the mechanical resistance against which the muscles can act. (pp. 527–32)

4 Individual muscle fibers resemble individual nerve cells in firing only if a stimulus is of threshold intensity, duration, or rate. (pp. 535–37)

5 Muscle contraction involves a ratchetlike mechanism in which actin and myosin microfilaments slide past one another. ATP provides the energy for this movement. (pp. 538–44)

6 Stimulatory neurotransmitter substances released by the neuronal axon cause a depolarization of the muscle membrane. If the depolarization is above threshold, an action potential is triggered over the surface of the fiber by the same combination of voltage-gated channels that is seen in nerve-cell membranes. The action potential activates the contraction process by means of Ca^{++} ions. (pp. 544–45)

7 The beating of cilia and flagella is believed to involve a ratchetlike mechanism in which the microtubules slide over one another. ATP provides the energy for this movement. (pp. 550–52)

OBJECTIVES

After studying this chapter and reflecting on it, you should be able to carry out the following objectives.

1 Describe the locomotion characteristic of an earthworm. In doing so, discuss the roles of the longitudinal and the circular muscles in effecting movement, and explain the importance of the hydrostatic skeleton; finally, show how segmentation is an advantage to an animal like an earthworm. (p. 528)

2 Distinguish between an exoskeleton and an endoskeleton, and list the advantages and limitations of each. Using diagrams such as those in Figure 20.6 (p. 531), explain how an appendage can be flexed by an insect and by a human. (pp. 530–32)

3 Using a diagram such as Figure 20.7 (p. 532), point out a Haversian system, a Haversian canal, the spaces where the bone cells are located, and the intercellular matrix. (pp. 532–33)

4 Give the difference between the two terms in each of the following pairs: appendicular skeleton–axial skeleton, origin of a muscle–insertion of a muscle, tendon–ligament. (p. 533)

5 Using a diagram such as Figure 20.9 (p. 534), identify smooth, skeletal (striated), and cardiac muscle. Contrast the three types of muscle with respect to the shape of the cell, the presence or absence of multiple nuclei in a cell, the presence or absence of striations, and the source of innervation (the somatic or the autonomic nervous system). (pp. 534–35)

6 For each of the following tissues, indicate whether the muscle cells are predominantly striated or smooth: iris of the eye, wall of an artery, leg muscle, abdominal muscle, tongue, wall of the small intestine, face muscle, wall of the esophagus. (pp. 534–35)

7 Using diagrams such as Figures 20.12 (p. 537) and 20.13 (p. 537), explain the terms simple twitch, tetanus, summation, and fatigue; compare the extent of shortening of a muscle undergoing simple twitches with that of a muscle in tetanus. Give the causes of fatigue in a muscle. (pp. 535–37)

8 Using diagrams such as Figures 20.14 (p. 539) and 20.16 (p. 541), identify a sarcomere, a bundle of muscle fibers, a myofibril, a Z line, an I band, an A band, an H zone, a thick filament, and a thin filament. (pp. 539–40)

9 Explain the Huxley sliding-filament theory of skeletal muscle contraction. In doing so, indicate the contribution to muscular contraction of each of the following: actin filament, myosin filament, myosin heads, regulatory proteins, Ca^{++}, ATP, creatine phosphate. (You will find examination of Figures 20.17–20.25 (pp. 542–47) helpful in meeting this objective.) (pp. 542–47)

10 Using a diagram such as Figure 20.22 (p. 545), identify a sarcomere, thick filaments, thin filaments, T tubules, the sarcoplasmic reticulum, and the cell membrane. Describe the relationship between T tubules and the cell membrane of a muscle cell, and explain

the role of the sarcoplasmic reticulum with respect to calcium ions and muscle contraction. (pp. 545–47)

11 Outline the events occurring between the time a nerve impulse reaches a neuromuscular junction and the time the muscle fiber contracts. (pp. 542–49)

12 Review Figure 5.25 (p. 131) to understand the general organization of cilia and flagella. Then, using Figures 20.32 (p. 551) and 20.33 (p. 552), describe the postulated mechanism for the movement of cilia and flagella, mentioning the similarities and differences to the mechanism for muscle contraction. (pp. 551–52)

KEY TERMS

The following terms are important in this chapter; you should become familiar with them.

hydrostatic skeleton (p. 528)
segmentation (p. 528)
exoskeleton (p. 531)
endoskeleton (p. 531)
Haversian system (p. 532)
ligament (p. 533)
tendon (p. 533)
origin of muscle (p. 533)
insertion of muscle (p. 533)
skeletal (voluntary, or striated) muscle (p. 533)
striations (p. 534)
red muscle (p. 534)
white muscle (p. 534)
myoglobin (p. 534)
smooth muscle (p. 534)
heart (cardiac) muscle (p. 535)
simple twitch (p. 537)
latent period (p. 537)
contraction period (p. 537)
relaxation period (p. 537)
summation (p. 537)
tetanus (p. 537)
tonus (p. 537)
fatigue (p. 537)
creatine phosphate (p. 538)
phosphagens (p. 538)
myoglobin (p. 538)
oxygen debt (p. 538)
actin (p. 538)
myosin (p. 538)
actomyosin (p. 539)
myofibril (p. 539)
I band (p. 540)
A band (p. 540)
H zone (p. 540)

Z line (p. 540)
sarcomere (p. 540)
thick filament (p. 540)
thin filament (p. 540)
cross bridges (p. 542)
sarcoplasmic reticulum (p. 545)
T system (p. 545)
tropomyosin (p. 546)
troponin complex (p. 546)
microfilament (p. 549)
microtubule (p. 551)
dynein (p. 551)

SUMMARY

Although plants do move by differential growth or turgor changes, the most elaborate mechanisms for producing movement are found among animals. Movement in animals depends on either microtubules or microfilaments. The *effectors* are the parts of the organism that do things, that carry out the organism's response to a stimulus. The principal effectors of movement in higher animals are the muscle cells.

Movement in many invertebrate animals is produced by the alternating contraction of the longitudinal and circular muscles against the incompressible fluids in the body cavity. The fluids function as a *hydrostatic skeleton*. The hydrostatic skeleton of the earthworm is particularly efficient; the body cavity is segmented, and each segment has its own muscles. Consequently, the worm is capable of localized movement, and burrows very effectively.

Animals with hard jointed skeletons Both arthropods and vertebrates have evolved paired locomotory appendages, a hard jointed skeleton, and elaborate musculature. Arthropods have an *exoskeleton*, a hard body covering with all muscles and organs inside it. Besides providing support, the exoskeleton functions as protective armor and as a waxy barrier to prevent water loss. Periodic molting of the exoskeleton is necessary for growth.

Vertebrates have an *endoskeleton*, an internal framework with the muscles outside. It is composed of bone and/or cartilage. *Cartilage* is firm, but not as hard or brittle as bone. It is found wherever firmness combined with flexibility is needed. *Bone* has a hard, relatively rigid matrix and provides structural support and protection.

Vertebrate skeletons are divided into two components, the *axial* skeleton and the *appendicular* skeleton. Some bones are connected by immovable joints; others are held together at movable joints by *ligaments*. Skeletal muscles, attached to the bones by *tendons*, contract and bend the skeleton at these joints. The movable bones behave like a lever system with the fulcrum at the joint. The action of any specific muscle depends on the position of its *origin* (the end attached to the stationary bone) and its *insertion* (the end attached to the moving bone) and on the type of joint between them. The muscles operate in antagonistic and synergistic groups.

Vertebrates possess three different types of muscle: skeletal, smooth, and cardiac. (This classification does not hold for many invertebrates.) The abundant *skeletal muscle* is responsible for most voluntary movement. Each muscle fiber is long and cylindrical, contains many nuclei, and is crossed by light and dark bands called *striations*. The fibers are usually bound together into bundles, and the bundles into muscles. There are two types of skeletal muscle: *red muscle* (or slow-twitch muscle) and *white muscle* (or fast-twitch muscle). They differ in structure and function. Skeletal muscle is innervated by the somatic nervous system.

Smooth muscle forms the muscle layers in the walls of the viscera and the blood vessels. The spindle-shaped cells interlace to form sheets of tissue, innervated by the autonomic nervous system. Smooth muscle is primarily responsible for movements in response to internal changes, while skeletal muscle is concerned with making adjustments to the external environment. These differences in action are reflected in many differences in the physiological characteristics of these two muscle types.

Cardiac muscle (or heart muscle) shows some characteristics of skeletal muscle and some of smooth muscle. Its fibers are striated, but it is innervated by the autonomic nervous system, and it acts like smooth muscle.

The physiology of skeletal muscle activity Individual muscle fibers contract only if they receive a stimulus of threshold intensity, duration, and rate. In vertebrates, individual muscle fibers seem to exhibit the all-or-none property. When a single threshold stimulus is applied to a muscle, there is a brief *latent period*, which is followed by the *contraction period*. This is immediately followed by the *relaxation period*. These three periods comprise a single *simple twitch* of the muscle.

When frequent stimuli are applied to a muscle, the muscle does not have time to relax between contractions. The result is *summation*. If the stimuli are very frequent, the muscle may not relax at all between successive stimulations; the resulting strong sustained reaction is called *tetanus*. If the frequent stimulation continues, the muscle may fatigue and be unable to sustain the contraction.

The energy for muscle contraction comes from ATP. Very little ATP is stored in muscle, but some high-energy phosphogens such as *creatine phosphate* are; these compounds pass their high-energy phosphate to ADP to form ATP. The energy to replenish the phosphogens and ATP comes from the complete oxidation of nutrients to CO_2 and water. During violent muscular activity, the energy demands are greater than can be met by respiration, because oxygen cannot be gotten to the tissues fast enough. Consequently, lactic acid fermentation takes place, and the muscle incurs an *oxygen debt*. The panting that occurs even after the activity is over supplies the oxygen needed to reconvert the lactic acid into pyruvic acid and to metabolize it.

Analysis of muscle shows that its contractile components are the proteins *actin* and *myosin*. Within each *myofibril* the thick myosin filaments and thin actin filaments are interdigitated, with the myosin located exclusively in the *A bands* and the actin primarily in the *I bands* but extending some distance into the A bands. The contractile unit (between the *Z lines*) is called the *sarcomere*. According to the Huxley sliding-filament theory, cross bridges from the myosin filaments hook onto the actin at specialized receptor sites and bend, pulling the actin along the myosin. The necessary energy comes from the hydrolysis of ATP by the myosin cross bridges. As the filaments slide past each other, the width of the *H zone* and I band decreases.

The control of contraction The membrane of the resting muscle fiber is polarized, with the outside charged positively in relation to the inside. Stimulatory neurotransmitter substances released by the neuronal axon reduce this polarization, triggering an action potential that sweeps across the muscle fiber. When the action potential penetrates into the interior of the fiber via the T-system, calcium ions are released by the *sarcoplasmic reticulum*. The Ca^{++} ions stimulate contraction by binding to the *troponin* complex and causing a conformational change, which moves the *tropomyosin* and exposes the myosin-binding sites of actin. The binding of each myosin cross bridge to actin, with the concomitant release of ADP and phosphate, initiates a conformational change in the cross bridge, whose bending—the power stroke —forces the filaments to slide past one another. The myosin head then dissociates from actin, and its ATP binding site is available. ATP then binds to myosin head, which is "cocked" for a new stroke. The filaments can now slide past each other by repeating the process of binding, power stroke, and release of the cross bridges. As long as Ca^{++} remains available, the muscle can continue to contract. When nervous stimulation ceases, the muscle relaxes because a calcium pump in the sarcoplasmic reticulum membrane moves the Ca^{++} back into the reticulum, and the myosin-binding sites are once again inhibited.

Smooth muscle contraction differs somewhat; Ca^{++} ions activate the myosin, through two intermediate enzymes, before ATP binds. The protein involved is calmodulin, rather than troponin.

Effectors of nonmuscular movement   Other types of eucaryotic cellular movement also depend on microfilaments, most of which also contain actin and a small amount of myosin. As in muscle, the complexing of actin and myosin activates the ATPase activity of the myosin, and contractions are stimulated by Ca^{++} ions. The same fundamental mechanism that is responsible for muscular movement appears to function in many other kinds of movement as well. The movements of microvilli, changes in cell shape, and cytoplasmic streaming all appear to be mediated by an actomyosin system.

Other movements in eucaryotic cells appear to be mediated by microtubules. Cilia and flagella contain a precise 9 + 2 arrangement of microtubules. Attached to the nine peripheral doublets are arms composed of the protein *dynein*, which has ATPase activity. The dynein arms may function as cross bridges between adjacent doublet microtubules, allowing them to slide over one another. Microtubules are also involved in the formation of the spindle in cell division, and in the movement of organelles within the neuronal axon.

QUESTIONS

Testing recall

The next two questions refer to the two bones in the drawing.

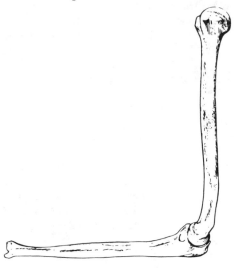

1   Draw in and label: ligaments, two antagonistic muscles (showing origins and insertions), tendons. (p. 531)

2   Do these bones form an endoskeleton, an exoskeleton, or a hydrostatic skeleton? (pp. 528–31)

For items 3–12, indicate whether the statement is true of skeletal, smooth, or cardiac muscle. A statement may be true of more than one kind of muscle.

3   crossed by alternating light and dark bands (p. 534)

4   occurs in two types, red and white (p. 534)

5   each cell contains a single nucleus (p. 534)

6   fibers bound together into bundles by connective tissue (p. 534)

7   innervated by the somatic nervous system (p. 534)

8   innervated by the autonomic nervous system (pp. 534–35)

9   can contract without nervous stimulation (p. 535)

10   capable of slow, sustained contraction (p. 535)

11 effects adjustment to the external environment (p. 534)

12 principal muscle type in insects (p. 535)

Use the kymograph record below to answer questions 13–16.

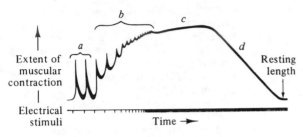

13 Referring to the diagram, state what is happening to the muscle in areas *a, b, c,* and *d*. (p. 537)

14 Why do several stimuli in rapid succession produce more contraction than a single stimulus of the same strength? (p. 537)

15 Are you using simple twitches or tetanic contractions when you write the answer to this question? (p. 537)

16 What causes fatigue? (p. 537)

17 Which of the following substances or processes are used by the cell to generate ATP for muscle contraction? (More than one may be correct.)

a glycolysis   *d* creatine phosphate
b Kreb's cycle   *e* myosin
c cell respiration   (p. 538)

18 Place the following seven events in the contraction of a muscle fiber in order.

a Transmitter substance diffuses across synaptic cleft at neuromuscular junction.
b Actin and myosin slide past each other.
c Permeability of the muscle membrane is altered.
d Impulse arrives at the junction between the motor neuron and the muscle fiber.
e Sarcoplasmic reticulum releases stored calcium ions.
f Action potential travels along membrane of the muscle cells and into the tubules.
g Troponin changes shape and moves the tropomyosin. (pp. 544–45)

Choose the one best answer.

Questions 19–24 refer to the following diagram of a portion of a muscle.

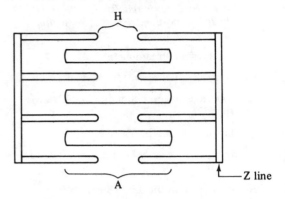

19 This diagram represents

a the sarcoplasmic reticulum.
b a sarcomere.
c the system of T tubules.
d an individual microfilament.
e a myofibril. (p. 541)

20 When a muscle contracts, the distance labeled A in the diagram

a remains unchanged.
b decreases.
c increases. (p. 542)

21 When a muscle contracts, the distance labeled H in the diagram

a remains unchanged.
b decreases.
c increases. (p. 542)

22 What would be the main substance remaining if you dissolved away the thin filaments?

a actin
b myosin (p. 542)

23 What would be the main substance remaining if you dissolved away the thick filaments?

a actin
b myosin (p. 542)

24 The energy for the movement of the cross bridges is supplied by

a ATP.   *d* actin.
b calcium ions.   *e* myosin.
c lactic acid. (p. 544)

For each of the following, indicate whether the movement is due mainly to: (1) *micro-tubules or* (2) *microfilaments. Also specify the probable mechanism of movement:* (a) *actomyosin mediated,* (b) *dynein-tubulin mediated, or* (c) *mediated by some other mechanism.*

25 cytoplasmic streaming (p. 550)

26 contraction of muscle (p. 539)

27 movement of mitochondria in the axon (p. 552)

28 beating of cilia (p. 551)

29 movement of ectoplasm through the endo-plasm (p. 550)

30 movement of the microvilli (p. 549)

31 changes in cell shape (p. 550)

Testing knowledge and understanding

Choose the one best answer.

32 Which of the following statements concern-ing the skeletal system is *false*?

   a Some bones are held together by liga-ments.
   b Bones help protect some internal organs.
   c Bones contain both nerve and blood cells.
   d Aside from nerve and blood cells there are no living cells in bone.
   e Many muscles are attached to bone.
                  (p. 533)

33 An important advantage of an endoskeleton over an exoskeleton is that the former

   a provides more structural support.
   b provides a structure against which muscles can work.
   c helps prevent water loss.
   d does not impose a severe limitation on overall size.
   e helps determine the shape of the animal.
                  (p. 531)

34 The following kymograph record resulted when a frog muscle was stimulated with a *single* stimulus. This is a record of

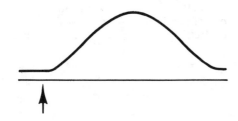

   a summation.   *d* simple twitch.
   b fatigue.   *e* subthreshold
   c tetanus.   stimulus. (p. 537)

35 Which of the following is *not* a difference between red and white varieties of skeletal muscle?

   a Red but not white muscle has a high myoglobin content.
   b White but not red muscle is found in muscle layers of the walls of the digestive tract, bladder, ducts, veins, and arteries.
   c Red muscle has a richer blood supply than does white.
   d White muscle contracts more rapidly than does red.
   e Red muscle is capable of even longer-term activity than is white.   (p. 534)

36 Vertebrate skeletal muscle is usually

   a under control of the sympathetic nervous system.
   b under control of the parasympathetic nervous system.
   c under control of the somatic (voluntary) nervous system.
   d dually innervated by sympathetic and parasympathetic neurons.
   e dually innervated by somatic and para-sympathetic neurons.   (p. 534)

37 If a single excised frog muscle is stimulated electrically and the responses are recorded on a kymograph drum, which one of the follow-ing *cannot* be demonstrated?

   a threshold   *d* fatigue
   b tetanus   *e* antagonism
   c summation   (p. 537)

38 A sustained muscular contraction is

   a fatigue.   *d* oxygen debt.
   b a twitch.   *e* the latent period.
   c tetanus.   (p. 537)

39 Muscular fatigue may occur with an accumu-lation of

   a ATP.   *d* myoglobin.
   b lactic acid.   *e* glucose.
   c oxyhemoglobin.   (p. 537)

40 Regeneration of ATP in the muscle cell utilizes stores of

   a myosin.
   b actin.
   c creatine phosphate.
   d NAD.
   e pyruvic acid.   (p. 538)

41 The energy levels in the muscle are directly restored after muscle contraction by

a cell respiration.
b lactic acid formation.
c ADP hydrolysis.
d protein breakdown.
e oxygen entry.   (p. 538)

42 A muscle has built up an oxygen debt. When there is enough oxygen for aerobic respiration to resume, all of the following will occur *except*

a lactic acid will be converted to pyruvic acid
b O_2 will be used up
c acetyl CoA will be converted to CO_2 and H_2O
d pyruvic acid will be converted to acetyl CoA
e an excess of NAD_{re} will accumulate
                                    (p. 538)

43 Once an impulse is transmitted across the neuromuscular junction by acetylcholine, continued transmission due to the same impulse is prevented by the breakdown of acetylcholine by

a ATPase.
b phosphatase.
c acetylcholinesterase.
d sodium-potassium pump.
e acetylase.   (see Ch. 18, pp. 473, 545)

44 The function of the T tubules in muscle contraction is to

a carry the impulse into the myofibrils of the muscle cell.
b release calcium ions.
c release sodium ions.
d split ATP.
e take up calcium ions.   (p. 545)

45 At the start of a muscle contraction, calcium ions are released from

a actin.
b the T tubule.
c the motor neuron.
d the sarcoplasmic reticulum.
e myosin.   (p. 545)

46 The role of calcium ions in muscle contraction is to

a facilitate the binding of ATP.
b trigger depolarization of the membrane of muscle fibers.
c bind to a regulatory protein associated with actin, allowing cross bridges with myosin to form.

d promote release of vesicles containing transmitter molecules.
e allow myosin to bind to myoglobin.
                                    (p. 545)

47 The reason that an A band within a sarcomere appears darker than adjacent I bands is that

a myofibrils are narrower in diameter at I bands.
b multiple nuclei tend to cluster within the A bands.
c the cell membrane is more opaque near A bands.
d A bands contain both actin and myosin, whereas I bands do not.   (pp. 542–45)

48 During muscle contraction one molecule of ATP is hydrolyzed each time

a a calcium ion is moved out of the sarcoplasmic reticulum into the cytoplasm.
b myosin binds to regulatory proteins.
c actin binds to regulatory proteins.
d a muscle fiber contracts by 30 percent of its length.
e a cross bridge between an actin molecule and a myosin molecule is made and broken.   (pp. 546–47)

49 Which of the following chemicals are necessary for muscle contractions?

a actin, myosin, calcium ions
b actin, myosin, ADP
c actin, ATP, myoglobin
d myosin, calcium phosphate, ATP
e myosin, calcium ions, myoglobin
                                    (pp. 538–39)

50 Which one of the following statements is *false*?

a In skeletal muscles, there is a regular alternation of light-colored I bands and dark-colored A bands.
b It is now thought that the A bands correspond to the lengths of thick myosin filaments.
c When a muscle contracts, the A bands become much shorter whereas the I bands remain roughly the same length.
d It is thought that cross bridges from the myosin filaments, acting like ratchets, provide the pull that slides the filaments together during contraction.
e The direct stimulant for contraction is thought to be release of intracellular calcium from cisternae of the sarcoplasmic reticulum.   (pp. 540–42)

51 When a skeletal muscle contracts, all of the following occur, with one exception. Which is the exception?

a The H zone almost disappears.
b The Z lines move closer together.
c The I bands diminish markedly in width.
d ATP is hydrolyzed to ADP + P$_i$.
e The A bands diminish markedly in width.
(p. 540)

52 To which of the following cellular components does calcium bind to initiate a muscular contraction in smooth muscle?

a actin   *d* myosin
b troponin   *e* calmodulin
c tropomyosin   (p. 549)

For further thought

1 The earthworm has exploited the hydrostatic skeleton to its fullest. How might the earthworm's movements differ if the worm were not segmented?

2 Why does a person pant and sweat after vigorous exercise?

3 Explain what happens when rigor mortis sets in after death.

4 In a chicken or turkey, the breast is white meat (white muscle) whereas the leg is dark meat (red muscle). In many migratory birds the situation is reversed: the breast meat is red muscle, the legs are white. Explain this difference.

ANSWERS

Testing recall

1

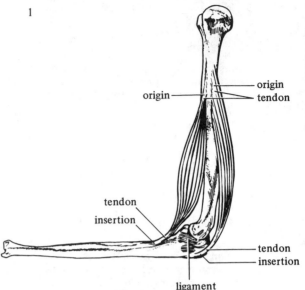

2 endoskeleton   8 smooth, cardiac
3 skeletal, cardiac   9 smooth, cardiac
4 skeletal   10 smooth
5 smooth, cardiac   11 skeletal
6 skeletal   12 skeletal
7 skeletal

13 *a* single twitch: latent period, contraction, relaxation
    b summation: muscle does not fully relax between contractions
    c tetanus: muscle does not relax at all between contractions; slow, sustained contraction
    d fatigue: the result of the buildup of lactic acid

14 The muscle does not fully relax between contractions.
15 tetanic contractions
16 buildup of lactic acid
17 *a, b, c, d*
18 *d, a, c, f, e, b*

19 *b*   24 *a*   28 1, *b*
20 *a*   25 2, *a*   29 2, *a*
21 *b*   26 2, *a*   30 2, *a*
22 *b*   27 1, *c*   31 2, *a*
23 *a*

Testing knowledge and understanding

32	*d*	38	*c*	43	*c*	48	*e*
33	*d*	39	*b*	44	*a*	49	*a*
34	*d*	40	*c*	45	*d*	50	*c*
35	*b*	41	*a*	46	*c*	51	*e*
36	*c*	42	*e*	47	*d*	52	*e*
37	*e*						

Chapter 21

ANIMAL BEHAVIOR

A GENERAL GUIDE TO THE READING

This chapter focuses on the mechanisms behind the behavior of organisms and applies what you have learned about control and effector mechanisms to the question of what, how, and why animals do what they do. As you read Chapter 21 in your text, you will want to concentrate on the following.

1 Fundamental components of behavior. You will want to pay particular attention to the material on sign stimuli and releasers, since understanding it is prerequisite to understanding much of the material that follows in this chapter and the next.

2 Motor programs. Motor programs were previously discussed in Chapter 18. Make sure you understand the difference between a fully innate motor program and a learned motor program.

3 Motivation and drives. Drives are discussed in some detail; you will want to read this carefully. Make sure you understand what a circadian rhythm is, and the role a clock can play in switching behavior patterns.

4 Learning and behavior. The relationship between instinctive behavior and learning is considered in some detail. You will need to know the differences between trial-and-error learning and classical conditioning, and the importance of cultural learning and imprinting on an animal's behavior.

5 Avian navigation. This section is provided to illustrate the interplay of sign stimuli, motor programs, drives, and innately guided learning on this particularly intriguing form of behavior.

KEY CONCEPTS

1 Behavior repertoires of animals are organized through an elaborate coordination of sensory organs, neural pathways and processing networks, the endocrine system, and effector organs. (p. 553)

2 Behavior cannot generally be classified as strictly innate or strictly learned; even the most rigidly automatic behavior depends on the environmental conditions under which it evolved, while most learning appears to be guided by innate mechanisms. (p. 554)

3 The sense organs of various animals have evolved to provide them with the range of sensory experience that suits their particular needs. (p. 555)

4 Releasers can initiate certain critical behavior responses automatically, thus bypassing the time-consuming and error-prone process of learning. (p. 557)

5 Visual, auditory, and olfactory feature detectors underlie much of the instinctive species recognition necessary in most animal communication. (p. 560)

6 Motor programs appear to be all-or-none innate responses to releasers; once begun, they proceed to completion with little or no need for further feedback. On the other hand, learned behavior can become stereotyped and largely automatic, taking on the characteristics of an innate motor program. (pp. 560–62)

7 A drive can alter an animal's threshold to stimuli, thus making a particular behavior more or less likely, or it can substitute entirely new programs for old ones. (p. 563)

8 Most complex innate behavior is probably organized into relatively simple subroutines, each with its own motor program, which start and stop in a pre-set ordered sequence, cued by signals from the environment. (p. 563)

9 All living things, whether individual cells or whole multicellular organisms, appear to have an internal sense of time—a biological clock. (pp. 566–67)

10 Instinctive behavior is modulated adaptively by drives, triggered and guided by external cues, and is subject only to the modifying influence of specific sensory feedback. (pp. 567–68)

11 The changing environmental and social conditions to which many species must adapt require the behavioral flexibility provided by learning. (p. 568)

12 Most animals have species-specific biases that channel their learning along paths that are adaptive for them; they are usually instinctively predisposed to recognize specific stimuli and to try specific sorts of behaviors in specific contexts. (p. 569)

13 Some animals have evolved the ability to pass novel and useful information from generation to generation, thereby saving others the risky and time-consuming exercise of rediscovering by trial and error the lessons of the past. (p. 573)

14 Many animals have incredible navigational abilities that allow them to find their way over great distances through unfamiliar territory. Their abilities arise from a variety of complex neurological and physiological systems that are still not well understood. (p. 575)

OBJECTIVES

After studying this chapter and reflecting on it, you should be able to carry out the following objectives.

1 Define the terms innate and instinct. (p. 554)

2 Describe, citing the experiments of Lorenz and Tinbergen, some of the characteristics of sign stimuli and releasers, and indicate their role in animal behavior. Explain why sign stimuli are biologically adaptive, and why they can be disadvantageous as well. (pp. 556–67)

3 Explain the role that feature-detector circuits play in releaser recognition and describe what is meant by heterogenous summation and supernormal stimuli. (pp. 557–58)

4 Differentiate between innate motor programs and learned motor programs. (pp. 560–62)

5 Discuss motivation and drives, and, using examples, explain how behavioral priorities are modulated. (pp. 563–65)

6 Explain what is meant by a biological clock and by circadian rhythms, and state the role the clock plays in effecting drives and changing behavioral programs. (pp. 566–67)

7 Distinguish between classical conditioning and trial-and-error learning. Explain, using a specific example, how species-specific biases help focus learning. (pp. 568–69)

8 Describe the processes of parental and sexual imprinting and explain what is meant by the critical period. (pp. 570–71)

9 Describe the process of cultural learning, citing examples of food learning and enemy recognition. Explain why cultural learning is adaptive for the animal. (pp. 573–75)

10 Discuss the two navigational strategies used by birds to navigate on long flights, indicating what is meant by a map sense and a compass sense. Then, describe the various cues that migratory birds and homing pigeons use to orient themselves on long flights. (pp. 575–80)

KEY TERMS

The following terms are important in this chapter; you should become familiar with them.

ethology (p. 553)
innate (p. 554)
instinct (p. 554)
sign stimulus (p. 556)
releasers (p. 556)
pheromones (p. 556)
supernormal stimuli (p. 558)
heterogeneous summation (p. 558)
fixed-action pattern (p. 560)
motor program (p. 560)
drives (p. 563)
circadian rhythms (p. 566)
classical conditioning (p. 568)
operant conditioning (p. 569)
trial-and-error learning (p. 569)
parental imprinting (p. 570)
critical period (p. 570)
sexual imprinting (p. 571)
cultural learning (p. 573)
map sense (p. 575)

SUMMARY

The behavior of an animal is organized through an elaborate coordination of sensory organs, neural pathways and processing networks, the endocrine system, and the effector organs. Behavior that depends largely on inherited mechanisms is usually referred to as *innate*. Behavior cannot generally be classified as strictly innate or strictly learned; even the most rigidly automatic behavior depends on the environmental conditions for which it evolved, while most learning appears to be guided by innate mechanisms. *Instinct* is the heritable, genetically specified neural circuitry that organizes and guides behavior.

Fundamental components of behavior Many animals have a range of sensory experience largely outside our own. The sense organs of various animals have evolved to provide them with the range of sensory experience that suits their particular needs.

An animal responds to a limited number of the thousands of stimuli its receptors are detecting. A simple signal that elicits a specific behavioral response is known as a *sign stimulus*. Experiments studying examples of rote behavior have shown that certain stimuli trigger a particular behavior because the organism possesses an exaggerated neural sensitivity to these stimuli: because sign stimuli are said to "release" specific behaviors, they are frequently called *releasers*. Depending on the sensory world of the organisms, behavior may be released by cues from many sensory modalities. Pheromones (odors), sounds, and colors may act as releasers. Releasers can be advantageous; they can initiate certain critical behavior responses automatically, thus bypassing the time-consuming and error-prone process of learning. However, they may be triggered by inappropriate stimuli.

Many releasers depend on feature-detector circuits. Exaggerated features are often superior to the natural stimulus in releasing a response; such stimuli are known as *supernormal stimuli*. Sometimes two or more releasers combine in such a way that the releasing values of the two stimuli add together to produce an increased probability of response. This process is called *heterogeneous summation*. Visual, auditory, and olfactory feature-detectors underlie much of the instinctive species recognition necessary in most animal communication.

Motor programs Some behaviors appear to be all-or-none responses to releasers; once begun, they proceed to completion with little or no need for further feedback. Formerly such behaviors were called *fixed-action patterns*; they are now referred to as *motor programs*.

It is often difficult to be sure whether or not a motor behavior has been learned if it is not actually exhibited at birth. In some cases, particularly among invertebrates, opportunities for learning are so slight that many behaviors seen only in adults must be regarded as innate. Even in vertebrates many behaviors that appear to be learned are actually innate. By contrast, many behaviors we know to be learned can become stereotyped and largely automatic and take on the appearance of fully innate motor programs. Even complex behaviors can be choreographed by instinct. Most complex innate behavior probably involves specific behavioral subroutines, cued by signals from the environment, which stop and start in a pre-set ordered sequence.

Motivation and drives   *Drives* are forces from within an animal that motivate it to do one thing now and another later. A drive may act by altering an animal's threshold to stimuli, thus making a particular behavior more or less likely. Current priorities, established by the shifting thresholds of response of various drives, can help the animal select among many behavioral possibilities. Performing the behavior with the highest priority will lower the urgency of the drive that motivated it, and behavior that was formerly less important will then become the animal's highest priority.

Drives may also act by bringing in or retiring entire behavioral programs. Animals often behave as though some sort of timer is repeatedly switching their behavior patterns from the set appropriate for one stage of their life cycle to the set appropriate for the next stage. All living things, whether individual cells or whole multicellular organisms, appear to have an internal timer—a clock. Because the period of animal clocks is approximately 24 hours, the cycles they control are called *circadian rhythms*. Although the basic period of the clock is innate, it is constantly being reset by the environmental cycle.

Learning and behavior   Purely instinctive behavior frequently lacks adaptive flexibility. The changing environmental and social conditions to which many species must adapt require the behavioral flexibility provided by learning. *Classical conditioning* is the association of a response with a stimulus with which it was not previously associated. In *trial-and-error* learning, animals learn by experiencing the consequences of their behavior and altering it accordingly. Most animals have species-specific biases that channel their learning along paths that are adaptive for them; they are usually instinctively predisposed to recognize specific stimuli and to try specific sorts of behaviors in specific contexts.

Some animals have evolved highly specialized learning programs. In *imprinting*, an animal learns to make a strong association with another organism or object. Imprinting occurs only during a short *critical period*. Various types of imprinting are known. *Parental imprinting* involves the recognition of the parent by the young; it occurs in mammals and birds. The critical period for parental imprinting is early and brief, and the learning is not reversible. During sexual imprinting, an animal learns to recognize its species. Various other types of imprinting are now known.

Some animals have evolved the ability to pass useful information from individual to individual and from generation to generation. Such *cultural learning* is important for passing information about food and in enemy recognition. In both cases, inborn predispositions and learning through experience appear to act together for the transfer of cultural information.

Avian navigation   Many animals can find their way over great distances through unfamiliar territory. Some birds are simply preprogrammed to fly a certain course. Others have a special *map sense*, which makes them behave as though they are aware of longitude and latitude. Migrating birds and homing pigeons have remarkable navigational abilities. Birds (and other animals) can tell compass directions by observing the sun and stars. Pigeons can use magnetic cues in addition to sun cues for orientation. Olfactory cues or measuring the field strength of the earth's magnetic fields may be involved in the map sense.

QUESTIONS

Testing recall

Select the correct term or terms to complete each statement.

1   Behavior that depends largely on inherited mechanisms is usually referred to as (innate, learned, instinctive, releasing) behavior. (p. 554)

2   Prey capture by toads is an example of a(n) (motor program, supernormal stimuli, food learning, heterogenous summation). (pp. 558, 560)

3   In humans, walking is an example of a(n) (innate motor program, learned motor program). (pp. 561–62)

4   The process by which newly hatched goslings follow their parents and recognize them is known as (parental imprinting, classical conditioning, trial-and-error learning, cultural learning). (p. 570)

5   A form of learning that ordinarily occurs quickly but only during a limited phase in the animal's life, and that is difficult to unlearn,

is called (conditioning, insight, habituation, imprinting). (p. 570)

6 The area of the brain that acts as the center for learned motor programs in higher vertebrates is the (cerebral cortex, hypothalamus, cerebellum, midbrain). (p. 562)

7 Circadian rhythms have a period of about a (day, week, month, year). (pp. 566–67)

8 Homing pigeons can use (the sun compass, the star compass, magnetic cues) as orientational cues. (pp. 577, 579)

Testing knowledge and understanding

Choose the one best answer.

9 Behavior can be modified by

a the nature of the sign stimulus.
b the season of the year and the time of day.
c the level of circulating hormones.
d previous experience.
e all the above   (pp. 567–75)

10 In an experiment, a single frog tadpole was raised in an isolated, soundproof box. As soon as the tadpole transformed into a frog, several adult males of the same species were put into the box for several days; they called continuously during this period and were then removed. Except for these males, the original frog was never exposed to another frog. At maturity, it produced a call typical of its species. From the information provided here, what can you say about the basis of this call?

a The call is entirely learned.
b The call is entirely innate.
c The call is partly learned, partly innate.
d The call is neither learned nor innate.
e None of the above can be concluded with certainty.   (p. 554)

11 All of the following statements characterize an instinct *except*

a It is stereotyped.
b Animals respond correctly to a stimulus the first time.
c It is often released by single stimuli.
d It is impossible to modify by learning.
e It is displayed widely in the population.
          (p. 554)

12 When ready to reproduce, the adult male stickleback fish develops a bright red belly, which he displays aggressively to other males when staking out his nesting territory. Cardboard models of the fish placed in a nesting male's territory also elicit an aggressive display if the lower half of the model is painted red. The red belly is an example of a(n)

a reflex.   *d* learned pattern.
b sign stimulus.   *e* taxis.
c instinct.   (p. 556)

13 The courting male stickleback fish responds most vigorously to an extra-large model that looks like the drawing. This preferred but unnatural stimulus is sometimes referred to as a

a releaser.
b supernormal stimulus.
c kinesis.
d displacement activity.
e misdirected effort.   (p. 558)

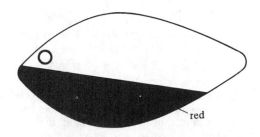

14 In human beings, certain perceptions of tactile stimuli release specific responses; for example, creeping things on the back of the hand release a shaking movement of the hand. This response is probably due to

a innate or instinctive behavior.
b imprinting.
c habituation.
d conditioning.
e insight learning.   (p. 554)

15 Which one of the following best exemplifies an innate motor program?

a a young child playing the piano
b a baby learning to walk
c food-choice behavior in rats
d human speech
e web-weaving by a spider   (p. 560)

16 A just-hatched gull chick pecks vigorously at the red spot on the parent's bill to elicit the feeding response. This red spot is an example of

a imprinting.
b trial-and-error learning.
c classical conditioning.
d a releaser.
e a super-normal stimulus.   (p. 556)

17 In many situations a behavior that is positively reinforced (for example by a reward) becomes more frequently practiced. This is known as

a instinct.
b classical conditioning.

c operant conditioning.
d imprinting.
e habituation. (p. 569)

Questions 18–21 describe experiments or observations that can be classified as examples of one of the forms of learning listed below. For each example, choose the corresponding form of learning. Answers may be used once, more than once, or not at all.

a cultural learning
b habituation
c classical conditioning
d trial and error
e imprinting

18 A duckling will follow any large moving object presented to it right after hatching. (p. 570)

19 Flatworms respond to an electric shock by contracting their body muscles. If the worms are exposed to 50 or 100 electric shocks and at the same time to a beam of light, the worms will learn to contract immediately upon presentation of the light beam even if no shock is present. (p. 568)

20 A hungry brown rat was placed in a closed metal box with a food slot and a food-releasing lever. The rat was allowed to poke around randomly until it accidentally tripped the lever and released a food pellet. The rat soon learned to trip the lever whenever it wanted food. (p. 569)

21 A young chick watches the mother hen peck at food, pick it up, drop it, and give the food call. The chick then pecks at the food. (pp. 573–74)

22 Ivan Pavlov discovered that dogs would learn to salivate upon hearing a sound if the sound was paired with the presentation of food over many trials. This type of learning is called

a habituation.
b classical conditioning.
c operant conditioning.
d cultural learning.
e imprinting. (p. 568)

23 A bird is hatched and raised indoors in complete isolation from other members of its own species. It becomes imprinted on the student who raised it. When the bird is released to the wild as an adult, which one of the following behaviors would you *not* expect it to have difficulty performing correctly?

a singing its species-typical song
b using star patterns in migratory orientation

c finding a mate
d flying (pp. 570–73)

24 Which one of the following statements about imprinting is *true*?

a Responsiveness to the releasing object is lost after the sensitive period is over.
b It occurs only in mammals.
c It is typically reversible, in both lab and natural situations.
d It occurs early in the animal's life. (pp. 570–73)

25 The chaffinch "learns" to sing when it is still a very young bird. Which of the following statements is correct?

a Instinct does not play any role.
b The chaffinch will sing its song without ever hearing another bird.
c The chaffinch will not sing its song unless it hears the song early in life.
d The chaffinch will sing the song of a number of other bird species if it hears the other song, rather than its own, during a critical phase early in life.
e The chaffinch will begin to sing its song if, early in life, it hears the song of another bird species. (p. 570)

26 Which of the following white-crowned sparrows would you expect to sing the most distorted song compared to normal birds of the same species?

a birds that have been hand-reared in isolation from all other birds but have had songs from other species played for them
b birds that have been reared by their parents but have been deafened before they start to sing
c birds that have been reared by their parents but placed in isolation from other birds before they have started to sing
d birds that have been reared in isolation from other birds but that have been exposed to tape recordings of songs from other white-crowned sparrows
e birds that have been reared in isolation from other birds but that have been exposed to tape-recordings of their parents' songs (p. 570)

27 Which of the following statements best describes circadian rhythms?

a They are of exactly 24-hour duration, in the absence of external cues.
b Once the "clock" is set in motion, it is not possible to reset it.
c They are not exactly 24-hour in duration, but are generally reset each day.
d All biological rhythms are circadian.

e Circadian rhythms that are set by light/dark cycles stop when the organism is kept in complete darkness. (pp. 566–67)

28 Which one of the following is *false* concerning circadian rhythms?

a The rhythmic activity pattern often persists for a long period even if the environment is kept constant.

b The rhythmic pattern can be reset to an altered pattern of light and dark cycles.

c When the "clock" is reset to coincide with a new environment all the physiological functions that have a circadian rhythm readjust simultaneously.

d The rhythm appears to be innate.

e The clock is not set precisely to a 24-hour day; it must be reset each day to keep it in synchrony with true time. (p. 567)

29 One which of the following flights would you expect passengers to experience the most severe case of jet lag?

a New York to San Francisco
b Cleveland to Miami
c Chicago to San Francisco
d San Francisco to New Orleans
e San Francisco to Chicago (p. 566)

30 Many animals can determine direction, that is, they have a "compass sense." Which one of the following cues is *not* usually used to determine direction?

a sun   c stars
b moon   d magnetism
                  (pp. 575–80)

31 Which statement is *false* concerning the navigation of homing pigeons?

a Homing pigeons are not the only animals that can return home when displaced.

b Pigeons with the phase of their internal clock shifted by six hours orient homewards on sunny days.

c Pigeons with their internal clock shifted by six hours orient homewards on overcast days.

d Magnets can disorient pigeons on overcast days. (pp. 577–78)

32 A well-trained group of homing pigeons is released (one at a time) *under clear, sunny skies while wearing magnets on their backs*. Which of the following results would you predict for the behavior of the group as a whole?

a They would be oriented in an approximately homeward direction.

b They would be randomly oriented.

c They would be oriented 90 degrees counterclockwise from the homeward direction.

d They would be oriented 90 degrees clockwise from the homeward direction.

e They would be oriented 180 degrees from the homeward direction. (p. 578)

33 Pigeons have a "compass sense" to determine direction and a "map sense" for correct orientation toward home from an unknown location. What is the major cue for the "map sense"?

a sun   d moon
b magnetic field   e we don't know
c stars   (pp. 579–80)

34 What is the most likely mechanism by which indigo buntings orient during their spring migration from the southern states into the northern states?

a the earth's magnetic field
b the sun and an internal sense of time
c odors produced by the northern temperate forest
d the moon
e the stars (p. 577)

Questions 35–37 refer to the following experiment. Several birds of the same species are kept in an outdoor cage and released under various circumstances. For each set of circumstances described in a question, choose the direction in which the released bird would be expected to fly. Answers may be used once, more than once, or not at all.

a north   d west
b south   e northeast
c east

35 In a test conducted at 9 A.M. it is found that the migratory direction of these birds is *south*. If a bird is then kept outdoors so that it can see the sun throughout the day and is retested six hours later, at 3 P.M., in which direction would you expect it to migrate? (p. 577)

36 The same bird is tested again, but immediately after its 9 A.M. test is put in a light-tight box for a day; during this time it cannot see the sun. After such treatment, the bird is tested at 3 P.M., under the sun. In which direction would you expect it to migrate? (p. 577)

37 During the experiment, several birds are born in the cage; these have never flown anywhere before. If one of these birds is exposed to the normal sky, in what direction would you expect it to go when released? Assume that these birds are native to the area in which the experiment is conducted. (p. 577)

ANSWERS

Testing recall

1 innate, instinctive
2 motor program, heterogeneous summation
3 learned motor program
4 parental imprinting
5 imprinting
6 cerebral cortex, cerebellum
7 day
8 sun compass, magnetic cues

Testing knowledge and understanding

9	e	17	c	24	d	31	b
10	e	18	e	25	c	32	a
11	d	19	c	26	b	33	e
12	b	20	d	27	c	34	e
13	b	21	a	28	c	35	b
14	a	22	b	29	a	36	b
15	e	23	d	30	b	37	b
16	d						

THE EVOLUTION OF BEHAVIOR

A GENERAL GUIDE TO THE READING

This chapter, which ends Part II, "The Biology of Organisms," emphasizes the concept that behavior is a biological attribute and evolves just as physical characteristics evolve. Try to keep this fundamental concept in mind as you do your studying. As you read Chapter 22 in your text, you will want to focus on the following topics.

1 Communication. You will find the discussion of animal communication (pp. 581–88) interesting and informative, and well illustrated by the diagrams and photographs. The section on communication in honey bees is relatively difficult; understanding Figure 22.6 (p. 586) will help you learn this material.

2 Niche and habitat. This section focuses on the ecological conditions that make different behavior patterns appropriate for different species, and thus drive behavioral evolution. The concept of the niche is an important one to learn, and will be referred to in subsequent chapters.

3 Population control. Population control is important in the evolution of sociality. The concept of territoriality is particularly important since it involves social behavior.

4 Dominance. You will want to read this section carefully since some important concepts are presented here.

5 Altruism. This is the first place that altruism is introduced; it will be mentioned further in Chapter 33. Try to understand how it is possible for altruistic genes to evolve. Be sure you understand what is meant by kin selection and reciprocal altruism.

6 Mating strategies and parental investment. These sections present important information that helps explain the mate-selection process and mating strategies. Read them carefully.

7 Social animals. You will find this section (pp. 601–710) fascinating; it applies what is currently known about the mechanisms and evolution of social behavior to the study of certain social species, including that of human beings.

KEY CONCEPTS

1 Most animal behavior is a product of the interplay of a small set of neural phenomena —releasers, motor programs, drives, and learning. These mechanisms and processes define in each species the range of behavioral possibilities that can evolve. (p. 581)

2 Natural selection can effect changes in a species behavior through differential survival: individuals whose behavioral traits improve their chance for survival and reproduction under existing circumstances will increase in the population. The set of traits that promote survival will therefore come to predominate in the species as long as the circumstances do not change. (p. 581)

3 Communication probably evolved first for reproduction, and from the mechanisms of sexual communication evolved the wide variety of signals that now communicate mood, intention, and other information necessary to maintain order and stability in social groups. (p. 582)

4 Much of the species-specificity of animal communication depends on the simultaneous presence of several cues at once; the multiplicity of cues reduces the potential for mistakes. (p. 583)

5 The nature and degree of a species' sociality largely determines the kinds of social signals it will need. Many ecological contingencies make different behavior patterns appropriate for different species, and thus drive behavioral evolution. (p. 585)

6 For a species to survive and remain stable it must avoid mating with other species, must not compete with another species for exactly the same food, and must not consume its entire food supply. Each species in an area must occupy its own niche. (pp. 588-89)

7 Social systems evolve primarily when groups of animals are better able than individuals to control resources like food and suitable habitats. (p. 589)

8 Niches, habitats, and behaviors of various species generally keep populations at or below the carrying capacity of the environment by means of opportunistic life histories, programmed dispersal, regulation by predators and disease, and territoriality. (p. 590)

9 In highly social species, the establishment of a dominance hierarchy reduces fighting and increases stability; every individual keeps its place, since it is aware of the probable outcome of a challenge to animals higher on the scale. (p. 595)

10 Many social animals exhibit altruistic behavior, behavior that may reduce the personal reproductive success of the individual exhibiting the behavior while increasing the reproductive success of others. (p. 595)

11 Primate social organization varies from one species to another, but some general ecological and behavioral trends are evident: most primates are arboreal herbivores and insectivores; they live in trees on generalist diets. (p. 604)

12 Fossil evidence indicates that our species evolved as a ground-dwelling hunter of animals and gatherer of vegetation, living in small groups on the plains of Africa. (p. 605)

13 Present evidence suggests that human language acquisition is the product of genetically specified neural circuitry in combination with the cultural environment, and that the drive to acquire language is under strong genetic control. (p. 610)

OBJECTIVES

After studying this chapter and reflecting on it, you should be able to carry out the following objectives.

1 Discuss, using specific examples, the roles that chemical, auditory, and visual signals play in sexual communication. Then explain how such signals reduce the possibility of mating errors. (pp. 585-88)

2 Using a diagram such as Figure 22.6 (p. 586), describe the honey bee's waggle dance, and the symbolic information it conveys. Given the position of a food source in relation to the hive, predict the angle at which a waggle dance would be performed. (pp. 585-87)

3 Carefully define the concept of the niche, explain how it differs from the habitat, and explain what is meant by the niche rule. Then discuss how the niche influences whether or not sociality evolves, and if it does, what kind. (pp. 589-90)

4 Describe in some detail the four means by which niches, habitats, and behavior of various species help control population growth. Explain how the distribution of critical resources determines whether territoriality is adaptive. (pp. 590-92)

5 Explain what is meant by a dominance hierarchy, and indicate the advantages of this system. (pp. 594-95)

6 Explain the evolutionary advantage to a population of having members who exhibit altruistic behavior. In doing so, show how the genes for altruistic behavior may increase in a population, even if the individual who has performed the altruistic act lowers his own reproductive potential. Also, contrast kin selection and reciprocity as explanations of altruistic behavior, and indicate which one of them is more generally applicable. (pp. 595–99)

7 Differentiate between monogamy and polygamy, and explain the role of parental investment in determining the evolution of these systems. (p. 599)

8 Indicate what is meant by sexual selection, describing the two types. (pp. 599–600)

9 Describe the social behavior seen in the honey bee colony, and discuss the advantages and disadvantages conferred by sociality. (pp. 602–4)

10 Describe the basic characteristics of primate social organization, and discuss how the social system of baboons confers stability on the group. (pp. 604–5)

11 Discuss, using information from the fossil record and studies of hunter-gatherer cultures living today, the evolution of human behavior. Give examples from infant behavior to show that humans too have some innate behavioral programming. (pp. 606–8)

KEY TERMS

The following terms are important in this chapter; you should become familiar with them.

behavioral ecology (p. 581)
sexual communication (p. 582)
social communication (p. 585)
niche (p. 589)
habitat (p. 589)
carrying capacity (p. 590)
opportunistic life history (p. 590)
territoriality (p. 591)
ritualization (p. 594)
dominance hierarchy (p. 594)
altruism (p. 595)
group selection (p. 596)
kin selection (p. 597)
reciprocal altruism (p. 599)
monogamy (p. 599)
polygamy (p. 599)
sexual selection (p. 599)
parental investment (p. 601)

angular gyrus (p. 609)
Wernicke's area (p. 610)
Broca's area (p. 610)

SUMMARY

Most animal behavior is a product of the interplay of a small set of neural phenomena—releasers, motor programs, drives, and learning. These mechanisms and processes define in each species the range of behavioral possibilities that can evolve. Natural selection can effect changes in a species behavior through differential survival: individuals whose behavioral traits improve their chance for survival and reproduction under existing circumstances will increase in the population. The set of traits that promote survival will therefore come to predominate in the species as long as the circumstances do not change. Just which traits these will be for the individual will depend on its environment, its food sources, and the nature of its interactions with other members of its species. *Behavioral ecology* seeks to understand how the dynamic interaction of animals with their environments and conspecifics brings about the evolution of behavior.

Communication Communication most probably evolved first for reproduction, and from the mechanisms of sexual communication evolved the wide variety of signals that now communicate mood, intention, and other information necessary to maintain order and stability in social groups.

The goal of most animal communication is survival and reproduction. Much of the behavior associated with mating is innate. The most primitive and widespread channel of communication is chemical; odor informs potential mates of the species, sex, reproductive readiness, and location of an appropriate mate. A number of species use visual and auditory signals. Male fireflies flash in a species-specific code to attract females. Crickets and frogs generate calling songs; as with fireflies the interval between pulses conveys specific information. In birds and mammals, different frequencies can be used to convey different messages.

Much of the species-specificity of animal communication depends on the simultaneous presence of several cues at once; the multiplicity of cues reduces the potential for mistakes. Sometimes the cues must also appear in a particular order. Advertising for a mate provides predatory organisms with the potential for deception; some animals specialize in luring other animals with the victims' own attractants.

The nature and degree of a species' sociality largely determines the kinds of social signals it will need. Social communication may use several behavioral channels, or may involve only one. Scout honey bees are able to communicate to workers in the hive the distance and direction of a food source through symbolic communication. The length and orientation of the waggle run convey information on the distance of the food source and its direction relative to the sun's position. The honey bee dance is entirely instinctive.

Postural and facial cues play a role in many bird and mammalian communication systems.

Behavioral ecology The behavior of animals is modified and adapted through natural selection to suit species' particular lifestyles. Another effect of natural selection on animal behavior has been the substitution of ritual, especially in courtship, for bright visual releasers, which for many species became increasingly costly as they attracted not only potential mates but predators. Many ecological contingencies make different behavior patterns appropriate for different species, and thus drive behavioral evolution.

For a species to survive and remain stable it must avoid mating with other species, must not compete with another species for exactly the same food (*niche rule*), and must not consume its entire food supply. Each species in an area must occupy its own niche. An animal's *niche* is its profession, or way of making a living; its *habitat* is the region that the given species occupies. Species occupy niches to which their behavior is well suited; their behavior is also well adapted to the nature of the habitat and its predators. Niche and habitat are also major determinants of the amount and degree of sociality that may evolve. Social systems evolve primarily when groups of animals are better able than individuals to control resources like food and suitable habitats, and perhaps defend against predators.

Niches, habitats, and behaviors of various species keep populations at or below the *carrying capacity* of the environment by means of "opportunistic" life histories, programmed dispersal, regulation by predators and disease, and territoriality. Opportunistic species are adapted for rapid reproduction during brief periods when resources are abundant; their population growth is generally limited by environmental factors. Other animals are programmed to disperse and seek new habitats when overcrowding occurs.

Predators and disease often function in maintaining a dynamic ecological balance, acting as important factors in keeping a population below the carrying capacity of the habitat.

Territoriality involves social organization. Individuals must compete to gain control of an area (*territory*) containing enough critical resources to permit bearing and raising young. Such a system of strictly defended areas limits population density by denying some members of the population access to certain resources. When individuals compete for or defend a territory, they employ a repertoire of species-specific social signals.

It is the distribution of critical resources like food and breeding areas that in large part determines whether territoriality is adaptive. The behavior of predators can also select for social living; predators may be seen sooner and attacked more effectively by a group than by an individual.

Many animal social systems revolve around ritualized contests that establish the likely winner of a fight with a minimum of risk. Such contests reduce the risk of injury and lessen the frequency of serious fighting. In highly social species such behavior can stabilize the group through the formation of a *dominance hierarchy*, or pecking order, in which every member knows which individuals it can defeat, and which can defeat it. Establishment of a hierarchy reduces fighting; every individual keeps its place, since it is aware of the probable outcome of a challenge to animals higher on the scale. Overt aggression in such species is generally reserved for mutual defense against rival groups and actually helps to stabilize the social bonds within the group. Aggression seems to be a programmed behavioral trait that provides social order and enhances the fitness of each individual.

Altruism is one individual's willingness to lower its own fitness—its personal reproductive potential—thereby raising another's. *Kin selection*, altruism that increases the fitness of kin, is one explanation for the evolution of altruistic behavior. Genes that cause one animal to be altruistic and forgo reproduction will still survive in the population if the resulting altruistic behavior sufficiently enhances the fitness of other individuals that carry the same genes. For such altruism to evolve, the donor and recipient must be close kin, and the benefit must be much larger than the sacrifice. In other instances *reciprocal altruism* may be an explanation, an altruistic act by one individual, not necessarily a relative, may increase the chance that the other individual will repay in kind.

Another set of behavioral alternatives subject to strong selection involves mate choice. *Sexual selection* is selection for characteristics necessary to attract and keep mates. There are two forms of sexual selection, one involving contests, the other female choice. The contest, or competition, form is widespread. Individuals of one sex, usually the males, compete for territories or position in a dominance hierarchy. The male's ability to attain a successful territory or survive seems to reflect high genetic quality. In other species, males do not engage in contests; instead, they perform ritualized displays, and females choose among them. In some cases, female-choice sexual selection may involve *sexual dimorphisms*—structures that differ between the sexes of a species.

Parental investment, the amount of time and energy adult animals spend raising offspring, varies radically from species to species. At one extreme are animals adapted to produce as many offspring as possible with minimal investment; at the other are animals that invest a great deal of time and effort to enhance the fitness of only a few offspring. The relative advantages of these investment alternatives differ not only among species, but between the sexes. In most mammals the males invest much less time than do the females in caring for the young. By contrast, for some species it has proven advantageous to have both parents care for the young. Most bird species form monogamous pair bonds and divide the labor relatively equally between the sexes.

Social animals Except for the primates, the social insects have the most complex and well-understood social system. The coordination engendered by the powerful force of kin selection and maintained by highly specialized behavioral programming has made the honey-bee colony a smoothly operating corporation of individuals that fully exploits the advantages of social living. Sociality makes possible the maintenance of the nest at a relatively precise temperature and humidity, and division of labor creates several subpopulations of efficient specialists. The protective cavity and large colony size permit effective group defense against predators, and the colony's numbers, combined with the dance communication system, facilitate the rapid recruitment of foragers, which enables bees to exploit new resources before other species locate them. The mechanism of social control during most stages of the life cycle of the honey bee is chemical.

Primate social organization varies from one species to another, but some general ecological and behavioral trends are evident. Most primates are arboreal herbivores and insectivores; they live in trees on generalist diets. About 6 million years ago baboons and early human ancestors shared the same niche, living on the African savanna and gathering seeds, corms and berries. Baboons still occupy this niche. Both the male and female baboons establish an elaborate hierarchy based on individual recognition and mutually accepted rank. Such a system benefits both high- and low-ranking individuals and confers great stability on a group.

Fossil evidence indicates that our species evolved as a ground-dwelling hunter of animals and gatherer of vegetation, living in small groups on the plains of Africa. About 1.5 million years ago, after 2 million years of rapid brain-growth, *Homo erectus* emerged as a group hunter, using stone weapons to kill large animals. Early *Homo sapiens* lived in hunter-gatherer societies, but the domestication of animals and plants made possible great changes in social structure. The result was the beginnings of civilization, characterized by cities, division of labor, private wealth, and large-scale central authority.

Possible insight into the evolution of human behavior may be derived from the study of primitive groups of hunter-gatherers living today. These people, such as the !Kung bush people of the Kalahari Desert, exhibit universal human behavioral characteristics: property rights, body adornment, incest taboos, sexual roles, rites of passage, intraspecific war, and belief in the supernatural.

Studies of human infants yields evidence that our species is provided with a repertoire of innate behaviors and reactions. Of particular interest is the acquisition of language, since human culture depends almost exclusively on language. Present evidence suggests that language is the product of genetically specified neural circuitry (located particularly in the left cerebral hemisphere) in combination with the cultural environment, and that the drive to acquire language is under strong genetic control.

QUESTIONS

Testing recall

Decide whether the following statements are true or false, and correct the false statements.

1   Natural selection can bring about changes in a species' behavior through differential survival; individuals whose behavioral traits improve their chances for survival and reproduction will tend to increase in the population. (p. 581)

2   The ultimate goal of most animal communication is to communicate mood, intention, and other information necessary to stabilize the social bonds within social groups. (p. 582)

3   Visual signals represent the most primitive and widespread channel of communication among animals. (p. 582)

4   Courtship sequences require a specific series of releasers that must appear in proper order. (p. 584)

5   Courtship sequences function both in bringing together the two sexes in the mating act and in avoiding mating errors. (p. 584)

6   Honey bees communicate both the distance of the food source and its direction relative to the sun through the orientation of the waggle run on the vertical comb in the hive. (p. 586)

7   An animal's niche is the region that the given species occupies. (p. 589)

8   Different species can occupy the same niche in different habitats. (p. 589)

9   Social behavior is often advantageous to animals that must compete for limited resources. (p. 590)

10   Bird songs may function in territorial defense and to attract females. (p. 591)

11   Dominance hierarchies are characterized by increased aggression among members of the group. (p. 595)

12   The mechanism of group selection provides a good explanation for the evolution of altruistic behavior. (p. 596)

13   A mating system in which one female mates with many males is called monogamy. (p. 599)

14   Sexual selection is selection for characteristics necessary to attract and keep mates. (p. 599)

15   All primates show a high degree of social organization, forming elaborate hierarchal systems. (p. 605)

16   Human beings probably evolved as hunter-gatherers on the plains of Africa. (p. 606)

17   The human brain is wired at birth for the circuitry for language acquisition. (pp. 609–10)

18   Linguistic meaning is generally processed in discrete areas in the right cerebral cortex. (p. 609)

19   Both spoken and written information is processed primarily in Broca's area of the brain. (p. 610)

Testing knowledge and understanding

Choose the one best answer.

20   Which one of the following communication signals is likely to evoke responses at the greatest distance?

   a   crickets singing a species-specific calling song
   b   wing beating by a mosquito
   c   pheromones released by a silkworm moth
   d   waggle dance of a honey bee
   e   song of a wood thrush defending his territory   (pp. 498, 582)

21   Wolves urinate to mark the edges of their territories. This behavior is an example of

   a   a reflex.
   b   using pheromones.
   c   kin selection.
   d   marking the niche.   (p. 582)

22   If mice are kept in a cage for several weeks and then are replaced by a new set of mice, the new mice become temporarily sterile, and put on weight. The type of communication involved is

   a   pheromones.   d   sign stimuli.
   b   hormones.   e   conditioning.
   c   releasers.   (p. 582)

Questions 23–25 refer to the following diagram showing locations of some bees' food sources (A through C) and their hive, as well as the position of the sun. Assume that the sun remains stationary.

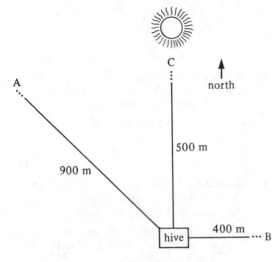

23 If a scout has returned to the hive from a successful flight to site A and performed its dance in a darkened vertical·comb, what would you expect its dance to look like?

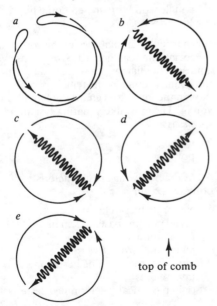

(p. 586)

24 If another bee returned to the same hive from site B and danced on a vertical comb, what would you expect its dance to look like?

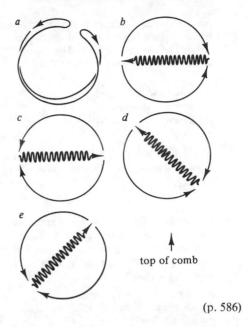

(p. 586)

25 If another bee returned to the same hive from site C and danced on a vertical comb, what would you expect its dance to look like?

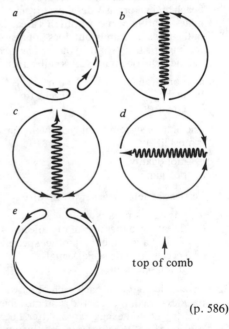

(p. 586)

26 Dung beetles of certain species dig burrows under a pile of dung and take the dung into the burrows to feed their larvae. When a female dung beetle finishes digging and stops bringing up soil from an underground burrow, the male seizes a piece of dung with his front legs and backs down into the burrow. When the tip of his abdomen strikes the

female's head, he drops his load and returns to the surface for more. In this case, the striking of the female's head by the male's abdomen is probably an example of a(n)

a sign stimulus.
b supernormal stimulus.
c sexual selection.
d kinesis.
e tropism. (p. 556)

27 The type of behavior seen in the above example would most likely be characterized as

a innate.
b learned. (p. 554)

28 A male dung beetle making a nuptial ball for a female to eat in relative safety underground exposes himself to considerable risk while reducing the risk to a potential mate. This is an example of

a aggressive behavior.
b altruistic behavior.
c nuptial love.
d kin selection.
e conditioning. (p. 595)

29 The male Alaskan fur seal secures an area of beach and vigorously defends it from all other males. Mating occurs within this area. The bull guards his family but does not assist in feeding or caring for the young. The area defended by one male seal is called the

a home range.   d nursery.
b niche.   e home turf.
c territory. (p. 591)

30 The mating system employed by the fur seal is most likely

a monogamy.
b polygamy. (p. 599)

31 Territorial defense generally involves

a physical combat.
b threats by a member of one species against a member of another species.
c species-specific social signals.
d group selection. (p. 591)

32 Two male dogs approach each other with hackles raised, teeth bared, and movements stiff-legged and exaggerated. After a period of time one of the dogs responds by putting his head down, tucking his tail between his legs, and bending his legs. This is an example of

a sexual dimorphism.
b ritualization.
c altruistic behavior.
d territoriality.
e niche rule. (p. 594)

33 When an intruder approaches the nest of a prairie warbler, the warbler drops from the nest to the ground and rapidly flutters its wings, distracting the observer from the nest. This is an example of

a altruistic behavior.
b learned behavior.
c kin selection.
d reciprocity.
e a and c are correct (pp. 595–99)

34 Which of the following factors has *not* been invoked by biologists as an explanation for the evolution of altruistic behavior?

a individual selfishness
b reciprocity
c kin selection
d mother love (pp. 595–99)

35 All of the following would be classified as illustrating types of social behavior, with one exception. Which is the exception?

a tricolor blackbirds foraging as a flock
b a monogamous male and female pair of Canada geese
c a flock of terns laying their eggs within a restricted rocky area on a small island
d a group of fish congregating in the last vestige of a drying pond
e a male wood thrush defending his territory (pp. 591–601)

36 In a particular animal species the young are precocial; they can help care for themselves almost immediately. The care and feeding of the young is done by the female parent, though the male does guard the female and young. The most likely mating system for this species is

a monogamy.
b polygamy. (p. 599)

37 All of the following are characteristic of primates *except*

a elaborate hierarchal systems
b social systems based on an individual recognition and rank
c males play protective roles
d most are tree-dwelling herbivores and insectivores
e rigid biologically determined caste system (pp. 602–5)

38 Structures that differ between the male and female of a species are called

a parental investments.
b sexual selections.
c sexual dimorphisms.
d ritualizations. (p. 600)

39 The area of the brain that acts as the processing center for the individuals' spoken and written expression is

a   Broca's area.
b   Wernicke's area.
c   the angular gyrus.
d   in the motor area.
e   in the midbrain.   (pp. 609–10)

For further thought

1 Several small songbirds in different families have evolved similar *seeeee* alarm calls. What mechanisms might be involved in the evolution of the similar alarm calls?

2 The behavior of Mexican jays has been studied extensively. These birds are communal breeders. Nests are built by pairs; there are usually two active nests in a flock. Most of the members bring food to the nestlings. The parents bring about half the food; the other half is brought by altruistic helpers—nonbreeding adults whose nests have failed, and immature birds. Suggest reasons for the evolution of this altruistic behavior.

ANSWERS

Testing recall

1   true
2   false—survival and reproduction
3   false—chemical
4   true
5   true
6   false—direction only
7   false—habitat
8   true
9   true
10 true
11 false—decreased
12 false—unsatisfactory
13 false—polygamy
14 true
15 false—most
16 true
17 true
18 false—left
19 false—Wernicke's area

Testing knowledge and understanding

20	c	25	c	30	b	35	d
21	b	26	a	31	c	36	b
22	a	27	a	32	b	37	e
23	c	28	b	33	e	38	c
24	c	29	c	34	d	39	b

CELLULAR REPRODUCTION

A GENERAL GUIDE TO THE READING

This chapter—the first in Part III, "The Perpetuation of Life"—focuses on the processes of cell division. Subsequent chapters will examine the inheritance and the functioning of the genetic material. As you read Chapter 23 in your text, you will want to give special attention to the following aspects of the topic.

1   The process of mitosis. Figure 23.11 (p. 622) graphically presents the stages of mitosis. Know these stages thoroughly. One mnemonic device for learning their order is the sentence "*I put my arm there*"; the first letter of each word is the first letter of a stage.

2   The structure of a late-prophase chromosome in mitosis. (p. 624) Notice that, as shown in Figure 23.6F (p. 618) and 23.8 (p. 620), the chromosome in late prophase is double-stranded, each strand being called a chromatid. The DNA for the duplicate strand is produced during the S stage of interphase, when the chromosomal material is replicated.

At the end of mitosis and cytokinesis, each chromosome in the newly formed cells is once again single-stranded.

3   The structure of the late-prophase chromosome in meiosis. Notice that the twin chromatids are not so obviously separated as they are in mitosis. The chromatids are held together by protein axis (see Fig. 23.20A, p. 632).

4   The process of meiosis. Students often have difficulty with this topic, so you will want to study pages 630–36 carefully. Figure 23.19 (p. 630) is central for understanding this process.

5   Differences between mitosis and meiosis. Compare metaphase of mitosis (Fig. 23.11:5, p. 623) with metaphase of meiosis (Fig. 23.19:4, p. 631), with special attention to the way the chromosomes line up totally independently during mitosis, but in pairs (synapsed) during meiosis. This difference is crucial. Other differences are summarized in Figure 23.26 (p. 636).

6 Synapsis and crossing over. You will need to read pages 631–34 carefully and study Figures 23.20 and 23.21 (pp. 632 and 633) in order to thoroughly understand this process. Notice that crossing over is not a rare event; it is a frequent and organized mechanism for producing genetic variation. You will need an understanding of crossing over to follow the material on linkage in Chapter 25.

7 The adaptive significance of sexual reproduction. Make sure you understand the section on this subject (pp. 636–40), since it lays important groundwork for Chapters 24 and 33.

KEY CONCEPTS

1 When a cell divides, it must make a complete copy of the genetic information in its nucleus and then, as it divides, give one complete set to each daughter cell. (pp. 616–17)

2 Mitosis produces new cells with exactly the same chromosomal endowment as the parent. (pp. 620–22)

3 Meiosis reduces the number of chromosomes by half so that when the egg and sperm unite in fertilization, the normal diploid number is restored. (pp. 628–30)

4 Sexual reproduction increases variation in the population by making possible genetic recombination. (pp. 636–37)

5 Unpredictably variable environments favor species that recombine sexually, whereas predictable, static environments favor species that are at least temporarily asexual. (p. 637)

OBJECTIVES

After studying this chapter and reflecting on it, you should be able to carry out the following objectives.

1 Using a diagram such as Figure 23.1 (p. 615), describe the transfer of information in a cell. (pp. 615–20)

2 Cite experimental evidence showing the essential role of the nucleus in the life of the cell and in the transmission of hereditary information. (pp. 625–28)

3 Using a diagram such as Figure 23.5 (p. 617), explain how cell division takes place in

procaryotic cells. Explain how the chromosomes of procaryotic cells differ from those of eucaryotic cells. (pp. 617–20)

4 Distinguish between mitosis and cytokinesis. (pp. 620–25, 626–28)

5 Describe the various stages in the cell cycle. In doing so, be sure to specify the stage in which genetic replication occurs and the stage in which cell division occurs, and to indicate which of the stages in the cell cycle comprise interphase. In addition, explain what is meant by G_1 arrest and by G_2 arrest, and for each name a type of cell that shows that type of arrest. (p. 621)

6 Explain the difference between a chromatid and a chromosome; for a cell in prophase that has six chromosomes, specify the number of chromatids and the number of centromeres. (pp. 618–20)

7 List the stages of mitosis and, using a diagram such as Figure 23.11 (p. 622), describe the principal events that occur during each stage. In doing so, explain how you would recognize a cell in prophase, metaphase, anaphase, or telophase; and how you would distinguish mitosis in a plant cell from that in an animal cell. (pp. 620–25)

8 Define cytokinesis, and show how it differs in algal cells, higher-plant cells, and animal cells; using a micrograph such as Figure 23.16 (p. 627) or Figure 23.18 (p. 629), point out the cell plate; and specify the organelles thought to be involved in the production of the cell plate. (pp. 626–29)

9 List the stages of meiosis and, using a diagram such as Figure 23.19 (p. 630), describe the principal events that occur during each stage. In doing so, define the terms haploid and diploid; explain what happens during synapsis, showing what is meant by a synaptonemal complex, and by a chiasma; specify how and when crossing over occurs; and indicate the significance of this process. (pp. 630–36)

10 Compare and contrast meiosis with mitosis. In doing so, explain the importance of each process in the life of the organism; compare the behavior of chromosomes in meiosis I and in mitosis (Figure 23.26, pp. 636–37, may be helpful); compare the behavior of chromosomes in meiosis II and in mitosis; and compare the end products of meiosis and mitosis. (pp. 636–37)

11 Discuss the advantages and disadvantages of sexual reproduction and the sexual recom-

bination the process allows. Describe the circumstances under which sexual reproduction is advantageous. Do the same for asexual reproduction. (pp. 636–37)

12   Describe the two methods by which variation is produced during sexual reproduction. (pp. 637–40)

13   Using a diagram such as Figure 23.30 (p. 640), compare and contrast spermatogenesis with oogenesis. (pp. 640–41)

14   Using diagrams such as those in Figure 23.32 (p. 642), describe the three representative life cycles. In doing so, explain how the products of meiosis differ in plants and in animals; and how the life cycle of a higher plant differs from that of an animal. (pp. 640–43)

KEY TERMS

The following terms are important in this chapter; you should become familiar with them.

binary fission (pp. 617–18)
nucleosome cores (p. 618)
histone proteins (p. 618)
centromere (pp. 619–20)
chromatid (p. 620)
mitosis (p. 620)
cytokinesis (p. 620)
centrioles (p. 620)
cell cycle (G_1, S, G_2, and M stages) (p. 621)
growth factors (p. 621)
prophase (p. 622)
metaphase (p. 624)
anaphase (p. 624)
kinetochore microtubules (p. 624)
polar microtubules (p. 624)
aster (p. 624)
spindle (p. 624)
telophase (p. 625)
cleavage furrow (p. 627)
cell plate (p. 627)
meiosis (p. 630)
synaptonemal complex (p. 632)
synapsis (p. 632)
crossing over (p. 632)
chiasmata (p. 632)
clone (p. 636)
alleles (p. 636)
linked (p. 639)
spermatogenesis (p. 640)
oogenesis (p. 641)
polar body (p. 641)

SUMMARY

The genes of an organism's DNA contain all the information necessary to determine the organism's characteristics and direct its activities. The entire set of genetic information must be duplicated and transferred when each new cell is formed.

Procaryotic cells divide by *binary fission*; the single circular chromosome replicates and a ring of new plasma membrane and wall material grows inward, separating the replicates.

Eucaryotic chromosomes differ from those of procaryotes in that they are linear, the DNA is wound on *nucleosome cores*, and the chromosomes occur in homologous pairs.

Mitotic cell division   Cell divisions in eucaryotic cells involves two processes: the division of the nucleus (*mitosis*) and the division of the cytoplasm (*cytokinesis*). Nuclear division entails, first, precise duplication of the genetic material and, second, distribution of a complete set of the material to each daughter cell.

Cells that are going to divide pass through a series of stages known as the *cell cycle*. After a cell has completed the division process, there is a gap in time, the G_1 stage, before replication of the genetic material begins. Next comes the S stage, during which new DNA is synthesized. Another time gap, the G_2, separates the end of replication from the onset of mitosis proper. Some animal cells become arrested in the G_1 stage and a few in G_2; these cells normally will not divide. The arrests appear to be due to failure to produce some control chemical. Production of these control substances depends on stimulation from certain specific peptide *growth factors* from blood serum.

During the G_1, S, and G_2 stages, the cell is nondividing; these three stages together constitute the *interphase* stage. During interphase the nucleus is visible, and one or more nucleoli are prominent. The chromosomes are long and thin and cannot be seen as distinct structures. After the cell has passed through G_1, S, and G_2 it enters mitosis. Mitosis is customarily divided into four stages:

1 *Prophase.* As the two *centrioles* move to the opposite sides of the nucleus, the chromosomes condense into visible threads. Each chromosome consists of *chromatids* held together by a *centromere*. Kinetochore microtubules radiate out from each centromere. The *polar* and *astral microtubules* appear near each centriole. The nuclear membrane and nucleoli disappear. Plant cells, unlike animal cells, lack centrioles and astral microtubules.

2 *Metaphase*. The centromere of each double-stranded chromosome attaches to a spindle microtubule along the equator of the cell. Metaphase ends when the centromere of each pair of twin chromatids splits.

3 *Anaphase*. The two new single-stranded chromosomes separate, one moving toward each pole. The kinetochore microtubules move poleward, shortening as they go, pulling the chromosomes apart. Cytokinesis often begins here.

4 *Telophase*. The chromosomes reach the poles, the nuclear membrane and nucleoli are reformed, cytokinesis is completed, and the chromosomes become longer and thinner.

Cytokinesis frequently accompanies mitosis. In animal cells, cytokinesis begins with the formation of a *cleavage furrow* running around the cell; the furrow deepens until it divides the cell in two. In plants, a *cell plate* forms in the center of the cytoplasm and enlarges until it cuts the cell in two.

Mitosis produces two new cells with exactly the same chromosomal endowment as the parent cell.

Meiotic cell division *Meiosis* is a special process of cell division that reduces the number of chromosomes by half so that when the egg and sperm unite in fertilization the normal number is restored. During the reduction division of meiosis, the chromosomal pairs are partitioned so that each gamete contains one of each type of chromosome. (It is *haploid*.) When the two gametes unite in fertilization, the resulting zygote is *diploid*, having received one chromosomal type from each parent.

Meiosis involves two successive divisional sequences, which produce four new haploid cells. The first division is the reduction division; the second separates the chromatids. Both divisional sequences can be divided into four stages. In meiosis I these are:

1 *First prophase*. The events are similar to the mitotic prophase except that in meiosis the homologous chromosomes move together and lie side by side in a process called *synapsis*. The twin chromatids are held tightly together by a pair of protein axes; the protein axes of the two homologous chromosomes then join by protein bridges to form a *synaptonemal complex*. Crossing over then takes place; chromatids are clipped and parts of separate chromatids are spliced together. When the synaptonemal complex breaks up, the *chiasmata* hold the spliced chromatids together. In late prophase, spindle microtubules

and kinetochore microtubules appear; however, the kinetochore microtubules go to only one pole.

2 *First metaphase*. The two chromosomes of each homologous pair attach to the same spindle microtubule along the equator of the cell. The centromeres do not split.

3 *First anaphase*. The two double-stranded chromosomes of each synaptic pair move to opposite poles.

4 *First telophase*. Two new nuclei are formed, each with half the chromosomes present in the parental nucleus. The nuclei are not identical.

A short period called *interkinesis* follows the first telophase. No replication of genetic material occurs during this stage. The second division sequence of meiosis is essentially mitotic; each double-chromatid chromosome attaches to a *separate* microtubule, the centromeres divide, and the new single-stranded chromosomes move to the poles. Four new haploid cells containing single-chromatid chromosomes are produced.

The first meiotic division then produces two haploid cells containing double-chromatid chromosomes. Each of these cells divides again in the second meiotic division, producing a total of four new haploid cells containing single-chromatid chromosomes.

Asexual reproduction produces *clones* of cells genetically identical to the parent cell, whereas sexual reproduction increases variation within the population by creating organisms with new combinations of characteristics. Each diploid cell undergoing meiosis can produce 2^n different chromosomal combinations, where n is the haploid number. Crossing over greatly increases the number of possible variants. Genetic variation increases the species' chances of surviving and reproducing in a changeable environment. Unpredictably variable environments favor species that recombine sexually, whereas predictable, static environments favor species that are at least temporarily asexual.

Higher animals are diploid. During reproduction, meiosis produces haploid gametes, which unite to produce the diploid zygote. *Gametes* are haploid cells specialized for sexual reproduction. In male animals, *spermatogenesis* produces four functional sperm cells from each diploid cell. In females, *oogenesis* produces only one mature ovum; the polar bodies are nonfunctional.

Meiosis in plants produces haploid cells called *spores* (stage 1), which often divide mitotically to form haploid multicellular plants (stage 2). Eventually these haploid plants produce gametes (stage 3) by mitosis. Two of the gametes unite to form the diploid zygote (stage 4), which develops into a diploid multicellular plant (stage 5), and the cycle is complete. Most multicellular plants have all five stages in their life cycles, but the relative importance of the stages varies greatly. In general, the haploid stages are dominant in the primitive plants and the diploid in the more advanced plants.

QUESTIONS

Testing recall

Match each event described below with the stage in mitosis in which it occurs. Answers may be used once, more than once, or not at all.

a	anaphase	*d*	prophase
b	interphase	*e*	telophase
c	metaphase		

1   Polar microtubules appear. (p. 622)

2   Chromosomes migrate to poles. (p. 624)

3   Centromeres divide. (p. 624)

4   Centrioles move to poles. (p. 622)

5   Chromosomes reach poles. (p. 625)

6   G_1, S, G_2 stages occur. (p. 620)

7   Nuclear membrane and nucleoli disappear. (p. 622)

8   Chromosomal replication occurs. (p. 622)

Indicate whether each event described below occurs in (a) mitosis, (b) the first division of meiosis, or (c) the second division of meiosis. Some of the events may occur in more than one type of cell division.

9   Twin-chromatid chromosomes move to poles. (p. 631)

10   Chromosomes shorten and thicken, are double-stranded. (pp. 620, 631, 635)

11   Centromeres divide. (pp. 620, 635)

12   Single-chromatid chromosomes move to poles. (pp. 620, 635)

13   Nuclear membrane and nucleolus disappear. (pp. 620, 631, 635)

14   Haploid cells are produced. (pp. 631, 635)

15   Crossing over occurs. (p. 631)

16   Cells genetically identical to the parent cell are produced. (p. 620)

17   Synaptonemal complex forms. (p. 631)

Decide whether the following statements are true or false, and correct the false statements.

18   If a cell in prophase of mitosis has 20 centromeres, it has 20 chromosomes. (p. 620)

19   Centromeres do not divide during meiosis I. (p. 634)

20   Duplication of the genetic material (that is, the synthesis of new genetic material) occurs during prophase. (p. 620)

21   Centrioles are essential for cell division in both higher plant and animal cells. (p. 620)

22   A cell in prophase I of meiosis has half as many chromosomes as a cell in prophase II. (pp. 631, 635)

23   Synaptonemal complexes form during both mitosis and meiosis. (p. 632)

24   Crossing over is a rare event in human chromosomes. (p. 632)

25   The vesicles of the cell plate formed during cytokinesis of a plant cell give rise to the middle lamella and the beginnings of the primary cell wall. (p. 627)

Testing knowledge and understanding

Choose the one best answer.

26   During the S phase of the cell cycle, the cell

    a   undergoes cytokinesis.
    b   undergoes meiosis.
    c   replicates its DNA.
    d   undergoes mitosis.
    e   enters interphase.   (p. 621)

27   If there are 12 single-chromatid chromosomes in a cell in G_1 of the cell cycle, what is the diploid number of chromosomes for the organism?

a	6	*d*	36
b	12	*e*	48
c	24		(p. 621)

28   Which one of the following occurs during mitosis?

    a   Homologous chromosomes synapse.
    b   Twin-chromatid chromosomes move to the poles.
    c   Crossing over takes place between homologous chromosomes.

d Chromosome number is reduced from diploid to haploid.
e Cells genetically identical to the parental cell are produced. (pp. 617–18)

b occurs in animal cells by a pinching-in of the plasma membrane.
c occurs in higher-plant cells by an inward growth of new wall and membrane.
d two of the above are correct (p. 627)

29 Cytokinesis

a always accompanies mitosis.

Questions 30–36 refer to the diagram at right of a diploid cell in prophase.

30 How many *chromatids* are present in this cell? (p. 620)

a 2   *b* 4   *c* 8

31 How many *chromosomes* are present in this cell? (p. 620)

a 1   *b* 2   *c* 4

32 Which diagram shows how the chromosomes would be arranged during metaphase of mitosis? (p. 624)

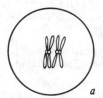 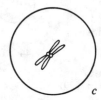  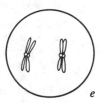

a ... *b* ... *c* ... *d* ... *e*

33 In G₁ after mitosis, which of the following would one daughter cell contain?

a two double-chromatid chromosomes
b two single-chromatid chromosomes
c one double-chromatid chromosome
d one single-chromatid chromosome (p. 621)

34 Which diagram shows how the chromosomes would be arranged during the first metaphase of meiosis? (p. 634)

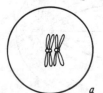

a ... *b* ... *c* ... *d* ... *e*

35 Which diagram shows how the chromosomes would be arranged at the start of the second division of meiosis following interkinesis? (p. 635)

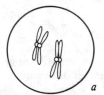

a ... *b* ... *c* ... *d*

36 Which diagram shows what the chromosomes in one daughter cell would look like at the end of meiosis? (p. 635)

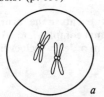 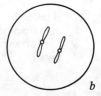 

a ... *b* ... *c* ... *d*

37 If there are 12 chromosomes in a cell that has just completed meiosis, what is the diploid number of chromosomes for that organism?

a 6
b 12
c 24
d 36
e 48   (p. 635)

38 Which one of the following occurs in meiosis but *not* in mitosis?

a Double-chromatid chromosomes move to the poles.
b Chromosomes shorten and thicken and are double-stranded.
c Single-chromatid chromosomes move to the poles.
d Nuclear membrane and nucleolus disappear.
e Centromeres divide.   (p. 635)

39 Which one of the following occurs in the first divisional sequence of meiosis?

a Diploid daughter cells are produced.
b Single-chromatid chromosomes move to the poles.
c Centromeres divide.
d Homologous chromosomes synapse.
e two of the above   (p. 629)

40 If a cell in the process of meiosis is haploid and its chromosomes are becoming individually visible, which stage is it in?

a first meiotic prophase
b first meiotic anaphase
c second meiotic prophase
d second meiotic telophase
e interkinesis   (p. 635)

41 At what stage of cell division does synapsis occur?

a anaphase of mitosis and meiosis
b prophase of mitosis and meiosis
c metaphase of mitosis and meiosis
d prophase of the first meiotic division
e metaphase of the second meiotic division
                                                     (p. 632)

42 Which one of the following statements is *false* concerning meiosis?

a DNA is replicated between each cell division.
b Each chromosome is double-stranded during prophase.
c Each chromosome pairs with its homolog during meiosis I.
d Cell division follows chromosome migration.
e Each chromosome may exchange a part of a chromosome with the equivalent part of a homologous chromosome.
                                                     (pp. 629–30)

43 D, E, F, and G are the four daughter cells resulting when cell A undergoes meiosis, as shown in the following diagram. If no crossing over has occurred, which cells listed below, if any, are genetically identical?

a cells B and C
b cells A and D
c cells D, E, F, and G
d cells D and E
e no two cells   (p. 635)

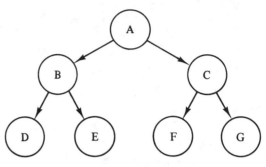

44 The drawing below shows a cell whose diploid chromosome number is four. This cell is in

a metaphase.
b anaphase of mitosis.
c first anaphase of meiosis.
d second anaphase of meiosis.
e telophase of mitosis.   (p. 634)

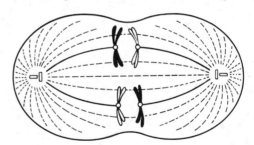

45 The above cell is from a(n)

a plant.
b animal.
c plant or animal.
d bacterium.
                                                     (p. 627)

46 From each primary oocyte that undergoes meiosis, the number of functional egg cells produced is

a one.
b two.
c three.
d four.
e eight.   (p. 641)

47 If the following events in the formation of the synaptonemal complex and crossing over were arranged in order, which event would be third?

a Homologous chromosomes are brought into perfect alignment and protein cross-bridges form.

b Recombination nodules appear.

c Axial proteins gather the DNA of the chromatids into paired loops.

d Chromatids from one chromosome exchange fragments with chromatids of the other chromosome.

e Chromosomes shorten and thicken; replication has already occurred. (p. 632)

48 Genetic recombination occurs during

a mitosis.

b first division of meiosis.

c second division of meiosis.

d crossing over.

e *b* and *d* are correct   (p. 632)

49 Each centromere sends kinetochore microtubules toward both poles during

a mitosis.

b first divisional sequence of meiosis.

c second divisional sequence of meiosis.

d *a* and *c* are correct

e *a, b* and *c* are correct   (pp. 624, 635)

50 Suppose the diploid chromosome number of a particular organism is 10. How many different chromosomal combinations could be produced by meiosis in this organism (i.e. how many different kinds of gametes could be formed)? Exclude combinations resulting from crossing over.

a 5   *d* 32
b 16   *e* 64
c 25   (pp. 631–34)

Questions 51–53 refer to the following diagram of a plant life cycle.

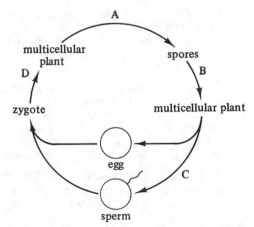

51 What type of cell division occurs during part D of the life cycle?

a mitosis   *b* meiosis   (p. 642)

52 What type of cell division occurs during part C of the life cycle?

a mitosis   *b* meiosis   (p. 642)

53 What type of cell division occurs during part A of the life cycle?

a mitosis   *b* meiosis   (p. 642)

For further thought

1 In a single sentence, how would you explain the basic significance of mitosis?

2 Which of the phases of mitosis do you think would require the greatest expenditure of energy? Explain your answer.

3 Suggest a hypothesis to explain the adaptive value of having genetic material packaged in chromosomes.

4 In a single sentence, state the fundamental difference between mitosis and meiosis in terms of chromosomal behavior.

5 Without looking at your notes or textbook, take a sheet of paper and make diagrams of mitosis and meiosis in an organism with three pairs of chromosomes. Make a list of the most important differences between mitosis and meiosis.

ANSWERS

Testing recall

1	*d*	7	*d*	13	*a, b, c*
2	*a*	8	*b*	14	*b, c*
3	*c*	9	*b*	15	*b*
4	*d*	10	*a, b, c*	16	*a*
5	*e*	11	*a, c*	17	*b*
6	*b*	12	*a, c*		

18 true
19 true
20 false—interphase
21 false—neither plant nor animal
22 false—twice as many
23 false—only during meiosis
24 false—frequent
25 true

Testing knowledge and understanding

26	*c*	33	*b*	40	*c*	47	*a*
27	*b*	34	*a*	41	*d*	48	*e*
28	*e*	35	*b*	42	*a*	49	*d*
29	*b*	36	*d*	43	*d*	50	*d*
30	*b*	37	*c*	44	*c*	51	*a*
31	*b*	38	*a*	45	*b*	52	*a*
32	*b*	39	*d*	46	*a*	53	*b*

Chapter 24

MENDELIAN GENETICS

A GENERAL GUIDE TO THE READING

This chapter and the next are concerned with the transmission of genetic information from generation to generation. Some beginning students experience difficulty with genetics because understanding it requires a skill in logic which they have not yet had the opportunity to develop. Both the chapters in the text and the corresponding material in the *Study Guide* are designed to assist you in developing this skill. Each includes a series of genetics problems which provide necessary practice. You will want to do these problems, since the best way to gain an understanding of genetics is to work with it. Suggestions on how to approach such problems effectively are presented in "Study Guide for Genetics Problems," on pages 198–200 of this *Study Guide*, and answers to the genetics problems in the text are on pages 200–3. In addition, as you read Chapters 24 and 25 in your text, you will find it useful to concentrate on the following topics.

1 Mendel's work. You will want to read carefully the material on Mendel (pp. 645–50), since our understanding of how heredity operates is based on his work.

2 Monohybrid and dihybrid crosses. Study carefully the text material on pages 645–59 since it is this material that you will need to know in order to do the genetics problems at the end of the chapter.

3 Gene interactions. Separate genes can interact with one another to influence phenotypic characteristics. This section (pp. 655–59) describes several different types of gene interactions, some of which will be tested in the genetics problems at the end of the chapter.

4 Penetrance and expressivity. These concepts, discussed on pages 659–62, are often confused by students; you will want to be able to distinguish between them.

KEY CONCEPTS

1 Somatic cells have pairs of homologous chromosomes, a pair consisting of one chromosome from each parent. Each gene is found in two copies, one on each chromosome of the homologous pair, at corresponding loci. (pp. 646–47)

2 In meiosis, each gamete receives a copy of only one chromosome from each homologous pair, and hence only one of the two alleles. (p. 647)

3 Genes do not alter one another; they remain distinct and segregate unchanged when meiosis occurs. (p. 646)

4 When two or more pairs of genes located on different chromosomes are involved in a cross, the members of one pair are inherited independently of the other. (p. 646)

5 Characteristics are often determined by many genes acting together. (p. 655)

6 The expression of a gene depends both on the other genes present and on the physical environment. All organisms are products of their inheritance *and* their environment. (pp. 661–62)

7 Scientists use statistical tests to determine whether the deviations they observe in their experimental results are significant or not. Tests of significance take into account both the amount of deviation and the sample size. (p. 663)

OBJECTIVES

After studying this chapter and reflecting on it, you should be able to carry out the following objectives.

1 Discuss Mendel's conclusions and relate them to the chromosomal theory of inheritance. (pp. 646–47)

2 Differentiate between the two terms in each of the following pairs: dominant–recessive, allele–gene, F_1–F_2, homozygous–heterozygous, phenotype–genotype, monohybrid–dihybrid. (pp. 645–48)

3 Use a Punnett square to find the expected genotypic ratios in a cross involving a single character (monohybrid). Compare the results when one allele is dominant over another with the results when there is partial dominance (intermediate inheritance). [Problems 1–7 in your text (p. 666) cover this material.] (pp. 649–50)

4 Explain how partial dominance differs from complete dominance. (pp. 646, 651)

5 Use a Punnett square or the multiplication of probabilities (as shown on p. 649) to find the characteristic phenotypic ratios in the F_2 of a cross involving two characters (dihybrid) in which the two genes are independent. Explain how gene interactions, such as those involving epistatic and modifier genes, could alter this ratio. [Problems 8–19 in your text (pp. 666–67) cover this material.] (pp. 656–58)

6 Using examples, differentiate between the terms penetrance and expressivity. (pp. 659–60)

7 Explain why statistical analysis is a useful genetic tool. (p. 663)

KEY TERMS

The following terms are important in this chapter; you should become familiar with them.

F_1 (p. 645)
F_2 (p. 645)
P (p. 645)
dominant character (p. 646)
recessive character (p. 646)
locus (p. 647)
allele (p. 647)
homozygous (p. 648)
heterozygous (p. 648)
genotype (p. 648)
phenotype (p. 648)
monohybrid cross (p. 649)
test cross (p. 650)
partial dominance (p. 651)
dihybrid cross (p. 653)
independent gene (p. 653)
complementary genes (p. 656)
epistasis (p. 656)
collaboration (p. 657)
modifier gene (p. 657)
penetrance (p. 660)
expressivity (p. 660)

SUMMARY

From his breeding experiments on the garden pea the Austrian monk Gregor Mendel concluded that each pea plant possesses two hereditary factors for each character, and that when gametes are formed the two factors segregate into separate gametes. Each new plant thus receives one factor for each character from each parent. The hereditary factors exist as distinct entities within the cell;

they do not blend or alter each other, and they segregate unchanged when gametes are formed. This theory is now supported by the occurrence of chromosomal segregation during meiosis.

Mendel was working with two different forms (*alleles*) of the gene for flower color. When both alleles were present, the gene for red flowers was expressed (*dominant*) while the gene for white was masked (*recessive*). The dominant allele is customarily represented by a capital letter, the recessive by a small letter. A diploid cell may be *homozygous* (have two doses of the same allele: *C/C, c/c*) or *heterozygous* (have one each of two different alleles: *C/c*). The term *genotype* refers to the possible genetic combinations, *phenotype* to the possible appearances. An easy way to figure out the possible genotypes produced in a cross is to construct a Punnett square.

Extensive investigation has demonstrated that Mendel's results have general validity. Whenever a monohybrid cross is made between two contrasting homozygous individuals, the expected genotype ratio in the second generation of offspring (the F_2) is 1 : 2 : 1. When dominance is involved, the expected phenotypic ratio is 3 : 1.

A test cross is performed to distinguish between homozygous dominant and heterozygous individuals. The unknown is crossed with a homozygous recessive individual. If all the progeny show the dominant phenotype, the unknown genotype is probably homozygous dominant; if any of the progeny show the recessive genotype, the unknown is heterozygous.

One allele is not always completely dominant over the other; in some cases heterozygous individuals show the effects of both alleles and are clearly distinguishable from both homozygous parents. Crosses between heterozygous individuals result in a phenotypic and genotypic ratio of 1 : 2 : 1. This type of inheritance is termed *partial dominance* or *intermediate inheritance*.

Mendel also made crosses involving two characteristics (a *dihybrid* cross). In his experiments Mendel found that the offspring consistently conformed to a 9 : 3 : 3 : 1 phenotypic ratio. This ratio is characteristic of the F_2 generation of a dihybrid cross (with dominance) in which the genes for the two characters are inherited *independently* of one another.

Some dihybrid crosses involve two independent genes that exert their phenotypic effect on the same character. The ratios from such crosses often differ from the basic ratio. *Complementary genes* are mutually dependent; neither one can exert its phenotypic effect unless the other one does also. *Epistasis* occurs when one gene masks the phenotypic effect of another entirely different gene. Sometimes two different genes interact to produce single-characteristic phenotypes that neither gene alone could produce; this is termed *collaboration*.

Probably no inherited characteristic is controlled by only one gene pair. Even when only one principal gene is involved, other genes act as *modifiers* to influence its expression.

Many characteristics vary in a continuous fashion. This is probably due to *multiple gene inheritance*, in which two or more separate genes affect the same character in an additive fashion.

The action of any gene can be fully understood only in terms of the overall genetic makeup of the individual organism. The expression of a gene depends on the other genes present and on the physical environment. *Penetrance* is the percentage of individuals carrying a gene that actually express it phenotypically. *Expressivity* is the intensity of expression.

In making genetic crosses, ratios are useful in predicting the expected results, but chance alone can cause deviations from the expected. Scientists use statistical analyses to determine whether the deviations they observe in their experimental results are significant. Tests of significance must take into account both the amount of deviation and the sample size, since marked departures from predicted results are not unusual in small samples. Experimental data may be evaluated by a variety of tests, all of which are simply ways to calculate the probability that the deviations are due to chance alone.

STUDY GUIDE FOR MENDELIAN GENETICS PROBLEMS

The best way to gain an understanding of genetics is to work with it. The problems at the end of Chapter 24 in your text illustrate the various patterns of inheritance the chapter discusses. The following information is intended to get you started on the problems. Answers are provided in this section, but you are strongly advised to work the problems *first* and check your answers later. Genetics problems are considerably easier when you start with the answers and work backward, but you will not learn how to do problems that way.

General Rules

1 Know your terms: dominant (capital letter), recessive (small letter); P, F_1, F_2; homozygous, heterozygous; genotype, phenotype.

2 Read the problem and assign appropriate symbols to the characteristics involved.

3 Set up a Punnett square correctly: female gametes go along the left side, male gametes across the top. Be sure to include all possible gametes.

♂ gametes

♀ gametes

4 Know how to use multiplication of probabilities instead of a Punnett square to solve problems. The basic principle is that the chance that a number of independent events will occur together equals the chance of the first event times the chance of each following event.

Example: The probability of getting heads when flipping a coin is $1/2$ or (0.5). If two pennies are flipped at the same time, the chance of getting two heads is the chance of the first being heads ($1/2$) times the chance of the second ($1/2$), or $1/2 \times 1/2 = 1/4$. The chance of getting five heads in a row is $1/2 \times 1/2 \times 1/2 \times 1/2 \times 1/2$, or $1/32$. (See problem 19 in your text.)

Types of problems

1 Monohybrid cross: cross involving one character

Example: If pole bean plants (*P*) are dominant over bush beans (*p*), what are the expected genotypes and phenotypes when two heterozygous individuals are mated?

P   female × male
         P/p   *P/p*

gametes: *P, p*   *P, p*

Punnett square:

	P	*p*
P	*P/P*	*P/p*
p	*P/p*	*p/p*

genotypic ratio: 1 *P/P* : 2 *P/p* : 1 *p/p*

phenotypic ratio: 3 pole : 1 bush

(Now try problems 1–3 in your text.)

2 Test cross: unknown organism crossed with a homozygous recessive

Example: If the genotype of a pole bean plant is in doubt (*P/−*), it is crossed with a bush bean plant.

P   *P/− × p/p*

a If − is dominant (*P*) the cross is *P/P × p/p* and all offspring are pole.

b If − is recessive (*p*) the cross is *P/p × p/p* and the offspring are half pole and half bush.

        P/p × *p/p*

gametes: *P, p*   *p*

Punnett square:

	p	*p*
P	*P/p*	*P/p*
p	*p/p*	*p/p*

genotypic ratio: 1 *P/p* : 1 *p/p*

phenotypic ratio: 1 pole : 1 bush

(See problem 2 in your text.)

3 Dihybrid cross: cross involving two characters

Example: Red pole beans heterozygous for color and shape were crossed with each other. The genes for color (*R* = red, *r* = white) and for shape (*P* = pole, *p* = bush) are on different chromosomes.

a Solving the problem using a Punnett square.

        female   male
parents: *R/r P/p*   ×   *R/r P/p*

gametes: *RP, Rp, rP, rp* × *RP, Rp, rP, rp*

Punnett square:

	RP	*Rp*	*rP*	*rp*
RP	*R/R P/P*	*R/R P/p*		
Rp				
rP				
rp				

Complete the Punnett square and determine the genotypic and phenotypic ratios. (Now try problem 8 in your text.)

b Solving the problem using probabilities.

It is very laborious to make Punnett squares for two or more pairs of genes, so multiplication of probabilities is often used instead. A convenient and rapid way to do problems where you are asked to calculate all the genotypic and phenotypic ratios is to use probabilities and the forkline method to do the crosses.

According to Mendel's second law, the members of one pair of alleles are inherited independently of the members of another. Consequently, the probability of each of the various genotypic combinations in the offspring is equal to the product of the genotypic probabilities of each pair of alleles.

To use the fork-line method, one calculates the genotypic probabilities of each pair of alleles separately, and then finds all possible genotypic combinations and multiplies their separate probabilities.

Example: R/r P/p × R/r P/p

1 Consider each pair of alleles separately.

Rr × Rr gives you 1/4 *R/R*, 2/4 *R/r*, 1/4 *r/r* (3 possible genotypes)

P/p × P/p gives you 1/4 *P/P*, 2/4 *P/p*, 1/4 *p/p* (3 possible genotypes)

(3 × 3 = 9 total possible genotypic combinations)

2 Now find all possible combinations.

Genotypes of offspring

$$1/4\ R/R \begin{cases} 1/4\ P/P = 1/16\ R/R\ P/P \\ 2/4\ P/p = 2/16\ R/R\ P/p \\ 1/4\ p/p = 1/16\ R/R\ p/p \end{cases}$$

$$2/4\ R/r \begin{cases} 1/4\ P/P = 2/16\ R/r\ P/P \\ 2/4\ P/p = 4/16\ R/r\ P/p \\ 1/4\ p/p = 2/16\ R/r\ p/p \end{cases}$$

$$1/4\ r/r \begin{cases} 1/4\ P/P = 1/16\ r/r\ P/P \\ 2/4\ P/p = 2/16\ r/r\ P/p \\ 1/4\ p/p = 1/16\ r/r\ p/p \end{cases}$$

If you make a cross involving three pairs of genes, you can simply add another column of "forks." (Start with a big piece of paper!)

Probabilities can be used to solve many genetics problems that would be tedious to solve with a Punnett square. (See example at the top of the next column.)

Example: A heterozygous red pole bean plant with smooth seeds (*R/r P/p S/s*) is crossed with a red pole bean plant with wrinkled seeds (*R/r P/P s/s*). What fraction of the offspring will have the genotype *R/r P/P s/s*?

The chance of *R/r* from the cross *R/r × R/r* is 1/2.

The chance of *P/P* from the cross *P/p × P/P* is 1/2.

The chance of *s/s* from the cross *S/s × s/s* is 1/2.

The chance of the combination *R/r P/P s/s* is equal to the product of their separate probabilities; 1/2 × 1/2 × 1/2 = 1/8. (Now try problem 19 in your text.)

Now do all the rest of the problems starting on page 666 of your textbook, and check your answers. Additional problems are given in the Questions section.

Answers to genetics problems, in text (pp. 666–68)

1 *a* All heterozygous, white

	w	w
W	W/w	W/w
W	W/w	W/w

b Genotype ratio: *W/w* : *w/w* is 1 : 1
Phenotype ratio: 1 white : 1 yellow

	w	w
W	W/w	W/w
w	w/w	w/w

c Genotype ratio: 1 *W/W* : 2 *W/w* : 1 *w/w*
Phenotype ratio: 3 white : 1 yellow

	W	w
W	W/W	W/w
w	W/w	w/w

2 *The cross is W/w × w/w* (as in 1*b*, above). The expected phenotype ratio is 1 white : 1 yellow. If there were 200 seeds, you would expect 100 to grow into white-fruited plants and 100 to grow into yellow-fruited. The chi-square is determined as follows:

	white	*yellow*
observed	110	90
expected (*e*)	100	100
deviation (*o-e*) = *d*	+10	−10
deviation squared (*d²*)	100	100
d²/e	100/100 = 1	100/100 = 1
$\chi^2 =$	1 +	1 = 2

Consulting the table on page 665 of your text, at one degree of freedom, the χ^2 value of 2.0 lies between 1.6 and 2.7, which means that a deviation of this magnitude would be expected between 10 and 20 percent of the time. The deviation is not significant; the deviation is a result of chance.

If there were 2000 seeds, we would expect 1000 white : 1000 yellow. The observed were 1100 white and 900 yellow. The chi-square would be determined as follows:

	white	yellow
observed	1100	900
expected (e)	1000	1000
deviation (o-e) = d	+100	−100
d^2	10000	10000
d^2/e	10000/1000 = 10.0	10000/1000 = 10.0
$\chi^2 =$	10 +	10 = 20

Our χ^2 value of 20 with one degree of freedom is highly statistically significant since 20 is greater than 6.64. Therefore, the deviation is not a result of chance.

3 Let B be for brown eyes and b be for blue eyes. The man must be b/b, but the woman is $B/-$. However, since her father is b/b, she must have a b allele. Thus, she is B/b. The cross is $b/b \times B/b$, and the Punnett square shows that 50 percent blue eyes are predicted.

	♂ b	b
♀ B	B/b brown	B/b brown
b	b/b blue	b/b blue

4 This time the man is $B/-$, and the woman is b/b. The cross is either $B/b \times b/b$, or $B/B \times b/b$. The first cross would yield 50 percent blue-eyed children as in question 2, and the second cross would yield all brown-eyed children as shown in the figure. Note that every child must receive a dominant B allele from his father. In this case the father is probably homozygous, but one cannot be certain that this is true. Furthermore, an eleventh brown-eyed child makes no difference. It cannot be proved beyond a doubt that a character is homozygous dominant. In this case there is a one in 2^{11} chance that the father is heterozygous, or some other factor may be prohibiting his b allele from being expressed.

	♂ B	B
♀ b	B/b brown	B/b brown
b	B/b brown	B/b brown

5 In this problem the brown-eyed man is B/b because his mother was b/b. His father is either B/B or B/b. His wife is b/b, so her brown-eyed parents must both be heterozygous, B/b. The blue-eyed son is, of course, b/b.

6 There are probably a number of explanations, but it is best to assume only one genetic locus at first. This is probably true since only

tail length is involved. The ratio is 3 : 6 : 2, which by inspection is similar to 1 : 2 : 1. This ratio appears as the expected genotype ratio of a monohybrid cross. This would mean that the parental short-tailed cats are heterozygous, T^1/T^2. Assuming then that T^1/T^1 yields a long tail and T^2/T^2 yields no tail, the Punnett square confirms this possible explanation. This assumes no dominance, with short tails a hybrid between none and long alleles.

	♂ T^1	T^2
♀ T^1	T^1/T^1 long	T^1/T^2 short
T^2	T^1/T^2 short	T^2/T^2 none

7 Again, assume that only a single locus is involved. The ratio of pups (haired : hairless : deformed) is 1 : 2 : ?. This is most like a monohybrid cross ratio, 1 : 2 : 1. This implies that the hairless dogs are monohybrids, H^1/H^2, which is substantiated by the number of hairless pups relative to haired pups, who would be H^1/H^1. The H^2 allele is then assumed to be lethal in the homozygous condition. A normal hairless cross, $H^1/H^1 \times H^1/H^2$, would yield $H^1/H^1 : H^1/H^2$ in the ratio 1 : 1, in agreement with the first statement.

8 a This is a dihybrid cross, as in problem 9. The genotypic ratio, $T/T\,S/S : T/T\,S/s : T/T\,s/s : T/t\,S/S : T/t\,S/s : T/t\,s/s : t/t\,S/S : t/t\,S/s : t/t\,s/s$, is 1 : 2 : 1 : 2 : 4 : 2 : 1 : 2 : 1. The phenotypes are tall, smooth; tall, wrinkled; short, smooth; short, wrinkled, in the ratios 9 : 3 : 3 : 1.

b This is a test cross, and the phenotypes will reflect the alleles present in the tall, wrinkled parent in equal numbers; the phenotypic ratio, tall, wrinkled : short, wrinkled, is 1 : 1. The genotypic ratio, $T/t\,s/s : t/t\,s/s$, is also 1 : 1. Note that there can be no homozygous dominant in a test cross.

c In this case each parent has only two possible gametes, tS and ts, and Ts and ts, respectively. The genotypic ratio, $T/t\ S/s : T/t\ s/s : t/t\ S/s : t/t\ s/s$, is $1 : 1 : 1 : 1$. The phenotypes are tall, smooth; tall, wrinkled; short, smooth; short, wrinkled, in the ratio $1 : 1 : 1 : 1$.

d Here each parent has only one possible gamete; hence there is only one possible offspring, $T/t\ S/s$, tall and smooth.

9 Let B be belted, b be no belt, F be fused, and f be normal. The cross is $b/b\ F/F \times B/B\ f/f$, yielding all double heterozygotes, $B/b\ F/f$, in the F_1 generation. The Punnett square shows the results of freely interbreeding these individuals, and the genotypic and phenotypic ratios are again, respectively, $1 : 2 : 1 : 2 : 4 : 2 : 1 : 2 : 1$ and $9 : 3 : 3 : 1$.

	BF	Bf	♂ bF	bf
BF	B/B F/F belted fused	B/B F/f belted fused	B/b F/F belted fused	B/b F/f belted fused
Bf ♀	B/B F/f belted fused	B/B f/f belted norm	B/b F/f belted fused	B/b f/f belted norm
bF	B/b F/F belted fused	B/b F/f belted fused	b/b F/F even fused	b/b F/f even fused
bf	B/b F/f belted fused	B/b f/f belted norm	b/b F/f even fused	b/b f/f even norm

10 Let S be green, s be striped, L be short, and l be long. The cross is $s/s\ l/l \times S/s\ L/l$. Since this is a test cross the phenotypes will reflect the alleles present in the green, short parent plant in equal numbers. (Assume no linkage.) The phenotypes are green, short; green, long; striped, short; striped, long, in the ratio $1 : 1 : 1 : 1$.

11 Let S be long-winged, s be vestigial-winged, H be hairless, and h be hairy. The parental cross is $s/s\ h/h \times S/S\ H/H$. Since the only gametes are s/h and S/H, the only F_1 offspring are $S/s\ H/h$, double heterozygotes. The Punnett square shows the results of the F_2 generation. (In the square "less" stands for "hairless," and "short" for "vestigial-winged.") The genotypic ratio, $S/S\ H/H : S/S\ H/h : S/S\ h/h : S/s\ H/H : S/s\ H/h : S/s\ h/h : s/s\ H/H : s/s\ H/h : s/s\ h/h$, is $1 : 2 : 1 : 2 : 4 : 2 : 1 : 2 : 1$. The phenotypes are long-winged, hairless; long-winged, hairy; vestigial-winged, hairless; vestigial-winged, hairy, in the ratio $9 : 3 : 3 : 1$. The F_1 phenotype is long-winged, hairless.

	SH	Sh	♂ sH	sh
SH	S/S H/H long, less	S/S H/h long, less	S/s H/H long, less	S/s H/h long, less
Sh ♀	S/S H/h long, less	S/S h/h long, hairy	S/s H/h long, less	S/s h/h long, hairy
sH	S/s H/H long, less	S/s H/h long, less	s/s H/H short, less	s/s H/h short, less
sh	S/s H/h long, less	S/s h/h long, hairy	s/s H/H short, less	s/s h/h short, hairy

12 The cross in this case is $s/s\ H/h \times S/s\ h/h$. The male can only contribute gametes with s/H and s/h while the female can contribute S/h and s/h. The Punnett square can be set up accordingly, indicating the following phenotypes: long-winged, hairless : short-winged, hairless : long-winged, hairy : short-winged, hairy—$1 : 1 : 1 : 1$.

	sH	♂ sh
Sh ♀	S/s H/h long, less	S/s h/h long, hairy
sh	s/s H/h short, less	s/s h/h short, hairy

13 Let S be barking, s be silent, D be erect, and d be drooping. The breeder wants $SSdd$. Any droop-eared dog must be pure and thus presents no problem. However, the breeder must use a test cross on his barkers since they may be heterozygous: $S/-\ d/d \times s/s\ -/-$. (The ears of the test dog are irrelevant.) Referring to problem 4, if the dog is homozygous, as the breeder wishes, all pups will be S/s, barkers, but if it is heterozygous, about 50 percent should be silent. However, as stated in problem 4, he can never be sure of having a homozygous dominant trait.

14 The cross is $A/-\ R/- \times A/-\ r/r$. If the male's second black allele were R, then there could be no yellow or cream offspring. Thus the allele is r. Similarly, if either second yellow allele were A, there could be no black or cream offspring. Thus they are both a. The cross was $A/a\ R/r \times A/a\ r/r$.

15 The cross is $A/-\ r/r \times a/a\ R/-$, and there are no black or cream offspring. Thus the second yellow allele of the male is not likely to be an a, which would make 50 percent black expected. The second black allele of the

female must be recessive, *r*, to produce yellow offspring. The cross was $A/A\ r/r \times a/a\ R/r$.

16 There are two possible crosses, $C/- i/i \times C/- I/-$ and $C/- i/i \times c/c\ -/-$. At least half the offspring of the first cross would receive an *I* allele and be colorless, but all the offspring are colored. Using the same argument in the second cross, neither inhibitory allele of the hen can be *I*. Finally, if the cock's second color allele were *c*, one would expect 50 percent white. Therefore, $C/C\ i/i \times c/c\ i/i$ is the most likely cross. The offspring are all $C/c\ i/i$.

17 The Punnett square shows the two gametes contributed by the parents. It shows that 75 percent of their offspring will be deaf. Without using a Punnett square one could note that 50 percent of the offspring would have the necessary *K* allele for hearing, but only 50 percent of these, and thus 25 percent of the total, would have it without the nasty *M* allele.

	♂ *kM*	*km*
♀ *Km*	*K/k M/m* deaf	*K/k m/m* normal
km	*k/k M/m* deaf	*k/k m/m* deaf

18 Assuming no linkage, each character, *Ks, Ls,* and *Ms*, assorts independently. The chances of any specific homozygous recessive appearing is $1/4$. The chance of all three appearing is therefore $(1/4)(1/4)(1/4)(1/4)$ or $1/64$.

19 The logic is the same here, where the parents are heterozygotes in five traits. In addition, the probability of any one heterozygous combination appearing is $1/2$. Thus, the chance of this particular offspring is $(1/4)(1/4)(1/2)(1/4)(1/2)$ or $1/256$.

QUESTIONS

Choose the one best answer.

1 In peas axial flowers are dominant over flowers borne terminally. What phenotypic ratios would you expect in offspring from a cross between a known heterozygous axial-flowered plant and one whose flowers are terminal?

a 3 axial:1 terminal   *d* all axial
b 3 terminal:1 axial   *e* none of the above
c 1 axial:1 terminal   (p. 649)

2 Cystic fibrosis is inherited as a simple recessive. Suppose a woman who carries the trait marries a normal man who does not carry it. What percent of their children would be expected to have the disease?

a 0 percent   *d* 50 percent
b 25 percent   *e* 75 percent
c 33 percent   (p. 649)

3 The light color variation in the peppered moth is inherited as a simple recessive characteristic. If a light moth is crossed with a dark moth which had a light parent, what percent of their offspring will be light?

a 25 percent   *d* 75 percent
b 33 percent   *e* 100 percent
c 50 percent   (p. 649)

4 A dominant gene *W* produces wire-haired texture in dogs; its recessive allele *w* produces smooth hair. A group of heterozygous wire-haired individuals are crossed and all of their F_1 progeny are then testcrossed. What is the expected phenotypic ratio among the testcross progeny?

a 3 wire-haired:1 smooth
b 1 wire-haired:1 smooth
c 3 smooth:1 wire-haired
d 100 percent wire-haired
e none of the above are correct
          (pp. 649–51)

5 In pigeons, the grizzle color pattern depends on a dominant autosomal gene *G*. A mating of two grizzle birds produced one nongrizzle youngster this year. If this pair of pigeons produces more youngsters next year, what percentage would be expected to be grizzles?

a 100 percent   *d* 25 percent
b 75 percent   *e* 0 percent
c 50 percent   (p. 649)

6 Among white human beings, when individuals with straight hair mate with those with curly hair, wavy-haired children are produced. If two individuals with wavy hair mate, what phenotypes and ratios would you predict among their offspring?

a 3 curly:1 wavy
b 1 curly:1 wavy:1 straight
c 1 straight:1 curly:2 wavy
d 3 wavy:1 straight
e 1 straight:2 curly:1 wavy   (p. 651)

7 In cocker spaniels, black color is due to a dominant gene *B*, and red color to its recessive allele *b*. Solid color is dependent on a dominant gene *S*, and white spotting on its recessive allele *s*. A solid red male was mated to a black-and-white female. They had five

puppies: one black, one red, one black-and-white, and two red-and-white. What were the genotypes of the parents?

a   male *b/b s/s* and female *B/B s/s*
b   male *b/b S/s* and female *B/b s/s*
c   male *b/b S/s* and female *B/b S/s*
d   male *B/b S/s* and female *B/b s/s*
e   male *B/b S/S* and female *B/b s/s*

(p. 653)

8   Suppose a solid red male cocker spaniel whose mother was spotted red was mated with a solid black female whose mother was also spotted red. What phenotypes and in what proportions might be expected from such a cross?

a   3 solid black:1 spotted black:3 solid red:
    1 spotted red
b   9 solid black:3 spotted black:3 solid red:
    1 spotted red
c   1 solid black:1 solid red:1 spotted black:
    1 solid red
d   3 solid black:1 spotted red
e   none of the above are correct   (p. 653)

9   Ignoring modifier genes, we can think of brown eyes in human beings as determined by a dominant allele *B* and blue eyes by a recessive allele *b*; free earlobes are determined by a dominant allele *F* and attached earlobes by a recessive allele *f*. A brown-eyed man with attached earlobes (whose mother was blue-eyed) marries a blue-eyed woman with free earlobes (whose father had attached earlobes). What phenotypes may be expected among their children?

a   Both blue- and brown-eyed children may be expected, but all will have free earlobes.
b   All four possible combinations of eye color and earlobe condition may be expected, in roughly equal frequencies.
c   All brown-eyed children will have attached earlobes, and all blue-eyed children will have free earlobes.
d   Both blue- and brown-eyed children may be expected, but all will have attached earlobes.
e   All brown-eyed children will have free earlobes, and all blue-eyed children will have attached earlobes.   (p. 653)

10   In a P cross, an *A/A B/B C/C* individual is paired with an *a/a b/b c/c* individual. Assuming no linkage, what will be the expected frequency of *A/A b/b C/c* individuals in the F_2 generation?

a   16/64   d   2/64
b   8/64   e   1/64
c   4/64   (p. 653)

11   A genetic disease known as Marfan's syndrome is caused by a dominant allele. In this disease the fingers and toes are excessively long. This and other skeletal defects are often accompanied by a misplaced eye lens and defects of the heart. Some individuals with this syndrome may have all these defects; others show only one or two of the defects yet may have children with all. Suppose a normal woman marries an affected man whose mother was normal. If this trait has 90 percent penetrance, what percentage of their children will be expected to express this trait?

a   22.5 percent   d   67.5 percent
b   45 percent   e   90 percent
c   50 percent   (p. 660)

12   In a diploid sexually reproducing organism with five pairs of chromosomes (I, II, III, IV, and V), what are the chances that one of its I, one of its III, and one of its V chromosomes were inherited from its paternal grandfather? (Ignore complications resulting from crossing over.) Be careful—this is a tricky question!

a   1/4   d   1/16
b   1/2   e   1/64
c   1/8

13   The allele for pea comb (*P*) in chickens is dominant to single comb (*p*), but the alleles black (*B*) and white (*B'*) for feather color show partial dominance, *B/B'* individuals having "blue" feathers. If birds heterozygous for both alleles are mated, what proportion of the offspring are expected to be white-feathered and pea-combed?

a   1/16   d   8/16
b   3/16   e   9/16
c   4/16   (p. 651)

For further thought

Answers to these questions are given in the Answer section. A chi-square table is provided below.

1   Mendel crossed pea plants with round seeds with plants with wrinkled seeds, and allowed the F_1 to self-fertilize. He collected the following data from the F_2:

round	5,474
wrinkled	1,850

Mendel hypothesized a 3:1 ratio. Is this data consistent with this hypothesis? Explain, using a chi-square test. (pp. 664–65)

2   A heterozygous burgundy flowered gladiolus was self-pollinated and produced 185 burgundy flowers and 71 white-flowered plants.

Are these results consistent with the hypothesis that flower color is controlled by one locus, with the allele for burgundy flowers dominant to the allele for white flowers? Explain your answer using a chi-square test and show all work. (pp. 664–65)

3 A student did a cross using fruit flies in which she expected a 9:3:3:1 phenotypic ratio. She ran a chi-square on her data and found that the chi-square value was 9.4. Does her data significantly deviate from the predicted ratio? (pp. 664–65)

4 A genetics student performed a cross in which he expected a 3:1 ratio. He ran a chi-square test on his data and found $\chi^2 = 4.1$. Does his data deviate significantly from the expected ratio? Explain. (pp. 664–65)

Probabilities for certain values of chi-square*

Degrees of freedom	P = 0.20 (1 in 5)	P = 0.10 (1 in 10)	P = 0.05 (1 in 20)	P = 0.01 (1 in 100)	P = 0.001 (1 in 1,000)
1	1.64	2.71	3.84	6.64	10.83
2	3.22	4.60	5.99	9.21	13.82
3	4.64	6.25	7.82	11.34	16.27
4	5.99	7.78	9.49	13.28	18.46
5	7.29	9.24	11.07	15.09	20.52

*Based on a larger table in R. A. Fisher, *Statistical Methods for Research Workers*, 10th ed., Oliver & Boyd, 1946.

ANSWERS

1 c   5 b   8 a   11 b
2 a   6 c   9 b   12 c
3 c   7 b   10 d   13 b
4 b

For further thought

1 $\chi^2 = 0.25$ Deviation due to chance; data is consistent with the hypothesis

2 $\chi^2 = 1.02$ Deviation due to chance; data is consistent with the hypothesis

3 Deviation at 3 degrees of freedom is larger than 6.3; therefore the deviation was significant at the 0.05 level

4 Deviation at one degree of freedom is greater than 3.8; deviation is significant at the 0.05 level

Chapter 25

NON-MENDELIAN PATTERNS OF INHERITANCE

A GENERAL GUIDE TO THE READING

This chapter continues the study of genetics but focuses on different, non-Mendelian patterns of inheritance. As in Chapter 24, a number of genetics problems are provided at the end of the chapter. You should plan to do all of the problems since the best way to gain an understanding of genetics is to do problems. Again, suggestions on how to approach such problems effectively are presented in the Study Guide for genetics problems on pages 209–11 of this *Study Guide*, and answers to the genetics problems in the text are provided on pages 211–12. In reading this chapter in the text, you should concentrate on the following topics.

1   Blood types. You will want to master this material (pp. 670–71), which illustrates the more general topic of multiple alleles. Tables 25.1–25.4 (pp. 670–71) are helpful.

2   Mutation. This concept, introduced on pages 672–73, will be discussed further in Chapter 26.

3   Sex and inheritance. Students often have difficulty with this topic. If you remember that "sex-linked" means "X-linked"—carried on the X chromosome—you will find it easier to understand sex-linked characters, as defined on page 676, and to avoid confusing them with sex-influenced characters, as defined on page 680.

4   Linkage. By and large, the most difficult section in this chapter is the one on linkage (pp. 680–86). For that reason some additional linkage problems are presented in "For Further Thought."

5   Chromosomal alterations. The section on this interesting topic (pp. 687–90) includes a box discussing specific examples of one of these alterations.

KEY CONCEPTS

1 A gene may exist in any number of allelic forms in a population, but a given individual can possess no more than two alleles. (pp. 669–72)

2 Mutational changes in genes occur at random; most of the mutations that have a phenotypic effect are deleterious. (p. 672)

3 Natural selection can act against a deleterious allele only if it causes some change in the organism's phenotype. Deleterious alleles that are recessive may be retained in the population for a long time. (p. 672)

4 Some alleles that are harmful when homozygous are beneficial when heterozygous. (pp. 672–73)

5 The inheritance patterns for characteristics controlled by genes on the X chromosomes are quite different from those for characteristics controlled by autosomal genes. (pp. 676–80)

6 Genes located on the same chromosome are linked together; they ordinarily remain together during meiosis. However, crossing over can occur between homologous chromosomes during synapsis, resulting in new linkages. (pp. 680–86)

7 Structural alterations to chromosomes sometimes occur, resulting in rearrangements of the genetic material; also changes in chromosomal number may occur, when the separation of chromosomes in cell division does not proceed normally. (pp. 687–89)

OBJECTIVES

After studying this chapter and reflecting on it, you should be able to carry out the following objectives.

1 Using the symbols shown in Table 25.3 (p. 671), list the possible genotypes of persons of each blood type (A, B, AB, and O) and explain the mode of inheritance. Then explain the importance of these blood types in giving blood transfusions; for example, explain why someone with type AB blood could not donate blood to a person with type O, while someone with type O blood could donate blood to someone with type AB. [Problems 13–14 in your text (p. 692) cover this material.]

2 Explain why most mutations are deleterious and how it is possible for a harmful recessive allele, even a lethal one, to persist in a population. (pp. 672–73)

3 Differentiate between the terms sex chromosome and autosome, and between the terms sex-linked and sex-influenced. (pp. 676–80)

4 Explain how sex is determined genetically in human beings. Then discuss the pattern of inheritance of sex-linked characters, showing why recessive sex-linked characters are expressed more often in males than in females. (pp. 675–76)

5 Use a Punnett square or the multiplication of probabilities to do problems involving sex-linkage, such as problems 1–5 (pp. 690–91) in your text. (pp. 676–78)

6 Explain the concept of linkage. Show how crossover frequencies are calculated and how they can be used to make chromosomal maps; then explain why the relative distances between genes on *Drosophila* chromosomal maps formulated from crossover data do not exactly match the corresponding distances on maps derived from salivary-chromosome studies, while the order of the genes on the chromosomes is the same in both. [Problems 8–11 in your text (pp. 691–92) cover this material.] (pp. 680–86)

7 Explain the process of nondisjunction and show how it affects the chromosomal composition of a cell. [Problem 12 in your text (p. 692) covers this material.] (p. 688)

8 Explain how translocations, deletions, duplications, and inversions alter chromosomal composition. Which of the above would you predict would have the most severe effect? The least severe effect? Explain.

9 Do all the problems on pages 690–92 of your text. Answers may be found on pages 211–12 of this *Study Guide*. (A word to the wise, however: Do the problems first, then check your answers. Genetics problems are easy if you look at the answers first and work backwards. You will learn more if you struggle along without reference to the answers.)

KEY TERMS

The following terms are important in this chapter; you should become familiar with them.

multiple alleles (p. 669)
mutation (p. 672)
sickle cell anemia (p. 673)
pleiotropic gene (p. 674)
sex chromosome (p. 675)
autosome (p. 675)
X chromosome (p. 675)
Y chromosome (p. 675)
sex-linked character (p. 676)
Barr body (p. 678)
dosage compensation (p. 679)
holandric (p. 680)
linked genes (p. 681)
crossing-over (p. 682)
polytene chromosomes (p. 684)
translocation (p. 687)
duplication (p. 687)
deletion (p. 687)
inversion (p. 688)
position effect (p. 688)
nondisjunction (p. 688)
trisomy (p. 688)
polyploid (p. 688)

SUMMARY

Multiple alleles Genes may exist in a number of allelic forms (*multiple alleles*). Each individual has only two alleles for a given trait, but other alleles may be present in the population. An example of multiple alleles in humans is the A-B-O blood series, which involves three alleles: I^A, I^B, and i. Both I^A and I^B are dominant over i, but neither I^A nor I^B is dominant over the other. Individuals with blood type A could be I^A/I^A or I^A/i; type B, I^B/I^B or I^B/i; type AB, I^A/I^B; and type O, ii. Individuals possessing the I^A gene will have antigen A on their red blood cells, those with I^B will have antigen B, those with ii will have no antigen. The two cellular antigens, A and B, react with certain antibodies, anti-A and anti-B, that may be present in the blood. Each antibody will clump red cells containing the corresponding antigen. The presence of these antigens and antibodies has important implications for blood transfusions.

Human Rh blood factors are another example of multiple alleles. Individuals can be divided into two phenotypic classes, Rh-positive and Rh-negative. Rh-positive individuals (genotype: *Rh/Rh* or *Rh/rh*) have the Rh antigen on their red blood cells; Rh-negative individuals (*rh/rh*) do not.

Mutations and deleterious genes A variety of influences can cause changes or *mutations* in the chemical structure of genes. Mutations occur constantly; most are deleterious. Natural selection can act against a deleterious gene only if it is expressed phenotypically. Harmful dominant genes can be eliminated rapidly by natural selection, but recessive genes are not expressed in the heterozygous state and thus cannot be easily eliminated.

An allele whose phenotype, when expressed, results in the death of the organism is called a *lethal*. Some alleles that are harmful when homozygous are beneficial when heterozygous. For example, individuals homozygous for the sickle-cell anemia gene suffer from this fatal disease. Heterozygous individuals do suffer from a mild anemia, but they have much higher than normal resistance to malaria. Thus the gene is beneficial in areas where malaria is common. Genes like this, with more than one effect, are said to be *pleiotropic*.

Marriages between closely related individuals are dangerous because they increase the chances of having children who are homozygous for deleterious traits.

Sex and inheritance In most higher organisms where the sexes are separate, the chromosomal endowments of males and females are different. One chromosomal pair, the *sex chromosomes*, differs in size and shape and determines the sex of the individual. All other chromosomes are called *autosomes*. In *Drosophila* and in human beings the females have two large *X chromosomes* whereas the males have an X and a smaller *Y chromosome*. The egg cells produced by meiosis are alike in chromosomal content, but the sperm cells are of two different types, one bearing an X, the other a Y. The sex is determined at the time of fertilization by the type of sperm fertilizing the egg. (The system is reversed in some organisms.)

The genes on the X chromosome are said to be *sex-linked*. Because females have two X chromosomes, they will always have two alleles for any sex-linked characteristic, whereas males will have only one (one on the X, none on the Y). Consequently, recessive sex-linked genes are always expressed phenotypically in the male but may be masked by the dominant allele in the female. In the male, all sex-linked characteristics are inherited from the mother. The Y chromosome apparently has few genes; these genes are said to be *holandric*.

Although females have two X chromosomes in each cell, only one is active; the other coils up into a tiny dark *Barr body* whose genes are inactive.

Many genes that control sexual characteristics are located on the autosomes of both sexes; these should not be confused with sex-linked characteristics.

Linkage Genes that are located on different chromosomes can segregate independently of one another during meiosis. Genes located on the same chromosome are *linked* together; they ordinarily remain together during meiosis. However, crossing over can occur between homologous chromosomes during synapsis in meiosis I; this results in new linkages.

The frequency of crossing over between any two linked genes will be proportional to the distance between them. The percentage of crossing over can be used to map gene locations. By convention, one unit of map distance is the distance within which crossing over occurs one percent of the time. Crossing over is the classical test of whether two characters are controlled by one gene or by two separate genes.

The giant chromosomes that occur in the salivary glands of many flies have been useful in determining visually the location of individual genes on the chromosomes. Such maps show that sequences of the genes determined by crossing-over frequencies are correct but the distances between the genes are not.

Chromosomal alterations Crossing over is one kind of chromosomal rearrangement; other alterations occur as well. In *translocation*, portions of two nonhomologous chromosomes are exchanged. Sometimes a portion of a chromosome breaks off and is lost, resulting in a *deletion*. Sometimes a piece breaks off one chromosome and fuses onto the end of its homologous chromosome, forming a *duplication*. Occasionally a portion of a chromosome breaks out, turns around, and fuses in its original position in reversed order, forming an *inversion*. This may alter gene expression because of the *position effect*.

Separation of the chromosomes during meiosis does not always occur normally; sometimes both members of one homologous pair move to the same pole. The result of this *nondisjunction* may be the production of a cell with an extra chromosome (*trisomy*). Occasionally all the chromosomes move to the same pole. If this cell then unites with another during fertilization, the resulting zygote has more than two sets of chromosomes and is said to be *polyploid*.

STUDY GUIDE FOR NON-MENDELIAN GENETICS PROBLEMS

Types of Problems

1 Sex linkage: Genes for sex-linked characteristics are carried on the X chromosome. Because females have two X chromosomes they always have two alleles for a characteristic on that chromosome, whereas males will have only one. Consequently, recessive sex-linked traits show up more often in males than in females.

Sometimes students find it helpful to use special symbols for sex-linked problems; X^C can represent a dominant sex-linked gene and X^c the recessive. The Y chromosome is designated Y.

Example: The disease hemophilia is sex-linked and recessive. A normal man marries a woman who is a carrier (heterozygous) for this trait. What phenotypes will the children probably have?

Let X^H = allele for normal blood
X^h = allele for hemophilia

The cross is ♀ X^H/X^h × ♂ X^H/Y.

Results: $1/2$ daughters will be normal.
$1/2$ will be carriers.
$1/2$ sons will be normal.
$1/2$ sons will have hemophilia.

(Now try problem 2 in your text.)

2 Pedigrees: Problems involving human pedigrees provide good practice, since they tie together many different aspects of inheritance. The following symbols are commonly used:

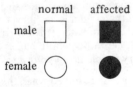

A horizontal line between two symbols indicates marriage, and the progeny are displayed below the parents:

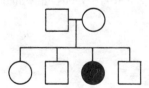

To do a pedigree problem, in which you are asked to determine the mode of inheritance of a given trait, first ascertain whether the trait in question is dominant or recessive. A dominant trait must occur in every generation, whereas a recessive trait can skip a generation. If you can find in the pedigree a family in which neither parent has the trait but one of the children does, you know the trait must be recessive. When a trait is identified as recessive, the next step is to decide whether it is autosomal or sex-linked, by analyzing the sex ratio of the affected individuals; remember that a recessive sex-linked trait is expressed more often in males than in females. Another clue is to look at the parents of the affected females. If a female offspring has the trait, it can be sex-linked only if the father expresses it. (For a female to express a sex-linked trait, she must get an allele from each parent; therefore the father would express the trait.) If a female has the trait and her father does not, the trait is autosomal recessive. Hence, in the pedigree above, the trait in question is autosomal recessive. Now look at the following pedigree, in which the affected trait is myopia (nearsightedness), and determine the mode of inheritance.

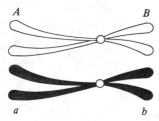

This trait too is autosomal recessive, as shown by the fact that a female has it even though neither parent expresses it. If the trait were sex-linked, her father would have expressed it. (Now try problem 7 in your text.)

3 Linkage: Genes are linked when they are on the same chromosome. In calculating crosses involving linked genes, special symbols are used. If the genes A and B are linked, the genotype of a heterozygous individual could be represented as AB/ab.

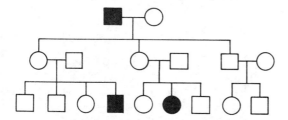

Linked genes are sometimes separated when homologous chromosomes change parts during synapsis in meiosis:

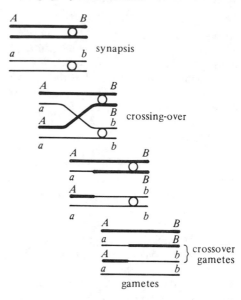

The normal and most common gametes are the noncrossovers (or "parentals"), AB and ab. The rare new gametes formed as a result of crossing over are the "recombinants"— Ab and aB. Linkage and crossing over can be detected in a test cross:

$$AB/ab \times ab/ab$$

	ab	# of offspring
Normal gamete AB	Ab/ab	45
Normal gamete ab	ab/ab	44
Crossover gamete aB	aB/ab	5
Crossover gamete Ab	Ab/ab	5

There are four phenotypes, but they are not present in equal proportions. (If A and B were not linked and were on different chromosomes the ratios of the phenotypes would be $1 : 1 : 1 : 1$.)

4 Chromosomal mapping: The distance between two genes can be calculated by means of the following procedure:

$$\frac{\text{number of crossover progeny}}{\text{total number of progeny}} \times 100 = \% \text{ crossing-over}$$

$$1\% \text{ crossing-over} = 1 \text{ map unit}$$

From our example above,

$$\frac{5+5}{100} \times 100 = 10\% = 10 \text{ map units}$$

(Now try problem 8 in your text.)

Now do all the rest of the problems starting on page 690 of your textbook, and check your answers. Additional problems are given in the Questions section of this *Study Guide*.

Answers to genetics problems in text (pp. 690–92)

1 The cross is $X^B/Y \times X^B/X^b$. As a result $1/2$ of the male offspring will receive the lethal allele and die. All the females will receive the X^B from their father and survive. Therefore, there will be twice as many female as male children.

2 Let X^C be normal and X^c be color-blind. The cross is $X^C/Y \times X^c/X^c$. Thus all males receive an X^c from their mother and will be color-blind. All the females will receive the dominant X^C from their father and be normal.

3 Yes, genetics can, and the man does have grounds. Since he is normal, X^C/Y, he will give all his daughters the dominant allele for the normal condition.

4 A lethal gene is probably involved. Since a disproportionate number of males survived, the allele is no doubt sex-linked. In fact, in accordance with the data given, $1/2$ of the female Z/W offspring died. If the lethal allele were on the W chromosome, all the female offspring would die. The only other possibility is that it is heterozygous in the male, Z^L/Z. The cross would be $Z^L/Z \times Z/W$, which would yield all viable males and 50 percent viable females ($1/3 : 2/3$).

5 Let B be for barred and b be for nonbarred. The male cannot be barred. If he is homozygous, all offspring will be barred, and if he is heterozygous, only $1/2$ of the females will be barred in the F_1. If the hen is barred, Z^B/W, and the cock is not, Z^b/Z^b, then all offspring cocks will be Z^B/Z^b, barred, and the hens will all be Z^b/W, nonbarred. This, of course, only works for the F_1 generation, since the cocks are now barred.

6 Let L be short hair and l be long hair and X^{B1} be yellow and X^{B2} black. The cross is $l/l\ X^{B2}/Y \times L/L\ X^{B1}/X^{B2}$. All the kittens will be L/l, short-haired. The color ratios are $X^{B1}/Y : X^{B2}/Y : X^{B1}/X^{B2} : X^{B2}/X^{B2}$—yellow male : black male : tortoise female : black female—$1 : 1 : 1 : 1$. If these freely interbreed, $1/4$ will be long-haired by the monohybrid ratio. Half of these will of course be males, and only $1/4$ of the males will be yellow since one out of the four female alleles is X^{B1}. Thus the chance of a long-haired yellow male is $(1/4)(1/2)(1/4)$ or $1/32$. The Punnett square illustrates this. Two out of 64 possibilities are such a cat. Each 4×4 block sector represents one of the four possible crosses. The two long-haired yellow males have thicker lines around them. (Note: This is a particularly difficult question—don't panic if you cannot do it!)

	LX^1	LY	lX^1	lY	♂ LX^2	LY	lX^2	lY
LX^1	$L/L\ X^1/X^1$ s, y, f	$L/L\ X^1/Y$ s, y, m	$L/l\ X^1/X^1$ s, y, f	$L/l\ X^1/Y$ s, y, m	$L/L\ X^2/X^1$ s, t, f	$L/L\ X^1/Y$ s, y, m	$L/l\ X^2/X^1$ s, t, f	$L/l\ X^1/Y$ s, y, m
LX^2	$L/L\ X^1/X^2$ s, t, f	$L/L\ X^2/Y$ s, b, m	$L/l\ X^1/X^2$ s, t, f	$L/l\ X^2/Y$ s, b, m	$L/L\ X^2/X^2$ s, b, f	$L/L\ X^2/Y$ s, b, m	$L/l\ X^2/X^2$ s, b, f	$L/l\ X^2/Y$ s, b, m
lX^1	$L/l\ X^1/X^1$ s, y, f	$L/l\ X^1/Y$ s, y, m	$l/l\ X^1/X^1$ l, y, f	**$l/l\ X^1/Y$ l, y, m**	$L/l\ X^2/X^1$ s, t, f	$L/l\ X^1/Y$ s, y, m	$l/l\ X^2/X^1$ l, t, f	**$l/l\ X^1/Y$ l, y, m**
lX^2	$L/l\ X^1/X^2$ s, t, f	$L/l\ X^2/Y$ s, b, f	$l/l\ X^1/X^2$ l, t, f	$l/l\ X^2/Y$ l, b, m	$L/l\ X^2/X^2$ s, b, f	$L/l\ X^2/Y$ s, b, m	$l/l\ X^2/X^2$ l, b, f	$l/l\ X^2/Y$ l, b, m
LX^2	$L/L\ X^1/X^2$ s, t, f	$L/L\ X^2/Y$ s, b, m	$L/l\ X^1/X^2$ s, t, f	$L/l\ X^2/Y$ s, b, m	$L/L\ X^2/X^2$ s, b, f	$L/L\ X^2/Y$ s, b, m	$L/l\ X^2/X^2$ s, b, f	$L/l\ X^2/Y$ s, b, m
LX^2	$L/L\ X^1/X^2$ s, t, f	$L/L\ X^2/Y$ s, b, m	$L/l\ X^1/X^2$ s, t, f	$L/l\ X^2/Y$ s, b, m	$L/L\ X^2/X^2$ s, b, f	$L/L\ X^2/Y$ s, b, m	$L/l\ X^2/X^2$ s, b, f	$L/l\ X^2/Y$ s, b, m
lX^2	$L/l\ X^1/X^2$ s, t, f	$L/l\ X^2/Y$ s, b, m	$l/l\ X^1/X^2$ l, t, f	$l/l\ X^2/Y$ l, b, m	$l/l\ X^2/X^2$ l, b, f	$L/l\ X^2/Y$ s, b, m	$l/l\ X^2/X^2$ l, b, f	$l/l\ X^2/Y$ l, b, m
lX^2	$L/l\ X^1/X^2$ s, t, f	$L/l\ X^2/Y$ s, b, m	$l/l\ X^1/X^2$ l, t, f	$l/l\ X^2/Y$ l, b, m	$L/l\ X^2/X^2$ s, b, f	$l/l\ X^2/Y$ s, b, m	$l/l\ X^2/X^2$ l, b, f	$l/l\ X^2/Y$ l, b, m

♀

7 Number the individuals in the diagram 1 through 15 from left to right and top to bottom.

   a The trait is not dominant autosomal because female 9 is deaf while neither parent is.

   b The trait could be recessive autosomal.

   c It is not sex-linked dominant for the same reason as *a*.

   d It is not sex-linked recessive, for all the sons of female 2 would have to be deaf and male 8 is not. Female 9 could not be deaf since her father, male 3, would have to be deaf for her to be homozygous recessive.

   e The trait is not holandric, for no female could have the trait. It would also have to appear in males 1 and 8.

8 Let *B* be gray, *b* be black, *V* be normal, and *v* be vestigial. The cross is

$$\frac{BV}{BV} \times \frac{bv}{bv}$$

yielding all

$$F_1 \quad \frac{BV}{bv}$$

The next cross

$$\frac{BV}{bv} \times \frac{bv}{bv}$$

a test cross, should segregate the characters into four equally occurring phenotypes if they are not linked. Since the phenotypes do not occur equally, the genes are linked and crossing over occurred

$$\frac{50 + 61}{600} \times 100 = 18.5\% \text{ crossing-over}$$

$$= 18.5 \text{ map units apart}$$

9 The genes are linked, since a ratio of 1 : 1 : 1 : 1 would be expected if the genes were inherited independently.

$$\frac{14 + 10}{200} \times 100 = 12\% \text{ crossing-over}$$

$$= 12 \text{ map units apart}$$

10 If the genes are 20 units apart, crossing over would occur 20 percent of the time. Hence 20 percent of the 1,000 offspring would be expected to show the crossover phenotypes. $(1000 \times .20 = 200)$

11 If *A* and *B* are 40 map units apart, and both *A* and *B* are 20 units from *C*, then *C* must lie halfway between them. Similarly, since both *C* and *B* are 10 units from *D*, *D* must lie halfway between them. The diagram below illus-

trates the conclusion, obtained through the incorporation of the various conditions in the order given in the text. The individual steps are lettered.

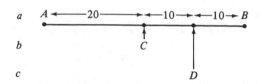

12

Genotype	Drosophila sex	Human sex	Number of Barr bodies
XO	male	female	none
XXX	female	female	two
XYY	male	male	none
XXXX	female	female	three
XXXY	female	male	two
XXXXY	female	male	three

13 The man is I^B/i since one of his parents is *ii*, which is actually irrelevant. The cross is $I^B/i \times I^A/I^B$, so that half of the offspring will have an I^A allele from the mother. They cannot be pure B type. The other half will get the I^B and will be B.

14 Every parent has one unknown allele and thus Shirley, who is *i/i*, may belong to either family. Jane, who is I^B/i, cannot be the daughter of the Joneses, where no I^B allele is present. Thus, a mixup did occur in her case, and she could belong to the Smiths. However, since Shirley may also be their daughter, a third family may be involved.

QUESTIONS

Testing knowledge and understanding

Choose the one best answer.

1 In an emergency, a person with blood type AB could receive blood from persons with blood type

   a A.   d AB.
   b B.   e all of the above
   c O.   (p. 670)

2 Knowledge of the blood-type genotypes of a certain couple leads us to say that if they were to have many children, the ratios of the children's blood types would be expected to approximate 1/2 type A and 1/2 type B. It follows that the blood types of the couple are

   a A and B.   d AB and A.
   b AB and AB.   e AB and O.
   c AB and B.   (p. 671)

3 In a paternity suit there were four men who
could have fathered the child. The child was
blood type O, the mother type B. The blood
types of the four men are listed below. Which
of the men can be eliminated from consideration on the basis of his blood type?

a A
b B
c AB
d O
e None of the men could be eliminated;
each one could have been the father.
(p. 671)

4 A particular sex-linked recessive disease of
human beings isn't usually fatal. Suppose
that, by chance, a boy with the disease lives
past puberty and marries a woman heterozygous for the trait. If they have a daughter,
what is the probability that she will have the
disease?

a 0 percent   *d* 75 percent
b 25 percent   *e* 100 percent
c 50 percent   (p. 677)

5 The rare trait ocular albinism (almost complete absence of eye pigment) is inherited as
a sex-linked recessive. A man with ocular
albinism marries a woman who neither has
this condition nor is a carrier. Which one of
the following is the best prediction concerning their offspring?

a All their sons will have ocular albinism,
and all their daughters will be carriers.
b All their children of both sexes will have
ocular albinism.
c About 50 percent of their sons will have
ocular albinism, and all their daughters
will be carriers.
d About 50 percent of their daughters will
have ocular albinism, but all their sons
will have normal eyes.
e None of their children will have ocular
albinism, but all their daughters will be
carriers.   (p. 677)

*Questions 6–8 refer to the following pedigree
for one type of deafness in human beings.
Squares symbolize males, circles females;
filled symbols designate deaf individuals; open
symbols individuals with normal hearing.*

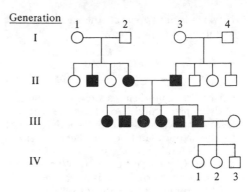

6 This type of deafness is probably inherited as

a an autosomal dominant.
b an autosomal recessive.
c a sex-linked dominant.
d a sex-linked recessive.
e a holandric.   (pp. 676–78)

7 The genotype of individual 4 in generation I
is probably

a D/D.   *d* X^D/Y.
b D/d.   *e* X^d/Y.
c d/d.   (pp. 676–78)

8 The genotype of individual 3 in generation IV
is probably

a D/D.   *d* X^D/Y.
b D/d.   *e* X^d/Y.
c d/d.   (pp. 676–78)

9 A certain man has the genotype *AB/ab*;
genes *A* and *B* are on one chromosome, and *a*
and *b* are on the homologous chromosome.
Suppose crossing-over occurs during a
meiotic division in this man's testis. With
regard to the two genes discussed here, how
many genetically different types of sperm
cells will result?

a 1   *d* 8
b 2   *e* 16
c 4   (pp. 681–82)

10 Suppose alleles *M* and *n* are linked on one
chromosome, and *m* and *N* are linked on the
homologous chromosome. Individuals
homozygous for *M* and *n* are mated with
individuals homozygous for *m* and *N*. Their
offspring are test-crossed (crossed with
homozygous recessive individuals), and the
following results are recorded:

Mn/mn	232
mN/mn	240
MN/mn	15
mn/mn	13

How many units apart are these genes on the chromosome?

a 2.5   d 12
b 5.1   e 50
c 5.6   (pp. 682–83)

11 Singer Arlo Guthrie has a 50 percent chance of dying prematurely from the same genetic disease that killed his father, Woody Guthrie. Neither Woody Guthrie's mother nor Arlo Guthrie's mother carries any allele for this disease (Huntington's Chorea). What type of inheritance pattern does this disease have? (It can be figured out from the information given.)

a autosomal dominant
b sex-linked dominant
c autosomal recessive
d sex-linked recessive   (pp. 678–81)

12 Pleiotropism describes

a a single gene having multiple effects.
b gene interaction of multiple alleles.
c a single trait being influenced by several genes.
d a trait that is not expressed for several generations.
e polygenic inheritance.   (p. 674)

13 The frequency of crossing over between linked genes A and B is 35 percent; between B and C, 10 percent; between C and D, 15 percent; between C and A, 25 percent; between D and B, 25 percent. The sequence of the genes on the chromosome is

a ACDB.   d ABCD.
b ACBD.   e ADCB.
c ABDC.   (p. 683)

14 In *Drosophila* a dominant allele P produces normal eye color, while the recessive allele p produces purple eye color. Another gene controls body color: the dominant allele C produces normal body color while the recessive allele c produces black body color. A female fruit fly heterozygous for the recessive alleles purple eye and black body is crossed with a purple-eyed black-bodied male. The following offspring are obtained:

both normal traits	151
purple eye, normal body	8
normal eye, black body	10
purple eye, black body	131

How do you explain these data?

a Genes C and P are linked and are 6 units apart on the chromosome.
b Genes C and p are linked and are 6 units apart on the chromosome.

c Genes C and P are on different chromosomes and assort independently of one another.
d Both *a* and *b* are correct.
e None of the above is correct.
                    (pp. 682–83)

15 Consider two linked autosomal genes. The dominant allele C of the first gene causes cataracts of the eye, whereas its recessive allele c produces normal eyes. The dominant allele of the second gene P causes polydactyly (presence of an extra finger on each hand), whereas its recessive allele p produces normal hands. A man with cataracts and normal hands marries a woman with polydactyly and normal eyes. Their son has both cataracts and polydactyly. The son marries a woman with neither trait. Assuming no crossing over, what is the probability that their first child will have both cataracts and polydactyly?

a 0 percent   d 75 percent
b 25 percent   e 100 percent
c 50 percent   (pp. 681–82, 660)

16 Gene A and gene B are known to be 10 units apart on the same chromosome. Individuals homozygous dominant for these genes are mated with homozygous recessives. The offspring are then test-crossed. If there are 1,000 offspring from the test cross, how many of the offspring would you expect to show the crossover phenotypes?

a 10   d 250
b 50   e 500
c 100   (pp. 682–84)

17 A person with XYY syndrome will have how many Barr bodies in his nuclei?

a 0   d 3
b 1   e either 1 or 2
c 2   (p. 678)

18 The exchange of parts between nonhomologous chromosomes is called

a inversion.   d transformation.
b translocation.   e duplication.
c transduction.   (pp. 687–88)

For further thought

[Answers to these questions are included at the end of the answer section of this chapter.]

1 Blood typing is often used as evidence in paternity cases in court. In a series of paternity cases, the mothers and their respective children had the blood types listed in the table below. For each, indicate the blood type(s) which, if found in the accused man, would exonerate him as the father. (p. 671)

MOTHER	CHILD	MAN EXONERATED IF HE BELONGS TO GROUP
A	O	
B	AB	
O	A	
AB	A	
O	O	
B	B	
A	B	
AB	AB	
A	A	
A	AB	
B	A	
B	O	
AB	B	

2   In pea plants, the genes for flower color and pollen-grain shape are on the same chromosome. Purple flowers (*P*) are dominant over red (*p*) and long pollen grains (*L*) dominant over round (*l*). Plants heterozygous for purple flowers and long pollen (*PL/pl*) are test-crossed with the homozygous recessive (*pl/pl*). The following offspring are produced:

purple, long	118
red, round	122
purple, round	32
red, long	28

How far apart are the genes for color and pollen-grain shape on the chromosome?
(pp. 682–83)

3   A tomato breeder wants to find out whether the genes for plant height and plant color are on the same chromosome. The allele for tall plants (*T*) is dominant over the dwarf allele (*t*), and the green allele (*G*) is dominant over the light-green allele (*g*). She crosses plants homozygous for tallness and green color with homozygous recessive plants. The progeny are then test-crossed, and the following results are observed:

tall, green	434
dwarf, light green	466
tall, light green	51
dwarf, green	49

Are the genes for height and color on the same chromosome? If so, how far apart are they on the chromosome?   (pp. 682–83)

4   In *Drosophila melanogaster* the genes for normal bristles and normal eye color are known to be about 20 units apart on the same chromosome. Individuals homozygous dominant for these genes are mated with homozygous recessive individuals. The F$_1$ progeny are then test-crossed. If there are 1,000 offspring from the test cross, how many of the offspring would you expect to show the crossover phenotypes?   (pp. 682–84)

5   The frequency of crossing-over between linked genes *A* and *B* is 6 percent; between *B* and *C*, 13 percent; between *C* and *D*, 18 percent; between *C* and *A*, 7 percent; between *D* and *B*, 5 percent. What is the sequence of these genes on the chromosome?
(pp. 683–84)

6   In watermelons, green color is dominant over striped and short length is dominant over long. The dominant allele for green (*G*) is linked to the dominant allele for short (*S*). In a test cross of heterozygous green short watermelons, how many types of gametes can each parent produce, assuming there is no crossing-over? Give the genotypes of the progeny of such a cross.   (pp. 681–82)

ANSWERS

1	e	6	b	11	a	15	a
2	e	7	b	12	a	16	c
3	c	8	b	13	e	17	a
4	c	9	c	14	a	18	b
5	e	10	c				

For further thought

1

MOTHER	CHILD	MAN EXONERATED IF HE BELONGS TO GROUP
A	O	AB
B	AB	B, O
O	A	B, O
AB	A	—
O	O	AB
B	B	—
A	B	A, O
AB	AB	O
A	A	—
A	AB	A, O
B	A	B, O
B	O	AB
AB	B	—

2   20 map units apart
3   same chromosome, 10 map units apart
4   200

5 *CABD* or *DBAC*

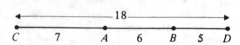

6   P   *GS/gs* × *gs/gs*

    gametes: *GS, gs*   *gs*

    F_1 *GS/gs, gs/gs*

THE STRUCTURE AND REPLICATION OF DNA

A GENERAL GUIDE TO THE READING

The last three chapters described how genes are transmitted from one generation to another and cited evidence for the linear arrangement of genes along the chromosome. This chapter discusses the chemical nature of the genetic material and how it replicates. It is important for you to thoroughly learn the material in this chapter since the next chapter builds upon this crucial information. As you read Chapter 26 in your text, you will want to concentrate on the following topics.

1 The flow of genetic information. Study carefully Figure 26.1 (p. 694) since it provides a helpful summary of the material presented in this chapter and the next. It will help you keep the new material in proper perspective.

2 Bacteriophage reproduction. This topic is presented early in the chapter, as part of an account that provides historical perspective, citing some of the experiments that showed

DNA to be the genetic substance. Figures 26.2 and 26.3 (p. 696) will help you learn and remember how bacteriophages replicate, which you will need to know for Chapters 28, 29, and 38.

3 The molecular structure of DNA. This topic is crucial; you must understand the section on it (pp. 697–701) before you continue reading, or you will be lost. You must learn the numbering convention for the five-carbon sugar, and be able to recognize the 3' and 5' ends of the molecule.

4 The replication of DNA. Make sure you understand this process thoroughly. Note that DNA is synthesized only in the 5' to 3' direction, and that one strand is copied continuously while the other is replicated in discontinuous segments. (pp. 701–2)

5 DNA repair. It is important for you to realize that there are special mechanisms to locate and correct errors that occur due to mutations and during replication. (pp. 706–8)

KEY CONCEPTS

1 DNA is the genetic material that makes up the genes; it contains all the information needed for the cell's growth and division into two similar cells. (pp. 694–97)

2 During replication, a faithful copy of the DNA is made. (pp. 701–3)

3 When DNA is replicated, the strands are copied in only one direction; the replicating enzymes move along one of the parental strands from 3' to 5', generating a complementary strand that goes from 5' to 3'; this means that one strand can be copied continuously, but the other must be replicated in discontinuous segments. (pp. 701–3)

4 Replication in eucaryotic cells is initiated at many independent sites on each chromosome simultaneously and proceeds in both directions away from the initiation sites. (p. 703)

5 Special mechanisms have evolved in both procaryotic and eucaryotic cells to locate and correct mutations and replication errors. (pp. 703–8)

OBJECTIVES

After studying this chapter and reflecting on it, you should be able to carry out the following objectives.

1 Cite two lines of experimental evidence to support the conclusion that DNA is the genetic material. (pp. 695–97)

2 Given a nucleotide, point out the phosphate group, the sugar group, the 1' carbon, 3' carbon, and 5' carbon of the sugar. Indicate whether the base is a purine or pyrimidine, and whether this nucleotide is from DNA or RNA. (pp. 697–99)

3 Name the four nitrogenous bases found in DNA, and indicate which are pyrimidines (single-ring structures) and which are purines (double-ring structures). Explain what the "D" in DNA stands for. (pp. 697–99)

4 Given a diagram of DNA such as that in Figure 26.7 (p. 698), point out the 5' and 3' ends of each chain, and an individual nucleotide and the three components of which it is made, the purine and pyrimidine bases, and the number of hydrogen bonds in a GC pair and an AT pair. (pp. 697–99)

5 Given a sequence of bases on one DNA strand, give the sequence on the complementary strand. (p. 700)

6 Show how the Watson-Crick model accounts for precise replication of genetic material, explain why one strand is copied continuously while the other is copied in discontinuous segments, and name the enzymes that catalyze DNA replication. (pp. 701–2)

7 Describe how repair enzymes act to correct mutations and errors that occur during replication. Give one example of a mutation that is actually created by the repair system. (Figures 26.11, p. 706, and 26.12, p. 707 may be helpful.) (pp. 703, 706–8)

KEY TERMS

The following terms are important in this chapter; you should become familiar with them.

replication (p. 693)
transformation (p. 695)
bacteriophage (p. 696)
adenine (p. 697)
guanine (p. 697)
purine (p. 697)
cytosine (p. 697)
thymine (p. 697)
pyrimidine (p. 697)
polymerase (p. 703)
misalignment deletion (p. 707)

SUMMARY

Experiments on bacterial transformations and radioactively labeled bacteriophage demonstrated that DNA, not proteins, constitutes the genetic material. DNA is composed of building blocks called *nucleotides*, which are made up of a five-carbon sugar, deoxyribose, attached by covalent bonding to a phosphate group and a nitrogenous base:

The phosphate group is bound to the 5' carbon of the sugar, while a hydroxyl is bound to the 3' carbon. There are four different nucleotides in DNA; they differ in their nitrogenous bases, which may be the double-ring purines, *adenine* and *guanine*, or the single-ring pyrimidines, *cytosine* and *thymine*.

The molecular structure of DNA Using information gained from chemical analysis, physical chemistry, and X-ray diffraction studies, James Watson and Francis Crick formulated a model of the DNA molecule. According to this model, the nucleotides are joined together by covalent bonds between the 3′ carbon of the sugar of one nucleotide and the phosphate group of the next nucleotide in the sequence; the nitrogenous bases are side groups of the chains. DNA molecules are usually double-chained structures, with the two chains held together by hydrogen bonds between adenine and thymine from opposite chains and between guanine and cytosine from opposite chains. The sequence of bases in one segment determines the complementary sequence in the other. The polarity of the two strands is opposite; one runs from 5′ to 3′, while the other goes from 3′ to 5′. The ladderlike double-chained molecule is coiled into a double helix, which is stabilized by hydrogen bonds.

The replication of DNA The Watson-Crick model of DNA explains how genetic replication can occur. The two chains of the DNA molecule are separated in places, and each chain acts as a mold or template for the synthesis of its new partner. The process produces two complete double-chained molecules, each identical in base sequence to the original double-chained molecule. The hydrogen bonds stabilizing the helical shape linking the strands of the two chains of DNA are broken and the chains separate. Each chain acts as a template for the synthesis of a new partner; complementary nucleotides are paired with nucleotides of each existing chain. Then the new nucleotides are covalently linked to form a chain. Each step is catalyzed by specific enzymes. The complementary strands of DNA are synthesized only in the 5′ to 3′ direction; one strand is copied continuously while the other is replicated in discontinuous segments.

DNA repair Special mechanisms have evolved in both procaryotes and eucaryotes to locate and correct mutations and replication errors. Enzymes find and bind to faulty or damaged sequences and clip out the flaws, and the intact complementary strand guides repair. The repair process is not perfect, however; some errors are missed, others cannot be detected, and still others may actually be created by the repair system. Nevertheless, the rate at which mutations accumulate is kept relatively low by the repair system.

KEY DIAGRAM

In the following diagram of a segment of a DNA molecule, color in the following, using a different color for each structure: (Reference: Figure 26.3, p. 698)

PHOSPHATE GROUPS

DEOXYRIBOSE GROUPS

HYDROGEN BONDS

Remembering the number of hydrogen bonds between an AT pair and a GC pair, and which bases are purines and which pyrimidines, color the following:

ADENINE

GUANINE

CYTOSINE

THYMINE

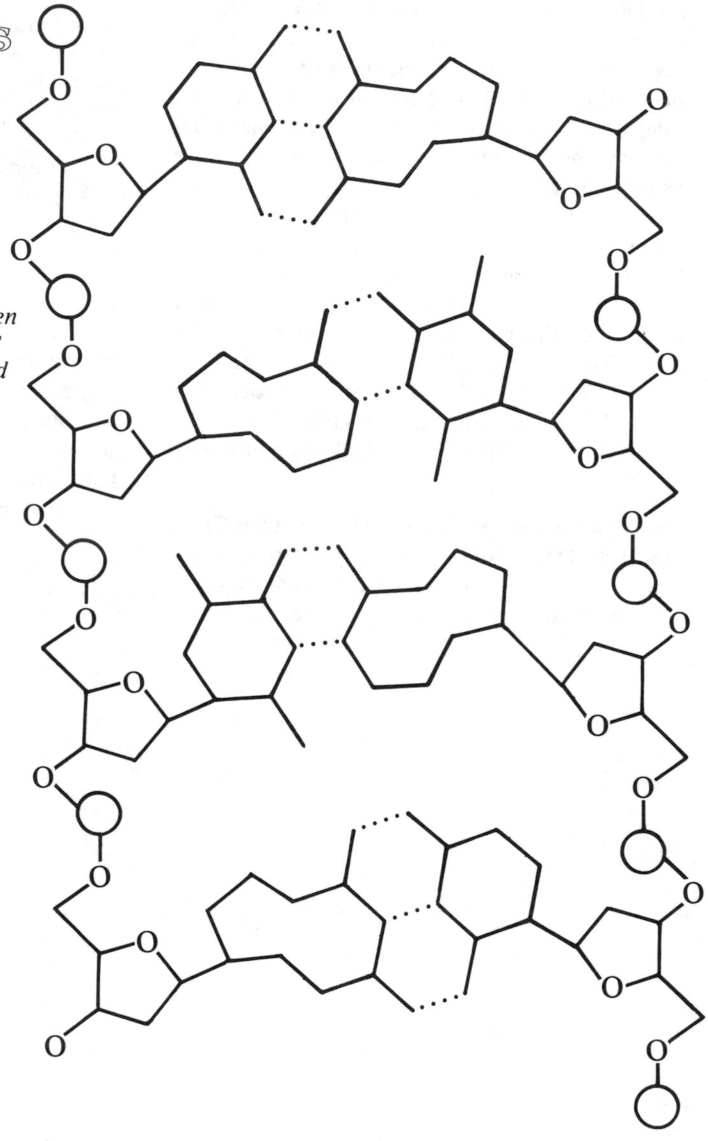

Ask yourself

Label the 3′ and 5′ ends of each strand.

QUESTIONS

Testing recall

Below are listed some of the important events that led to our present knowledge of the nature of the gene and its action. Match each

of these with the investigators associated with it.

a   Feulgen
b   Chargaff
c   Griffith
d   Hershey and Chase
e   Meselson and Stahl
f   Watson and Crick
g   Wilkins and Franklin

1 proposal of the double-helix model of DNA (p. 699)

2 transformation of bacteria by material extracted from heat-killed virulent cells (pp. 695–96)

3 infection of bacteria using radioactively labeled bacteriophage (p. 696)

4 discovery that in any DNA the amount of A equals the amount of T and the amount of G equals the amount of C (p. 700)

5 X-ray diffraction studies of DNA (p. 699)

6 use of ^{15}N-labeled DNA to obtain evidence in support of the Watson-Crick mechanism of DNA replication (p. 702)

Match each item below with the appropriate term. A term may be used more than once or not at all.

a sugar-phosphate groups
b purine(s)
c pyrimidine(s)
d covalent bonds
e hydrogen bonds

7 backbone of the DNA molecule (pp. 697–99)

8 forces between the two polynucleotide chains (p. 699)

9 single-ring nitrogenous bases (p. 697)

10 double-ring nitrogenous bases (p. 697)

11 adenine and guanine (p. 697)

12 cytosine and thymine (p. 697)

Testing knowledge and understanding

Choose the one best answer.

13 Which of the following is *not* compatible with the concept that DNA is the genetic material?

a DNA content of nuclei is constant in the cells of any one species but is only half the usual amount in the nuclei of the gametes.
b Each species has equal amounts of adenine, thymine, guanine, and cytosine.
c DNA content of the nuclei doubles before division.
d Only the DNA of a bacteriophage enters a new host bacterium. (pp. 695–99)

14 A nucleic acid is composed of a chain of nucleotides. Nucleotides themselves are made of three components. Which of these components could be removed from a nucleotide

that is part of a nucleic acid chain without breaking the chain?

a sugar
b phosphate
c nitrogenous base (pp. 695–99)

15 The structure of DNA as proposed by Watson and Crick *depended* on all of the following observations *except*

a that DNA is capable of replicating itself precisely.
b that DNA base sequences vary from organism to organism.
c that DNA contains nitrogenous bases, sugars, and phosphates.
d X-ray periodicities of 3.4 nm, 2 nm, and 0.34 nm.
e Chargaff's rules of A=T and G=C. (pp. 697–99)

16 A parent molecule of DNA containing only radioactive nitrogen ^{15}N is placed in an environment containing only ^{14}N. After four replications, how many DNA molecules would still contain some ^{15}N? (Assume no crossing-over.)

a 2   d 8
b 4   e 16
c 6   (p. 702)

17 If a segment of nucleic acid is CATCATTAC, the complementary DNA strand is

a CATTACTAC.   d GTAGTAATG.
b CAUCAUUAC.   e CUACUACAT.
c GUAGUAAUG.   (p. 701)

18 The DNA of a certain organism has guanine as 30 percent of its bases. What percentage of its bases would be adenine?

a 10   d 40
b 20   e 50
c 30   (p. 700)

19 Suppose that you provide an actively dividing culture of *E. coli* bacteria with radioactive thymine. What would you expect to happen if a cell replicated its DNA and divided once in the presence of the radioactive base?

a One of the daughter cells, but not the other, would have radioactive DNA.
b Neither daughter cell would have radioactive DNA.
c All four bases of the DNA would be radioactive.
d Radioactive thymine would pair with nonradioactive guanine.
e DNA in both daughter cells would be radioactive. (p. 702)

20 The diagram below represents a student's view of DNA synthesis occurring in an animal cell. Arrows represent newly synthesized DNA. This diagram is:

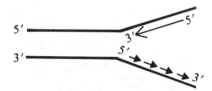 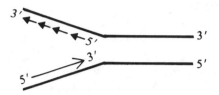

a   correct as shown.
b   incorrect because DNA synthesis in animal cells is unidirectional.
c   incorrect because DNA synthesis proceeds in a 3′ → 5′ direction.
d   incorrect because Okazaki fragments are synthesized on both strands.
e   incorrect because DNA synthesis is proceeding in the wrong direction on two of the strands. (pp. 704–5)

21 Which one of the following statements best describes the replication of DNA?

a   The double helix is cleaved at the phosphate bonds and new nucleotides pair with the appropriate sugars.
b   The hydrogen bonds between the bases are broken and new nucleotides pair with complementary bases on the old strands.
c   The hydrogen bonds between the bases are broken and these bases are replaced by new bases.
d   New bases are added at the 5′ carbon on the sugar so that DNA synthesis occurs in a 3′ to 5′ direction.
e   DNA is cleaved by DNA polymerases and new nucleotides are assembled by ligases. (p. 701)

22 A mutation that *cannot* be detected by the repair enzymes is

a   substitution of a guanine for an adenine.
b   fusion of two thymines.
c   misalignment between a series of adenines and thymines.

d   a cytosine that has been modified to form a thymine.
e   conversion of cytosine to uracil. (pp. 703, 706–8)

For further thought

1   If nucleic acids are synthesized in a 5′ to 3′ direction, what problem does this present during DNA replication? What are Okazaki fragments? Why is the enzyme DNA ligase important in DNA replication?

2   Explain why the DNA repair enzymes can detect and remove a mutation in which cytosine has been converted into uracil, but are not able to detect a mutation in which cytosine has been converted into thymine.

ANSWERS

Testing recall

1	f	4	b	7	a	10	b
2	c	5	g	8	e	11	b
3	d	6	e	9	c	12	c

Testing knowledge and understanding

13	b	16	a	19	e	21	b
14	c	17	d	20	a	22	d
15	e	18	b				

Chapter 27

TRANSCRIPTION AND TRANSLATION

A GENERAL GUIDE TO THE READING

The last chapter described how the DNA in chromosomes was replicated; in this chapter we see how the information in DNA is used to produce proteins. This is a very important chapter; as you read it you will want to concentrate on the following topics.

1 Transcription. You will want to understand how transcription works and be able to correctly transcribe a sequence of DNA. Remember that RNA is always synthesized in a 5' to 3' direction. You should note the differences between transcription in procaryotes and eucaryotes and learn what introns and exons are. Figure 27.5 (p. 714) is an excellent summary of messenger RNA processing in eucaryotes.

2 Translation. You will want to learn thoroughly the process of translation, and how to use the genetic code (Table 27.1, p. 717). A useful summary of information flow is found on pages 723–24.

3 Mutation. This concept, introduced in Chapter 26 (see p. 706), is enlarged upon here (pp. 724–26).

KEY CONCEPTS

1 The information encoded in DNA is used to produce both the proteins that form cellular structure and the enzymes that direct cellular metabolism; these determine the phenotypic characteristics of the organism. (p. 709)

2 During transcription, the information in DNA is faithfully copied into RNA. (pp. 709–15)

3 The sequence of the bases in DNA determines the sequence in which amino acids will be linked in protein synthesis. (pp. 715–21)

4 In eucaryotes, specific regions within the transcript must be removed in the nucleus to create a functional mRNA molecule. (pp. 713–15)

5 Translation of the information contained in DNA into functional proteins is indirect, involving several types of RNA. (pp. 716–22)

6 The genetic code consists of triplet coding units; a combination of three nucleotides specifies one amino acid. (pp. 716–17)

7 Mutations are alterations in DNA that result in new alleles. (pp. 724–26)

OBJECTIVES

After studying this chapter and reflecting on it, you should be able to carry out the following objectives.

1 State three ways in which RNA is different from DNA; name the three types of RNA and indicate where each is synthesized; and describe the function of each type of RNA. (pp. 710–11)

2 Given a sequence of bases on one DNA strand, write the sequence of mRNA that might be transcribed from this DNA. Indicate the 5′ and 3′ ends of the transcript. State the role of the promoter, RNA polymerase, the start signal and the termination signal in the process. (pp. 710–17)

3 Explain how transcription in procaryotes differs from that in eucaryotes. Figure 27.5 (p. 714) may be helpful. (pp. 711–15)

4 Describe in some detail the process of translation; be sure to show how the sequence of bases in DNA determines the sequence in which amino acids are linked in protein synthesis. (pp. 715–17)

5 Using the genetic code in Table 27.1 (p. 717), be able to give the amino acid sequence for any given mRNA.

6 Sketch a typical tRNA molecule, indicating the location of the anticodon and the position at which an amino acid becomes attached. (p. 721)

7 Describe in some detail how the genetic code is read by the ribosome. In doing so, explain the terms codon, anticodon, initiation codon, and termination codon. (pp. 721–23)

8 Define mutation, give three examples of different types of mutations, and indicate how mutations can be induced. Then, explain the relationship between mutagenicity and carcinogenicity. (pp. 724–26)

9 Discuss three possible definitions of the word "gene." (p. 726)

KEY TERMS

The following terms are important in this chapter; you should become familiar with them.

transcription (p. 710)
messenger RNA (mRNA) (p. 710)
uracil (p. 710)
ribosomal RNA (rRNA) (p. 711)
transfer RNA (tRNA) (p. 711)
RNA polymerase (p. 711)
promoter (p. 711)
consensus sequences (p. 712)
hairpin loop (p. 713)
primary transcript (p. 713)
exon (p. 713)
intron (p. 713)
translation (p. 715)
codon (p. 715)
anticodon (p. 721)
P-site (p. 723)
A-site (p. 723)
addition mutation (p. 724)
deletion mutation (p. 724)
base substitution (p. 724)
transposition (p. 725)
mutagenic agent (p. 725)

SUMMARY

The information encoded in DNA is used to produce both the proteins that form cellular structure and the enzymes that direct cellular metabolism. First, the information in the genes is transcribed into ribonucleic acid (RNA); second, the transcripts are translated into protein.

RNA differs from DNA in three ways: (1) The sugar in RNA is ribose whereas that in DNA is deoxyribose. (2) RNA has uracil where DNA has thymine. (3) RNA is usually single stranded where DNA is double stranded. There are three types of RNA: messenger RNA (mRNA), ribosomal RNA (rRNA), and transfer RNA (tRNA). All three are synthesized from DNA.

Transcription During transcription, the RNA polymerase complex binds to the DNA and opens up the helix. The polymerase complex moves from the 3′ end to the 5′ end of one of the DNA strands, and synthesizes a complementary RNA strand of the opposite polarity (5′ to 3′). In procaryotes, specific control sequences mark the beginning and end of a gene. The *promoter* is the area where the polymerase binds; transcription begins at a start signal and ends at a termination signal; a *hairpin loop* then forms and the mRNA leaves the DNA.

In eucaryotes, the transcript is capped and tagged with a polyadenine tail to form the *primary transcript*. Specific regions within the transcript must then be removed in the nucleus to create a functional mRNA molecule. The intervening sequences that are removed are called *introns*; the regions that remain and are used for protein synthesis are called *exons*. Specific base sequences mark the beginning and ends of the introns so removal is precise. The functional mRNA leaves the nucleus and moves into the cytosol where it becomes associated with the ribosomes.

Translation During translation, the sequence of bases in mRNA is used to order the amino acids to form a polypeptide. The coding unit or *codon* in nucleic acids is three nucleotides long. All but three of the 64 possible triplet codons code for one of the 20 amino acids; these three are termination codons, marking the end of the message. Most amino acids are represented by more than one codon.

The mRNA acts as a template for synthesis of polypeptide chains. As the ribosomes move along the mRNA, they read the codons, starting at the 5' end. Amino acids to be incorporated into the polypeptide chains are picked up by molecules of transfer RNA specific for each of the 20 amino acids. The tRNAs bring their amino acids to the ribosome as it moves along the mRNA. Each tRNA attaches to the mRNA at the point where a triplet of mRNA bases (a codon) is complementary to the exposed triplet (*anticodon*) on the tRNA. This ordering of the tRNAs along the mRNA automatically orders the amino acids, which are then linked by peptide bonds. This process is called *translation*; the nucleic acid message has been translated into an amino acid sequence.

Synthesis of the polypeptide chain proceeds one amino acid at a time in an orderly sequence as the ribosomes move along the mRNA. As each tRNA donates its amino acid to the growing polypeptide chain, it uncouples from the mRNA and moves away to pick up another load. When a ribosome reaches a termination codon, it releases the completed polypeptide chain.

In this process, the DNA of the gene determines the mRNA, which determines the protein enzymes, which control chemical reactions, which produce the characteristics of the organism.

Mutations Mutations are alterations in the DNA that change its information content and thus produce new alleles. Several types of mutations are possible. The *addition* or *deletion* of nucleotides in DNA often results in the production of inactive enzymes because of frame shifts in the translation process. In *base substitution* (point mutation), one nucleotide is exchanged for another. Base-substitution mutations are not as serious as additions or deletions because only one codon (and hence one amino acid) is involved, and often the change in the codon does not result in a change in the amino acid. *Transposition* mutations occur when DNA from one part of the genome is inserted into the middle of another.

High-energy radiations and a variety of chemicals can cause genetic mutations. Ionizing radiations sometimes induce point mutations, but frequently produce large deletions of genetic material. Some mutagenic chemicals convert one base into another. There is a strong relationship between the mutagenicity of a chemical and its cancer-inducing activity (carcinogenicity).

Definitions of the gene A gene can be defined in many ways. Using evidence from the mold *Neurospore*, George W. Beadle and Edward L. Tatum proposed the one gene-one enzyme hypothesis, which states that each gene directs the synthesis of one enzyme that controls a chemical reaction of the cell; the reactions, in turn, determine the phenotypic characteristics. Because some proteins are composed of two or more chemically different polypeptide chains, each determined by its own gene, and because some genes act as templates for tRNA and rRNA synthesis, it would be more accurate to define the gene as the length of DNA that codes for one functional product, whether those products are RNAs to be used directly, enzymes, structural proteins, or polypeptide chains that must combine with other chains to form functional products.

QUESTIONS

Testing recall

Questions 1–11 refer to the following: Indicate whether each of the following structural or functional features is true for

a DNA only
b RNA only
c both DNA and RNA
d neither DNA nor RNA

1 is usually single stranded (p. 710)

2 contains pyrimidines (p. 710)

3 contains deoxyribose (p. 710)

4   is coiled in a double helix (p. 679)

5   contains thymine (p. 710)

6   contains cytosine (p. 710)

7   contains uracil (p. 710)

8   brings amino acids to the ribosome (p. 711)

9   is present in the ribosome (p. 711)

10 is involved in transcription (p. 710)

11 is involved in translation (p. 711)

12 Fill in the blanks.

DNA (strand 1)   T G T _ _ _   _ _ _
DNA (strand 2)   _ _ A   C _ _   _ _ _
mRNA (from
   strand 2)   U _ _   _ C A   _ _ _
tRNA anticodons   _ _ _   _ _ _   G C A

Testing knowledge and understanding

13 One strand of a DNA molecule has the
    sequence of bases
    3'   TACCTTCAGCGT   5'.

a   What is the sequence of bases on the
    complementary strand of DNA? (p. 701)
b   What is the sequence of bases on the
    strand of mRNA that is synthesized from
    the original strand? (p. 710)
c   Name the organelle where the synthesis of
    mRNA takes place. (p. 710)
d   How many different codons are there in
    this strand? (p. 715)
e   What are the anticodons for the mRNA
    transcribed from this segment of DNA?
    (p. 721)
f   Name the organelle where the codon and
    anticodon couplings take place. (pp.
    722–23)
g   Using the genetic code below and the
    sequence of codons on the mRNA strand
    (b, above), determine the sequence of
    amino acids coded for by this strand.
    (pp. 710, 715–17)

The genetic code (messenger RNA)

First base in the codon	Second base in the codon				Third base in the codon
	U	C	A	G	
U	Phenylalanine	Serine	Tyrosine	Cysteine	U
	Phenylalanine	Serine	Tyrosine	Cysteine	C
	Leucine	Serine	Termination	Termination	A
	Leucine	Serine	Termination	Tryptophan	G
C	Leucine	Proline	Histidine	Arginine	U
	Leucine	Proline	Histidine	Arginine	C
	Leucine	Proline	Glutamine	Arginine	A
	Leucine	Proline	Glutamine	Arginine	G
A	Isoleucine	Threonine	Asparagine	Serine	U
	Isoleucine	Threonine	Asparagine	Serine	C
	Isoleucine	Threonine	Lysine	Arginine	A
	Methionine	Threonine	Lysine	Arginine	G
G	Valine	Alanine	Aspartic acid	Glycine	U
	Valine	Alanine	Aspartic acid	Glycine	C
	Valine	Alanine	Glutamic acid	Glycine	A
	Valine	Alanine	Glutamic acid	Glycine	G

14 Given the following strand of DNA in a
    bacterial cell and its polarity

        3'   TATTACCATCCCGGGATT   5'

a   Replicate the strand.
b   If the original strand is the coding strand,
    transcribe this strand and indicate the 3'
    and 5' ends of your transcript.
c   Using the genetic code above, translate
    the mRNA and give the correct sequence
    of amino acids coded for the strand.
    (Don't forget that translation begins with
    the initiator codon.)

Choose the one best answer.

15 If mRNA is transcribed from a chain of DNA
    with the base sequence 3'   CATTAG   5',
    the mRNA will have the sequence

a   GUAAUC.   d   TUCCUA.
b   GTAATC.   e   GTUUTC.
c   TGCCGA.   (p. 710)

16 DNA, but not RNA, contains

a   adenine.   d   cytosine.
b   guanine.   e   uracil.
c   thymine.   (p. 710)

17   Transfer RNA functions in

a   carrying RNA from the ribosomes to
      mRNA.
b   attaching RNA to the ribosomes.
c   joining proteins to form the ribosomes.
d   carrying mRNA from the nucleus to the
      cytoplasm.
e   carrying amino acids to the correct site on
      the mRNA.   (p. 721)

*Questions 18–20 refer to the following
segment of the base sequence of a gene:
3'   ACGTGCCCGGAT   5'.*

18   How many amino acids would the polypep-
     tide encoded by this gene segment have?

a   1   *d*   12
b   3   *e*   36
c   4   (p. 715)

19   The *second* codon on the mRNA derived
     from the segment will be

a   TAC.   *d*   CAC.
b   ACG.   *e*   UGC.
c   ATG.   (p. 715)

20   The anticodon of the tRNA for the *first*
     codon will be

a   TCG.   *d*   UGC.
b   ACG.   *e*   GCA.
c   TCG.   (p. 721)

*Questions 21–24 refer to the following
situation:
The following is a short mRNA sequence.
Use this sequence, and the genetic code
above, in question 13, to answer the next
four questions*

5'   AUGCCCUACUAC   3'

21   The sequence of the template strand of DNA
     of the gene that codes for this message will be
     (assume 3' end is at left side in the answers)

a   AUGCCCAACUAC.
b   TACGGGATGATG.
c   UTGCCCUUCTAC.
d   UACGGGUUGAUG.
e   ATGCCCTACTAC.   (p. 710)

22   The anticodon in the tRNA that attaches to
     the first codon will be

a   UAC.   *d*   GUA.
b   TAC.   *e*   ATG.
c   AUG.   (p. 721)

23   The protein coded for by this message will
     have _____ amino acids.

a   1   *d*   5
b   3   *e*   12
c   4   (p. 715)

24   The second amino acid in the polypeptide
     chain will be (use table at left)

a   lysine.   *d*   aspartic acid.
b   proline.   *e*   phenylalanine.
c   glycine.   (p. 717)

25   The diagram represents tRNA which recog-
     nizes and binds a particular amino acid (in
     this instance, valine). Which one of the fol-
     lowing triplets of bases on the *DNA* strand,
     or template, codes for this amino acid?

a   TGG   *d*   ACC
b   GUG   *e*   CAT
c   GUA   (p. 721)

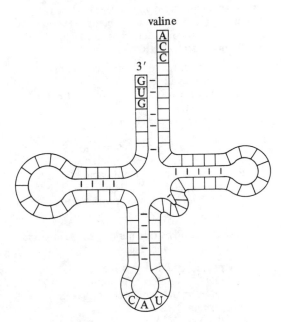

26   According to current ideas about the DNA
     genetic code, which one of the following
     statements is *false*?

a   The codon is three nucleotides long.
b   Every possible triplet codes for some
      amino acid.
c   The code is redundant (i.e. it contains
      "synonyms").
d   The code is read in a regular sequence,
      beginning at the 5' end.
e   The code is nonoverlapping.
                                  (pp. 715–17)

27 Which one of the following statements is *false*?

  a The nucleolus is a specialized region of a chromosome where tRNA is synthesized.
  b In base substitution mutations only a single nucleotide of a gene is altered.
  c Molecules of mRNA are synthesized on the ribosomes from nucleotides brought by tRNA.
  d Some amino acids are specified by several "synonymous" codons.   (p. 710)

28 Though a gene codes ultimately for all aspects of a protein's structure, it codes *directly* only for

  a primary structure.
  b secondary structure.
  c tertiary structure.
  d quaternary structure.
                    (p. 724; see also Chapter 3, p. 66)

29 As proteins are being synthesized, tRNA molecules are constantly being released from the site of amino acid incorporation. What happens to these tRNA molecules?

  a They return to the nucleus and bind to DNA again.
  b They are used to code for synthesis of a protein.
  c They immediately bind to another mRNA.
  d They pick up another amino acid of the same type that they had before.
  e They pick up an amino acid of another type, specifically the amino acid coded for by the codon next to the one to which they originally bound.   (pp. 721–22)

30 The direction of transfer of genetic information in most living things is

  a protein → DNA → mRNA.
  b DNA → mRNA → protein.
  c DNA → tRNA → protein.
  d protein → tRNA → DNA.
  e RNA → DNA → mRNA → protein.
                                        (pp. 723–24)

31 A certain gene codes for a polypeptide that is 126 amino acids long. The portion of the gene that codes for this polypeptide is probably how many nucleotides long?

  a 42   d 378
  b 126   e 504
  c 252
                                        (p. 715)

32 Suppose a gene has the DNA nucleotide sequence

  1 2 3 4 5 6 7 8 9 10 11 12 13 14 15 16 17
  C T G G C A T G C T T C G G A A A

(No real gene could be this short, but for our purposes this will suffice.) Which one of the following mutations would probably produce the greatest change in the activity of the protein for which this gene codes?

  a substitution of A for G in position 3
  b deletion of the C at position 5
  c deletion of the A at position 16
  d addition of a G between positions 14 and 15   (pp. 724–26)

33 An *intron* is

  a a foreign RNA sequence inserted in the normal message for a protein.
  b an RNA sequence that is edited from a transcript before translation.
  c a DNA sequence that is used to link a *plasmid* with a foreign DNA.
  d a DNA sequence that codes for the protein product of the gene.
  e a DNA sequence that is not transcribed.
                                        (p. 713)

For further thought

1 Given the following strand of DNA in a bacterial cell and its polarity

     5'   CTATGCTGGATACAT   3'

  a Replicate the strand.
  b Knowing that the DNA template strand is read from 3' to 5', and that mRNA is synthesized 5' to 3', determine which strand is the template strand and *transcribe* the proper DNA strand, if the direction of transcription is from right to left. (Be sure to indicate the 3' and 5' ends of your transcript.)
  c Using the code on page 717 of your text, *translate* the above strand of mRNA.

2 A nonsense mutation is one in which the substitution of one nucleotide for another results in a triplet that does not code for any amino acid. A missense mutation is one in which such a substitution results in a triplet that codes for a different amino acid. What effect would each of these have on protein synthesis? Which of them would be expected to have the more severe effect on enzyme activity?

3 Active beef insulin is a small protein composed of 51 amino acids. How many nucleotide units does it take to code for this protein? Using the diagram of beef insulin on page 61 of your text and the genetic code on page 717, determine the nucleotide sequence in the DNA that codes for the shorter chain.

4   DNA has a ladderlike structure, with the sugar-phosphate uprights held together by covalent bonds and the cross rungs consisting of nitrogenous bases joined by hydrogen bonds. How is the presence of these two types of bonds biologically significant?

5   What would be the effect on protein synthesis of changing a single nucleotide in the anti-codon of a tRNA molecule?

ANSWERS

Testing recall

1	b	4	a	7	b	10	c
2	c	5	a	8	b	11	b
3	a	6	c	9	b		

12	DNA	T G T	G C A	C G T
	DNA	A C A	C G T	G C A
	mRNA	U G U	G C A	C G U
	tRNA	A C A	C G U	G C A

Testing knowledge and understanding

13 a   A T G G A A G T C G C A
    b   A U G G A A G U C G C A
    c   nucleus
    d   4
    e   U A C, C U U, C A G, C G U
    f   ribosome
    g   methionine, glutamic acid, valine, alanine

14 a   5′   A T A A T G G T A G G G C C C T A A   3′
    b   5′   A U A A U G G U A G G G C C C U A A   3′
    c   methionine-valine-glycine-proline

15	a	20	b	25	e	30	b
16	c	21	b	26	b	31	d
17	e	22	a	27	c	32	b
18	c	23	c	28	a	33	b
19	b	24	b	29	d		

EXTRACHROMOSOMAL INHERITANCE

A GENERAL GUIDE TO THE READING

Chapter 28 focuses on other sources of genetic information in cells—organelle DNA, plasmids, and viruses—and describes the important features of recombinant DNA technology. As you read this chapter, you will want to focus on the following topics.

1 Organelle heredity. The endosymbiont hypothesis was first introduced in Chapter 5 (see page 134). This section provides additional evidence for this hypothesis. Note the similarities between transcription and translation in organelles and that in procaryotic cells.

2 Plasmids. You will need to know what plasmids are, and their basic properties, since plasmids are such an invaluable tool in recombinant DNA technology. The material on the sex factor is difficult and you will need to read it carefully. Figure 28.5 (p. 732) is helpful.

3 The bacteriophage reproductive cycle. You will need to learn the lytic and lysogenic cycles, both for this chapter and for Chapter 38. Figure 28.8 (p. 736) is a useful summary of the various viral reproductive strategies; pay special attention to the differences between DNA and RNA viruses.

4 Recombinant DNA. The discussion of this topic (pp. 738–40) is up-to-date and relevant. You will want to read the section thoroughly, as this area is growing rapidly and new information is accumulating at an amazing rate. You will no doubt see reports in magazines and newspapers of exciting new advances through use of this technology.

KEY CONCEPTS

1 Some cytoplasmic structures are capable of self-replication; they have their own DNA which is transcribed and translated. (pp. 730–33)

2 Recombination may occur in bacteria through the movement of genetic material from one

cell to another during conjugation. (pp. 730–33)

3   The DNA of temperate viruses can exist in a vegetative state, replicating independently and eventually destroying the cell, or it can exist in the integrated state, as provirus, functioning and replicating as a portion of the bacterial chromosome. (pp. 735–37)

4   Nonepisomal plasmids play a major role in the recombinant DNA technique for inserting genes from one kind of organism into cells of another kind of organism. (p. 738)

5   Gene cloning makes it possible to use host cells as chemical factories to produce substances of medical importance, and also to study the sequencing and activity of genes from eucaryotic cells. (pp. 739–40)

OBJECTIVES

After studying this chapter and reflecting on it, you should be able to carry out the following objectives.

1   Compare and contrast the chromosomes and ribosomes of organelles with the chromosomes and ribosomes of procaryotic cells and eucaryotic cells. Then, compare the processes of transcription and translation in organelles with that in procaryotic cells, noting the similarities and differences. (pp. 727–30)

2   Explain what a plasmid is, and describe how the sex factor can be transferred from one cell to another, using both F^+ and *Hfr* cells. Differentiate between episomes and nonepisomal plasmids. (pp. 730–33)

3   Using a diagram such as Figure 28.7 (p. 734), describe the lytic and lysogenic modes of bacteriophage replication. Be sure to include in your description the following terms: virulent, temperate, lysogenic cells, lytic cycle, provirus, and vegetative virus. (pp. 735–37)

4   Compare and contrast the mode of reproduction of a DNA bacteriophage virus with that of a typical RNA virus and a retrovirus. See Figure 28.8 (p. 736). (p. 735)

5   Using a diagram such as Figure 28.9 (p. 737), describe the process of transduction by bacteriophages and mention some implications and uses of this process. (p. 737)

6   Using a diagram such as Figure 28.11, describe the steps involved in inserting genes from one kind of organism into cells of another kind of organism. Include the following in your description: restriction endonuclease, "sticky" ends, plasmid, ligase, and transformation. (pp. 738–39)

7   Describe in some detail the process of gene cloning. Indicate the advantages that this process has over that of artificial transformation. Then, explain how gene cloning can be used to sequence genes from eucaryotic cells. (pp. 739–40)

KEY TERMS

The following terms are important in this chapter; you should become familiar with them.

plasmid (p. 730)
conjugation (p. 730)
episomes (p. 733)
nonepisomal plasmids (p. 735)
lytic cycle (p. 735)
viroids (p. 735)
retrovirus (p. 735)
temperate viruses (p. 735)
lysogenic cycle (p. 736)
provirus (p. 737)
transduction (p. 737)
recombinant DNA (p. 738)
artificial transformation (p. 738)
restriction endonucleases (p. 738)
gene cloning (p. 739)
interferon (p. 739)

SUMMARY

Such cytoplasmic organelles as chloroplasts and mitochondria have their own DNA and replicate themselves independently of the nucleus. The DNA in these organelles is usually circular and transcription and translation are similar to those in procaryotes.

Plasmids   Procaryotic and yeast cells often contain, in addition to their main circular chromosome, other small circular pieces of DNA called *plasmids*. These are free in the cytoplasm and replicate independently of chromosomes.

Bacteria are able to exchange genetic material through the process of *conjugation*. Two bacterial cells lie very close to one another, and a cytoplasmic bridge forms between them. Genetic material can pass through this bridge. Conjugation can occur only between cells of different mating types.

The cells of one mating type contain a *sex factor* (or F⁺ factor), a type of plasmid. When the sex factor is free in the cytoplasm, it can easily be transferred from cell to cell by conjugation. At times, however, the sex factor becomes inserted into the circular bacterial chromosome. When conjugation occurs, synthesis of a linear chromosome on the template of the circular chromosome begins and the new linear chromosome moves through the conjugation bridge into the recipient cell. Conjugation usually stops before the entire chromosome is transferred. Since the chromosome moves at a steady rate, disrupting conjugation at measured intervals permits mapping of the chromosomal genes. The sex factor is sometimes free in the cytoplasm; at other times it is integrated into the bacterial chromosome. Plasmids that exhibit this dual behavior are called *episomes*. Some plasmids are never integrated into the main chromosome; they are always autonomous. Sometimes these plasmids possess genes for antibiotic resistance.

Viral reproduction When virulent DNA bacteriophage viruses reproduce, the bacteriophage attaches to the bacterial cell wall and injects its DNA into the cell, leaving the protein coat outside. Once inside, the viral DNA takes control of the bacterial cell's metabolic machinery and puts it to work making new viral DNA and proteins. Later the bacterial cell lyses, releasing the new virions which may attack new cells and start the *lytic cycle* over again. RNA viruses use different reproductive strategies: Some RNA viruses have a region that codes for a special enzyme, RNA replicase, that catalyzes replication of the RNA. Other RNA viruses, the *retroviruses*, carry an enzyme, RNA transcriptase, that catalyzes the formation of a DNA copy (cDNA) of the RNA; this copy is then incorporated into a host chromosome. *Viroids* have tiny circles of naked RNA and can infect a variety of plant cells.

However, the DNA of some viruses (*temperate viruses*) can be present in the cell in an inactive state; integrated into the bacterial chromosome as a *provirus*, it functions and replicates as an additional part of the bacterial chromosome (in the *lysogenic cycle*). Under unfavorable conditions the provirus may leave the chromosome and begin the lytic cycle. Sometimes fragments of the bacterial chromosome become enclosed within the new viral coats; such viruses will inject both viral and bacterial DNA into the cells they infect, and the introduced genes can undergo recombination with the host's genes. The virus has thus carried genes from one bacterial cell to another, a process called *transduction*.

Recombinant DNA Nonepisomal plasmids can be used to transfer genes from the cells of one organism to the cells of another. Plasmids are isolated, the desired genetic material is added, and these modified plasmids can be picked up by bacterial cells, which thereby acquire the foreign genes. The transformed bacterial cells can then grow and divide rapidly, creating limitless numbers of bacteria containing the new genetic material.

Gene cloning is also used to introduce foreign genes into bacteria. In this process mRNA is isolated from cells specialized to produce the desired product and reverse transcriptase is used to make cDNA from the mRNA template. DNA polymerases then replicate the DNA strand, the new gene is inserted into a plasmid, and the plasmid is introduced into a bacterial host. The transformed bacterial cells grow and divide, producing large numbers of bacteria that may synthesize the desired product. Gene cloning makes it possible not only to use host cells as chemical factories to produce substances of medical importance, but also to study the sequencing and activity of genes from eucaryotic cells.

QUESTIONS

Testing recall

Decide whether each of the following questions is true or false; if false, correct the statement.

1 The mitochondrial chromosome contains all the genes necessary for all the products required by the mitochondria for normal functioning. (p. 728)

2 The ribosomes of chloroplasts more closely resemble the ribosomes of bacteria than they do the ribosomes of a eucaryotic cell. (p. 728)

3 Conjugation between an F⁺ *E. coli* and an F⁻ cell results in the F⁻ cell becoming F⁺. (p. 731)

4 The sex factor in *E. coli* is an episome. (p. 733)

5 In bacterial conjugation, usually only part of a chromosome is transferred from the donor cell to the recipient cell. (p. 732)

6 The DNA of a phage in the lytic cycle is integrated into the host cell's chromosome. (p. 735)

7 Retroviruses carry an enzyme, reverse transcriptase, that catalyzes the synthesis of cDNA from the RNA template. (p. 735)

8 All RNA viruses are virulent and cause lysis of the host cell. (p. 735)

9 During transduction, a plasmid is passed from one bacterium to another. (p. 737)

10 Ligases are enzymes used to cut DNA at specific sites, creating sticky ends. (p. 738)

11 Gene cloning, using radioactive cDNA, can be used to map genes. (pp. 739–40)

12 The enzyme reverse transcriptase is used to catalyze RNA replication. (pp. 736–37)

Testing knowledge and understanding

Choose the one best answer.

13 Transcription in the organelles of eucaryotic cells

   a is identical to that in eucaryotic cells.
   b resembles that in procaryotic cells.
   c requires the addition of a polyadenine tail.
   d requires the enzyme reverse transcriptase.
   (p. 728)

14 When a virulent phage attacks a bacterial cell,

   a only the DNA is injected into the cell.
   b both the DNA and the protein coat enter the cell.
   c a plasmid is injected into the cell.
   d the DNA becomes integrated into the host cell chromosome.
   e the cell enters the lysogenic cycle. (p. 735)

15 In a conjugation between an F⁻ bacterium and an *Hfr* bacterium, the number of genes transferred to the F⁻ cell is directly proportional to

   a the distance the genes are apart.
   b the point on the chromosome which first enters the F⁻ cell.
   c the length of time of the mating.
   d the length of the *Hfr* chromosome.
   e the number of bacterial chromosomes.
   (pp. 731–32)

16 During conjugation between an F⁻ bacterium and an *Hfr* bacterium, the sex factor always

   a remains in the *Hfr* cell.
   b moves into the F⁻ cell.
   c attaches to the bacterial chromosome at the same location in all strains.
   d is found at the rear end of the linear chromosome. (pp. 731–32)

17 A plasmid is

   a a fragment of a bacterial chromosome.
   b a portion of a bacteriophage.
   c a self-replicating non-chromosomal circle of DNA.
   d a virus that infects bacteria.
   e formed from the DNA of a lysogenic virus. (pp. 730, 733)

18 Bacteria that harbor a provirus are said to be

   a lytic.   *d* virulent.
   b lysogenic.   *e* transformed.
   c F⁺.   (p. 737)

19 The retroviruses

   a synthesize cDNA from the viral RNA.
   b replicate their RNA using the enzyme RNA replicase.
   c always cause lysis of the host cell.
   d consist of single-stranded DNA surrounded by a protein coat. (p. 735)

20 The process by which a virus transfers genes from one bacterial cell to another is called

   a transformation.   *d* transduction.
   b translation.   *e* translocation.
   c transcription.   (p. 737)

21 Restriction endonucleases are useful in recombinant DNA technique because they

   a cut DNA at specific sites.
   b restrict the number of nucleotides that can be removed at one time.
   c restore the bonds in the DNA backbone.
   d synthesize cDNA from mRNA.
   e can be used to locate genes for mapping.
   (pp. 738–40)

22 A genetic engineer prepares DNA fragments from two species and mixes them together. Two of the many fragments are shown below. Which one of the following statements is correct?

	TTCC
ATGC	

   a No sticky ends were produced.
   b The two fragments shown above will join by complementary base pairing.
   c The two fragments were prepared by two different restriction endonucleases.
   d A single restriction endonuclease was used to cut at different locations in the two types of DNA. (pp. 738–39)

23 One problem with using the recombinant DNA technique to insert genes from one organism into cells of another kind of organism such as bacteria is that

   a the two organisms often use different genetic codes.
   b the foreign genes generally contain introns which bacteria are unable to remove.

c restriction endonucleases do not cleave plasmid DNA in the same way they cleave eucaryotic DNA.

d ligases often remove bases at one end of the gene. (pp. 738–39)

24 If you were to arrange the following steps in gene cloning in order, which step would be third?

a A cDNA copy of mRNA is made using reverse transcriptase and the DNA strand is replicated.

b The plasmid is inserted into the bacterial host cell.

c mRNA molecules coding for the desired product are isolated.

d Restriction endonucleases are used to cut DNA, producing sticky ends.

e The plasmid DNA and the cloned gene are joined and sealed by ligase.
(pp. 738–39)

25 Suppose a pharmaceutical company planned to clone the gene for human phosphoribosyl transferase, an important enzyme. Which of the following steps would *not* be involved in this process?

a joining of sticky ends by the enzyme ligase

b breaking human DNA by a restriction endonuclease

c breaking human DNA into individual nucleotides

d transforming an *E. coli* with recombinant plasmid / human DNA

e joining human DNA to a vector such as a plasmid (pp. 738–39)

For further thought

1 In recombinant DNA technology, why is it necessary to attach a foreign gene to a plasmid? Why not just introduce the foreign gene directly into the bacterium?

2 Discuss some of the benefits and potential dangers of recombinant DNA technology.

3 Assume that you are working for a large pharmaceutical company in their research division. Your goal is to produce interferon in large quantity by genetic engineering techniques. You are provided with a DNA fragment, containing the interferon gene, which had been isolated using restriction endonuclease A. You know that the interferon gene is flanked on both ends by recognition sites for B, another restriction endonuclease. An *E. coli* plasmid with single B recognition site is available. What would you do?

ANSWERS

Testing recall

1 false—some of the genes
2 true
3 true
4 true
5 true
6 false—never integrated
7 true
8 false—some RNA viruses
9 false—a virus carries bacterial genes from one bacterium to another
10 false—restriction endonucleases
11 true
12 false—RNA replicase

Testing knowledge and understanding

13	*b*	17	*c*	20	*d*	23	*b*
14	*a*	18	*b*	21	*a*	24	*d*
15	*c*	19	*a*	22	*c*	25	*c*
16	*d*						

CONTROL OF GENE EXPRESSION

A GENERAL GUIDE TO THE READING

This chapter describes the control mechanisms that determine for each cell when and how it will act on its inherited genetic instructions. Many students find this material difficult, since the control mechanisms are often complex. You will find Chapter 29 easier if you read it slowly and carefully and allot plenty of time to studying it. You will want to concentrate on the following topics.

1 Control of gene expression in bacteria. Focus your attention on the Jacob-Monod model and learn the different parts of the model and how it works. Figure 29.1 (p. 743) is particularly helpful in learning the material. Make sure you understand the Jacob-Monod model before you go on to gene repression; otherwise you will find the latter confusing. Figures 29.1 (p. 743), 29.2 (p. 745), and 29.3 (p. 747) are key to understanding this section.

2 Control mechanisms in eucaryotic cells. Note that controls in eucaryotic cells, discussed on pages 747–56, are far more complex than those in procaryotic cells. Your focus should be on histone and nonhistone proteins, repetitive DNA, euchromatin, heterochromatin, lampbrush chromosomes, and chromosomal puffs as evidences of eucaryotic control mechanisms at work. Figure 29.5 (p. 749) is a useful summary of the material on the organization of eucaryotic chromosomes.

3 The section on highly repetitive DNA and pseudogenes is one which you will want to spend some time on since concepts and terms are introduced which will be used in the next chapter. Time spent learning this material now will aid your understanding of gene evolution. Figure 29.5 (p. 749) is helpful.

4 Cancer. The section on cancer is interesting to most students. The material on the multi-step hypothesis and environmental causes of

cancer is up-to-date and fascinating. You will want to read the material on oncogenes carefully since this material is rather complex. The outline in the middle of page 761 nicely summarizes this material.

KEY CONCEPTS

1   Although every cell in the body of a multicellular organism has identical genetic information, individual cells have different structural and functional characteristics. Various control mechanisms determine when and how each cell will act on its inherited genetic instructions. (pp. 741–56)

2   Each gene codes for only one kind of messenger RNA; regulators determine if and when each gene will be transcribed. (pp. 743–47)

3   Most chromosomal DNA of eucaryotes is never transcribed and only about 1 percent of the eucaryotic DNA that codes for mRNA will be translated. In procaryotic cells approximately 90 percent of the DNA is translated. (pp. 749–51)

4   Gene activation in eucaryotes involves the unwinding of particular sections of the nucleosomes, followed by activation of a specific gene or groups of genes in the unwound regions. (pp. 748, 751)

5   In procaryotic cells, most control of gene transcription is by inhibition of specific genes, in contrast to eucaryotic cells where most control is by activation of specific genes. (p. 749)

6   There are numerous points other than transcription where control can be exerted on the flow of information. In some cases a gene's expression can be controlled at every step from before transcription until after translation. (pp. 755–56)

7   The single most distinctive feature of cancer cells is their unrestrained proliferation; normal cell-control mechanisms no longer work. (p. 756)

8   The conversion of a normal cell into a cancer cell involves several changes: loss-of-fixed-number-of-division control, loss of contact inhibition, loss of anchorage dependence, and, sometimes, cell surface changes. (pp. 757–59)

9   Some of the genetic changes that induce a cell to become cancerous involve the same oncogenes found in the cancer-inducing retroviruses. (pp. 760–62)

OBJECTIVES

After studying this chapter and reflecting on it, you should be able to carry out the following objectives.

1   Differentiate between inducible, constitutive, and repressible enzymes. (pp. 742–45)

2   Describe the Jacob-Monod model of gene induction. In doing so, explain the roles of the inducer, the operator, the promoter, the repressor protein, the regular gene, and the structural genes. Using a diagram such as Figure 29.1 (p. 743), show how expression of the *lac* operon of *E. coli* is regulated. (pp. 742–43)

3   Using Figure 29.2 (p. 745), describe the functioning of a repressible operon and explain how it differs from an inducible operon. (pp. 743–45)

4   Differentiate between positive control and negative control of gene transcription, and give two examples of each. Then, using Figure 29.3 (p. 747), describe how the cAMP-CAP complex facilitates transcription. (pp. 745–47)

5   Explain why the models for the control of gene transcription in bacteria are not directly applicable to eucaryotic cells. (pp. 747–51)

6   Review Figures 5.13 (p. 113) and 23.6 B and C (p. 618) to refresh your memory about the structure of nucleosomes. Point out the relationship between histone proteins and the nonhistone proteins and DNA. Specify a function proposed for the nonhistone proteins. (pp. 747–49)

7   Differentiate among highly repetitive DNA, moderately repetitive DNA, and single-copy DNA, and give the approximate proportions of each type in the eucaryotic genome. Explain why there are so many copies of certain genes whereas one copy is sufficient for other genes. (pp. 749–51)

8   Give the percentage of the eucaryotic genome that is both transcribed and translated and contrast that with the percentage of the procaryotic genome that is transcribed and translated. (p. 751)

9   Explain what lampbrush chromosomes are and how they are formed; describe the structure of the giant chromosomes of *Drosophila*, and summarize the evidence indicating that the chromosomal puffs seen in them are sites of intense transcriptional activity. Differentiate between euchromatin and heterochromatin. (pp. 751–54)

10 Define gene amplification and relate this process to the formation of the nucleolus. (pp. 754–55)

11 Explain what is meant by "post-transcriptional control" and give three examples of points where cellular control can be applied in the path of information flow from the genes to their phenotypic expression. (pp. 755–56)

12 Give three principal differences between cancer cells and normal cells, and show how these differences are related to the tendency of cancer cells to metastasize. (pp. 757–59)

13 Describe the multistep hypothesis and relate this hypothesis to the changes that a normal cell must undergo to be converted into a cancer cell. (pp. 759–60)

14 Give the relationship between oncogenes and proto-oncogenes and discuss the ways in which these genes might function to convert a normal cell into a cancer cell. (pp. 761–62)

15 State the relationship among radiation, carcinogenic chemicals, and cancer. Review the Ames test (p. 755) and relate the mutagenicity of a chemical to its carcinogenicity. (p. 763)

KEY TERMS

The following terms are important in this chapter; you should become familiar with them.

inducible enzyme (p. 742)
constitutive enzyme (p. 742)
structural genes (p. 742)
regulator gene (p. 742)
repressor protein (p. 742)
operator (p. 742)
operon (p. 742)
promoter (p. 742)
corepressor (p. 743)
repressible enzymes (p. 745)
negative control (p. 745)
positive control (p. 745)
cAMP-CAP complex (p. 747)
chromatin (p. 748)
histone proteins (p. 748)
nonhistone proteins (p. 748)
DNA hybridization (p. 749)
highly repetitive DNA (p. 749)
pseudogene (p. 749)
moderately repetitive DNA (p. 750)
single-copy DNA (p. 750)
euchromatin (p. 751)
heterochromatin (p. 751)

lampbrush chromosomes (p. 753)
chromosomal puffs (p. 753)
gene amplification (p. 754)
contact inhibition (p. 756)
somatic cell hybridization (p. 757)
multistep hypothesis (p. 758)
oncogenes (p. 760)
proto-oncogenes (p. 760)

SUMMARY

Although every cell in the body of a multicellular organism has identical genetic information, individual cells have different structural and functional characteristics. Only a small percentage of the genetic material is active at any one time. Various control mechanisms determine when and how each cell will act on its genetic instructions; these control mechanisms involve chemicals that bind directly or indirectly to DNA or mRNA.

Control of gene expression in bacteria Bacterial cells have many systems that control gene transcription; the best understood are gene induction, gene repression, induction of positive control (the cAMP-CAP system), and repression of positive control.

In the gene induction system, the genes specifying particular enzymes are inactive until turned on by an inducer substance. Enzymes which are produced only when an inducer is present are called *inducible enzymes*, as opposed to *constitutive enzymes*, whose production is continuous. The *Jacob-Monod model of gene induction* in bacteria proposes that three parts of the chromosome are involved in controlling transcription of the *structural genes*: the repressor gene and the operator and promoter regions. The *regulator gene* codes for the *repressor protein*. When the repressor binds to the *operator*, it blocks the *promoter's* binding sites for RNA polymerase and thus prevents transcription of the structural genes. According to this model, transcription of the structural genes occurs only when the substrate is present: the substrate acts as an *inducer* and binds to and inactivates the repressor; now the RNA polymerase can bind to the promoter region and initiate transcription of the structural genes. These, transcribed as a unit, determine production of polycistronic mRNA, that is, mRNA coding for more than one gene product. The mRNA is then translated, and the enzymes are synthesized.

In a gene repression system, the operator is always turned on and enzymes are automatically synthesized unless the operator is turned off by a repressor protein. Here the repressor protein encoded by the regulator gene is initially inactive; only if it binds to a *corepressor molecule* (often the end product of a biochemical pathway) can it bind to the operator and block transcription of the structural genes.

Both of the above systems are examples of negative control; a repressor binds to an operator and turns off transcription. Though negative control is the most common way of regulating gene expression in bacteria, some systems are regulated by *positive control*. For example, the binding of cAMP to the inactive CAP complex induces a conformational change that allows the complex to bind to the promoter and facilitates the binding of RNA polymerase, speeding up transcription. Another positive control sequence involves a normally active control protein that is turned off by the binding of a small molecule, so the control protein can no longer help with the binding of RNA polymerase.

Control of gene expression in eucaryotes Control mechanisms in eucaryotes differ from those in bacteria. Eucaryotes have many more genes than procaryotes, their DNA has introns which must be removed from the mRNA transcript, and the DNA of eucaryotic chromosomes is not naked but wrapped on histone-protein cores to form *chromatin*. In eucaryotes the *histone* proteins make up the nucleosome cores and appear to play a largely structural role whereas the *nonhistone* proteins may play a more regulatory role, acting as selective agents in gene regulation. Some may bind to control regions in the DNA to cause unraveling of the nucleosomes. Nonhistone proteins may also be important in activating specific genes or groups of genes. Activation, rather than inhibition of transcription, is the rule in eucaryotes.

The technique of *DNA hybridization* has shown the organization of eucaryotic chromosomes to be complex. About 10 percent of the eucaryotic genome contains base sequences that are repeated thousands of times. The function of much of this *highly repetitive DNA* is not well understood, but one class consists of long, tandemly repeated units that code for four of the ribosomal RNAs. Because a single copy of this sequence would be inadequate to meet the needs of the cell for ribosomal RNA, most eucaryotes have about 25,000 copies of it. Another 20 percent of the genome consists of *moderately repetitive DNA*; each base sequence is repeated hundreds of times. The genes for the other three ribosomal RNAs are not only tandemly repeated hundreds of times but also this

portion of the chromosome can be replicated independently of the rest of the chromosome to form a polytene region, which makes up the nucleolus. The remaining 70 percent of the genome consists of *single-copy* sequences. Most chromosomal DNA is never transcribed and only about 1 percent of the eucaryotic DNA that codes for mRNA will be translated, in contrast to procaryotic cells where 90 percent is translated.

Control of gene transcription in eucaryotic cells Gene activation in eucaryotes involves the unwinding of particular sections of the nucleosomes, followed by activation of a specific gene or groups of genes in the unwound regions.

When stained, eucaryotic chromosomes can be differentiated into *euchromatic* and *heterochromatic* regions. The euchromatic regions contain active genes, heterochromatic regions are inactive. Even though eucaryotic genes are not organized as operons, the synthesis of some groups of enzymes can be controlled as a unit. The mechanisms for coordinating activation and repression are still unknown.

Lampbrush chromosomes and *chromosomal puffs* are visible evidences of gene activity. In both, the DNA of active regions of the chromosomes loops out laterally from the main chromosome, exposing the maximum surface for transcription. The pattern of chromosomal puffs varies with the developmental stage and with changes in the extranuclear environment, showing that changes in the cytoplasm can alter gene activity.

Gene amplification is another transcriptional control mechanism. In this process, many parallel copies of rRNA genes are produced from moderately repetitive DNA.

There are numerous points other than transcription where control can be exerted on the flow of information. In some cases a gene's expression can be controlled at every step from before transcription until after translation.

Cancer: A failure of normal cellular controls The single most distinctive feature of cancer cells is their unrestrained proliferation. Cancer cells almost always have an abnormal set of chromosomes; there may be extra chromosomes, or specific chromosomal rearrangements.

Cancer cells also show many differences related to cell surfaces. They are more spherical in shape than normal cells and are anchorage-independent. Cancer cells typically have an abnormal cell surface, having fewer glycoproteins and glycolipids. They do not show normal *contact inhibition* of cell division, and cannot recognize other cells of their own tissue type. The surfaces of cancer cells characteristically bear antigens not found on normal cells.

The conversion of a normal cell into a cancer cell involves several changes: loss-of-fixed-number-of-division control, loss of contact inhibition, loss of anchorage dependence, and, sometimes, cell surface changes. Evidence indicates that each change is the result of a separate genetic event. According to the *multistep hypothesis*, two to four independent genetic events are required to induce a cell to become cancerous.

Oncogenes are genes that cause cancer. Sometimes the oncogene is brought into the cell by a retrovirus. Other retroviruses do not carry oncogenes; these can cause cancer only when the cDNA from the retrovirus inserts near a *proto-oncogene* in the host genome and is able to convert the proto-oncogene into an oncogene. Studies of human and nonviral animal tumors have shown that some of the genetic changes that induce the cell to become cancerous involve the same oncogenes found in the cancer-inducing retroviruses.

Some of the nonviral oncogenes appear to have arisen through mutation in a proto-oncogene, others may involve translocation of a proto-oncogene to a position near an active control region (or vice versa). The modes of action of most oncogenes remain to be discovered.

Mutations and translocations that can play a role in inducing cancers can be caused by external agents—radiation and mutagenic chemicals. However, it is difficult to link particular chemical or radiation sources to cancer in humans.

KEY DIAGRAM

The following coloring exercise is designed to help you learn the similarities and differences in the structure of an inducible operon, such as the lac *operon, and a repressible operon. First color in the titles, and then the appropriate structure, noting the similarities and differences as you do so. (Reference: Figs. 29.1A, p. 743, and 29.2A, p. 745)*

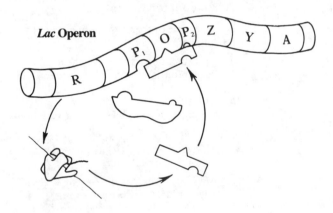

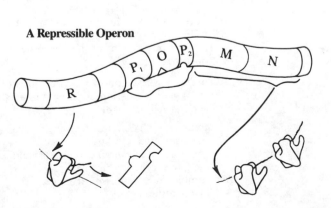

REGULATOR GENE

OPERATOR GENE

PROMOTER SEQUENCES

STRUCTURAL GENES

RNA POLYMERASE

mRNA

REPRESSOR PROTEIN

RIBOSOMES

Ask yourself

1 In which one of the two models are the structural genes automatically transcribed until they are turned off? In which are they normally repressed until the operator is turned on?

2 Can you redraw each of the above drawings to show what happens when lactose is present in A, and the end product is present in B?

QUESTIONS

Testing recall

For each of the processes listed in questions 1–6, write yes if it is associated with increased gene transcription and no if it is not.

1 binding of the inducer to the repressor (pp. 742–43)

2 binding of the RNA polymerase to the promoter (pp. 742–43)

3 binding of cAMP-CAP complex to the operator (p. 747)

4 puffing of the chromosomes (p. 753)

5 formation of lampbrush chromosomes (p. 753)

6 binding of the repressor to the operator (p. 742)

For each of the items listed in questions 7–15, write C if it is more characteristic of cancer cells and N if it is more characteristic of normal cells.

7 extra chromosomes (p. 757)

8 anchorage dependent (p. 757)

9 unrestrained proliferation (p. 756)

10 abnormal cell surface (p. 758)

11 recognition of other cells of their own tissue type (p. 758)

12 spherical shape (p. 758)

13 contact inhibition of growth and cell division (p. 758)

14 excess glucose consumption, with increased lactic acid production (p. 758)

15 secretion of large amounts of protein-digesting enzymes (p. 759)

Testing knowledge and understanding

Choose the one best answer.

The next five questions refer to the diagram of the *lac* operon from *E. coli*, shown below. Each letter may be used more than once.

DNA strand

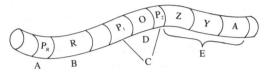

16 Which letter indicates the location of a gene or genes that code for constitutive enzyme(s)? (p. 742)

17 Which letter indicates the binding site for the repressor protein? (p. 742)

18 Which letter indicates the binding site for RNA polymerase? (p. 742)

19 Which letter indicates the structural genes? (p. 742)

20 Which letter indicates the gene that codes for the repressor protein? (p. 742)

21 In the lactose operon system in *E. coli*, the repressor is

a a product of a structural gene locus.
b bound to the promoter sequence.
c lactose.
d a protein.
e a short length of DNA.   (pp. 742–43)

22 Which of the following statements concerning the regulator gene (R) associated with the *lac* operon is correct?

a mRNA is transcribed from the R gene whether lactose is present or not.
b mRNA is only transcribed from the R gene when lactose is present.
c mRNA is only transcribed from the R gene when lactose is not present.
d Lactose inhibits the translation of R gene mRNA.
e Lactose binds to the promoter of the *lac* operon.   (pp. 742–43)

23 According to the Jacob-Monod (*lac* operon) model of gene regulation, inducer substances in bacterial cells probably

a combine with operator regions, activating the associated operons.
b combine with structural genes, stimulating them to synthesize messenger RNA.
c combine with repressor proteins, inactivating them.

d combine with promoter regions, activating RNA polymerase.
e combine with nucleoli, triggering production of more ribosomes. (pp. 742–43)

24 The promoter region of a bacterial operon

a codes for repressor proteins.
b codes for inducer substances.
c codes for corepressor substances.
d is a binding site for inducers.
e is a binding site for RNA polymerase. (pp. 742–43)

25 The regulator gene of a bacterial operon

a codes for inducer substances.
b codes for repressor proteins.
c acts as an on-off switch for the structural genes.
d is a binding site for RNA polymerase.
e is a binding site for inducers. (pp. 742–43)

26 The sugar lactose induces synthesis of the enzyme lactase. What happens when an *E. coli* (bacterial) cell runs out of lactose?

a Repressor protein binds to the operator.
b Repressor protein binds to the promoter.
c RNA polymerase attaches to the promoter.
d RNA polymerase attaches to the repressor. (p. 742)

27 The *lac* operon is an example of

a translational control.
b post-transcriptional control.
c replicational control.
d transcriptional control.
e positive control. (p. 742)

28 In the *lac* operon, RNA polymerase

a binds to the promoter.
b binds to the operator.
c binds to the regulator gene.
d is synthesized by the regulator gene.
e binds to the regulator protein. (p. 743)

29 In bacteria, the structural genes can be turned off when

a the end-product of a reaction combines with the repressor protein and activates it.
b the substrate combines with a repressor protein and inactivates it.
c the substrate combines with the promoter.
d the substrate combines with its enzyme and produces a repressor molecule.
e the cAMP-CAP complex binds to the promoter. (pp. 742–45)

30 In the tryptophan operon of *E. coli* the end-product of the biochemical pathway, tryptophan, binds to the repressor protein which then binds to the

a promoter to inhibit transcription.
b promoter to accelerate transcription.
c operator to inhibit transcription.
d operator to accelerate transcription.
e repressor gene to accelerate transcription. (p. 743)

31 Nucleosomes

a are composed of DNA and histones.
b are small particles found in only the nuclei of plant cells.
c disappear during transcription.
d play a role in coiling and uncoiling the chromosomes.
e two of the above are correct (p. 748)

32 All of the following are *true* about transcriptional control in eucaryotes *except*

a Most control mechanisms involve activation of transcription rather than inhibition.
b Much of the mRNA transcript must be excised before translation.
c The DNA wound around the histone-protein core must be unraveled before transcription can take place.
d Some base sequences are repeated thousands of times to form highly repetitive DNA.
e Almost 90 percent of the genome is transcribed and translated. (pp. 745–47)

33 Difficulty in extending the bacterial operon model to developing eucaryotes lies in

a differences in processing the transcriptional message.
b animals preferring to use glucose instead of lactose.
c genes for related function, or for different polypeptides assembled in the individualized protein, being located on different chromosomes.
d not all operons being the inducible kind.
e eucaryotes using different RNA polymerase than bacteria. (pp. 747–48)

34 If an insect is given the hormone ecdysone, certain regions of the insect's chromosomes will soon exhibit puffing. What important process is taking place in the puffs?

a mRNA synthesis *d* enzyme synthesis
b rRNA synthesis *e* histone synthesis
c DNA replication (p. 754)

35 Regions of active gene transcription are found in

   a lampbrush chromosomes.
   b heterochromatic regions.
   c pseudogenes.
   d Barr bodies.
   e polyribosomes.   (pp. 751–53)

36 All of the following changes are involved in converting a normal cell into a cancer cell *except*

   a loss or reduction of contact inhibition.
   b loss of fixed-number-of-division control.
   c loss of anchorage dependence.
   d decrease in tissue-type affinity.
   e loss of vascularization.   (p. 759)

37 Human cancer cells differ from normal cells in that cancer cells

   a have fixed-number-of-division control.
   b frequently possess extra chromosomes.
   c are anchorage-dependent.
   d show contact inhibition.
   e do not produce lactic acid.   (pp. 757–59)

38 Genes that cause cancer are referred to as

   a oncogenes.   *d* operator genes.
   b pseudogenes.   *e* promoter genes.
   c structural genes.   (p. 760)

39 All of the following may play a role in triggering cancer *except*

   a mutations.
   b retroviruses.
   c gene amplification.
   d carcinogenic chemicals.
   e proto-oncogenes.   (p. 762)

For further thought

1 Cancer cells seem to lack the normal control mechanisms for growth and cell division. Scientists hope that by studying the control of normal cell growth and division they will gain some understanding of why these controls are not exercised in cancer cells. At which points in the flow of information from gene to phenotype do normal controls operate? At what points do these controls appear to be lacking in cancer cells?

2 Radiation is extensively used to treat cancerous tumors. Explain why radiation can destroy cancer cells while allowing normal cells to continue to function.

3 In what ways do aberrations of the cell surface contribute to the properties of cancer cells?

4 Differentiate among inducible, repressible, and constitutive enzymes and describe the Jacob-Monod operon model for substrate induction. Include in your answer the role of the inducer, operator, regulator, promoter, repressor protein, and structural genes. Explain how gene repression differs from substrate induction.

ANSWERS

Testing recall

1 yes
2 yes
3 yes
4 yes
5 yes
6 no

7	C	10	C	13	N
8	N	11	N	14	C
9	C	12	C	15	C

Testing knowledge and understanding

16	B	22	*a*	28	*a*	34	*a*
17	D	23	*c*	29	*a*	35	*a*
18	C	24	*e*	30	*c*	36	*e*
19	E	25	*b*	31	*e*	37	*b*
20	B	26	*a*	32	*e*	38	*a*
21	*d*	27	*d*	33	*c*	39	*c*

IMMUNOLOGY AND THE MECHANISMS OF GENETIC VARIABILITY

A GENERAL GUIDE TO THE READING

This chapter begins with a discussion of the immune response and goes on to consider the genetic basis of antibody diversity. This background as well as the background provided by material in Chapter 29 is used to introduce the topic of eucaryotic gene evolution. Many students find the material in this chapter difficult, since the immune system is complex and the ideas presented here challenging. You will find Chapter 30 easier if you proceed slowly and carefully and allot plenty of time to studying it. The payoff for concentrated study is great; you will emerge from this reading with a sound basis for understanding the molecular basis of the immune response to many human pathogens. Two particularly interesting and devastating human diseases—sleeping sickness and AIDS—are presented in this chapter in "Exploring Further" sections. You will want to concentrate on the following topics.

1 The immune response. Begin your study by learning the boldface terms on pages 766–67. Once you know the terms, go on with your reading. Notice that although T lymphocytes and B lymphocytes are both important for the immune response, they function in quite different ways. You should focus on their different modes of action and the interactions between the two. Figures 30.1–30.7 (pp. 766–71) are central to an understanding of this area.

2 Monoclonal antibodies. The box on this subject (pp. 772–73) will prove valuable reading, since monoclonal antibodies are widely used for both basic research and in medical technology; you will read about their use frequently.

3 The antibody molecule and the genetic basis of antibody diversity. You will want to read this section very carefully as it is difficult.

Figures 30.11 (p. 776) and 30.12 (p. 778) are crucial to a firm understanding of this material. We suggest that you do not move on to the next section until you have gained a basic understanding of the nature and generation of antibodies.

4 Mechanisms of genetic variability. This section builds on the preceding one, focusing on mechanisms of gene and exon duplication. Again, the figures that accompany this section (Figures 30.13–30.16) will help you.

5 Evolution and the genome. This section focuses on the evolutionary consequences of duplication, asking how evolution might make use of existent genes to create proteins with novel and adaptive properties. This section is an exciting area of biological debate; it is difficult, but rewarding.

KEY CONCEPTS

1 An important defense against disease in vertebrate animals is their ability to manufacture antibodies that can inactivate or destroy foreign substances. (pp. 764–69)

2 When stimulated, the B lymphocytes manufacture and secrete antibodies resulting in the humoral immune response, which is particularly effective against pathogenic cells, viruses, and toxins. (p. 766)

3 Stimulated T lymphocytes are responsible for the cell-mediated immune response, which primarily defends against large cellular targets such as internal parasites, cancer cells, and virus-infected cells. (p. 766)

4 When antigen molecules bind to membrane-mounted antibodies of a virgin B cell, the lymphocyte enlarges and proliferates, producing plasma cells, which secrete antibodies, and memory cells. (pp. 766–68)

5 The balance among the three types of T cells —cytotoxic T cell, helper T cells, and suppressor T cells—determines the overall level of the immune response. (p. 769)

6 Each lymphocyte, whether of the B or T type, has only one type of membrane-mounted antibody or T-cell receptor. An antigen can bind only to those lymphocytes whose surface antibodies are specific for its antigenic determinant regions. (p. 770)

7 Antibodies are globulin proteins that consist of four polypeptide chains: two identical "heavy" chains and two identical shorter "light" chains; the chains are linked by disulfide bonds. Most of the antibody molecule is constant in its amino-acid sequence and structure; the two binding sites for the antigens are at the ends of the variable portion of the molecule. (pp. 775–76)

8 The vertebrate genome is able to produce an almost infinite variety of antibodies using a strategy of exon selection from only three genes—one for the heavy chains and one for each of the two classes of light chains. (pp. 776–78)

9 Many proteins seem to be encoded by genes with very similar sequences, suggesting that gene duplication may be a major force in evolution. (pp. 779–83)

10 Eucaryotic gene evolution depends in part on gene duplication followed by changes in base sequence that give rise to functionally different proteins. (pp. 783–86)

OBJECTIVES

After studying this chapter and reflecting on it, you should be able to carry out the following objectives.

1 Differentiate between T lymphocytes and B lymphocytes with respect to a) the site of maturation (thymus vs. bone marrow), b) the type of immune response (humoral vs. cell-mediated) elicited, c) the type of antigens to which each responds. (pp. 765–66)

2 Explain what happens when an organism is exposed to an antigen such as a bacterial cell and is stimulated to produce antibodies, making clear the role of the B lymphocytes, plasma cells, memory cells, antibodies, macrophages, K lymphocytes, mast cells, and the complement system. Figures 30.2 (p. 767) and 30.4 (p. 768) may be helpful in meeting this objective. (pp. 766–68)

3 Describe how the cell-mediated immune response acts to defend the body against antigens such as those found on cancer cells or virus-infected cells. Be sure to include the roles of memory cells, cytotoxic T cells, helper T cells, suppressor T cells, interleukins, lymphokines, macrophages, and mast cells. (pp. 768–69)

4 Using a diagram such as Figure 30.7 (p. 771), explain the process of clonal selection. Differentiate between antigenic determinants and haptens. (pp. 770–71)

5 Explain how the body is able to recognize "self" from "non-self" and discuss the role of the antigens produced by the major histocompatibility complex (MHC) in the recognition of infected or transplated cells. Figure 30.8 (p. 774) may be helpful in meeting this objective. (pp. 771–74)

6 Using diagrams such as Figure 30.9 (p. 775) and 30.11 (p. 776), point out the light chains, the heavy chains, the four disulfide bonds that link the chains, the constant regions, the variable regions, the carboxyl and amino ends of the chains, and the antibody binding sites. (pp. 775–76)

7 Explain how the vertebrate genome is able to produce an almost infinite variety of antibodies using only three genes. (pp. 776–78)

8 Describe how genes and exons could be duplicated by breakage and fusion, unequal crossing over, reverse transcription, and transposition. Figures 30.13–30.16 (pp. 780–83) may be useful in meeting this objective. (pp. 779–83)

9 Cite two examples providing evidence for the duplication of genes coding for enzymes or structural proteins, and explain how gene duplication can lead to the evolution of new functional genes with novel properties. (p. 784)

10 Using Figure 30.17 (p. 785), describe the exon recombination model, indicating how it could lead to the formation of a new gene that would code for a functionally different product. (p. 785)

KEY TERMS

The following terms are important in this chapter; you should become familiar with them.

antigen (p. 764)
lymphocytes (p. 765)
humoral immune response (p. 766)
cell-mediated immune response (p. 766)
B lymphocyte (p. 766)
plasma cells (p. 767)
memory cells (p. 767)
macrophages (p. 767)
K lymphocytes (p. 768)
complement system (p. 768)
mast cell (p. 768)
T lymphocyte (p. 768)
helper T cell (p. 769)
interleukins (p. 769)
lymphokines (p. 769)
suppressor T cell (p. 769)
antigenic determinants (p. 770)
haptens (p. 770)
clonal selection (p. 770)
autoimmune disease (pp. 771, 773)
major histocompatibility complex (MHC antigens) (p. 773)
light chains ⎫ parts of antibody
heavy chains ⎰ molecule (p. 775)
gene duplication (p. 779)
reverse transcription (p. 780)
transposition (p. 780)
transposons (p. 781)
exon recombination (p. 784)

SUMMARY

Vertebrate animals have an immune system that directs the manufacture of highly specific *antibodies* to inactivate or destroy invading *antigens*, which are large molecules (proteins or polysaccharides) ordinarily foreign to the organism's body. The cells that respond to the antigen are the *lymphocytes*, of which there are two types: the *B cells*, which mature in the bone marrow, and the *T cells*, which mature in the thymus. When stimulated, the B cells manufacture and secrete antibodies resulting in the *humoral immune response*, which is particularly effective against pathogenic cells, viruses, and toxins. The T cells are responsible for the *cell-mediated immune response*, which primarily defends against larger targets such as internal parasites, cancer cells, and virus-infected cells.

The humoral immune response When antigen molecules bind to membrane-mounted antibodies of a virgin B cell, the lymphocyte enlarges and proliferates. Some of the resulting cells are *memory cells*, others become specialized as *plasma cells*, which secrete antibodies. Each antibody molecule can bind to two antigen molecules; hence the antibodies cause agglutination of the antigen-bearing bacteria or viruses. This agglutination aids in the destruction of the antigens by (1) ingestion by *macrophages*, (2) destruction by *K lymphocytes* (also called *killer cells*), or (3) lysis mediated by *complement-system proteins*. In addition, circulating antibodies can become attached by their tails to mast cells. When an antigen binds to such an antibody, the mast cell is induced to release histamine, causing the nearby blood vessels to become leaky and thus enabling other elements of the immune system to reach the site of the histamine release.

The cell-mediated immune response The antibody-like receptors of T lymphocytes are not secreted but remain attached to the lymphocyte. Each virgin T lymphocyte has membrane-mounted receptors for two different antigens. When antigens are encountered on an antigen-bearing cell, the T cell divides to produce memory cells and three types of activated T cells: *cytotoxic T cells*, which bind to the antigens on the cells and kill the cells; *helper T cells*, which release local chemical mediators such as *interleukins* and *lymphokines* that activate other cells of the immune response; and *suppressor T cells*, which when bound to the appropriate antigens inhibit the other cells involved in the immune response. The balance between activation and inhibition determines the overall level of the immune response.

Only certain regions of the large antigen molecules serve as *antigenic determinants*, sites of interaction with the receptors on lymphocytes. Isolated antigenic determinants, or *haptens*, can also be bound by antibodies and stimulate an immune response if there has been a prior exposure to this antigen. Each lymphocyte, whether of the B or T type, has only one type of membrane-mounted antibody or T-cell receptor. An antigen can bind only to those lymphocytes whose surface antibodies are specific for its antigenic determinant regions. The binding of an antigen by lymphocyte receptors facilitates cross-linking of the receptors and this appears to induce the lymphocytes to begin dividing. The proliferation of the particular lymphocytes that react with a specific antigen is called *clonal selection*.

The immune system of an organism is able to distinguish between "self" and "not-self." However, sometimes this ability is impaired and an *auto-immune disease* results. The antigens involved in these reactions are generally glycoproteins. Some of these glycoproteins are involved in cell-type recognition; others, which are encoded by genes of the major histocompatibility complex (MHC), appear on the surface of most cells of the body. The MHC antigens of different individuals generally have only a few antigenic determinants in common. The MHC antigens play a major role in transplant rejections. T cells bind most effectively when there is both a familiar (self) MHC antigen and a foreign antigen available on a cell surface. A single T-cell receptor binds to both antigens simultaneously. This system seems to allow T cells to recognize when one of an organism's own cells is infected. Cytotoxic T cells act directly, by binding to and attacking infected and foreign cells, but helper and suppressor T cells act by modulating the activity of B cells. B cells ingest and process the antigen, attach it to a protein, and insert it back into the membrane. With the processed antigen in the membrane, the T cells with receptors specific to it can bind simultaneously to this antigen and the MHC antigen, and the T cell can begin to secrete local chemical mediators, promoting the immune response.

The genetic basis of antibody diversity Antibodies are globulin proteins that consist of four polypeptide chains: two identical "heavy" chains and two identical shorter "light" chains; the chains are linked by disulfide bonds. Most of the antibody molecule is constant in its amino-acid sequence and structure; the two binding sites for the antigens are at the ends of the variable portion of the molecule.

The vertebrate genome is able to produce an almost infinite variety of antibodies using a strategy of exon selection from only three genes—one for the heavy chains and one for each of the two classes of light chains. The original heavy-chain antibody gene contains approximately 240 exons, while the protein product coded for by this gene is encoded by only seven or eight exons. During fetal development, large regions of the gene are randomly excised and the remaining ends are spliced together. The remaining exons and introns are transcribed, but the superfluous transcribed exons are removed during RNA processing. Only one exon of each exon group survives processing and appears in the mature mRNA. The two steps of exon removal permit a B lymphocyte to produce a unique kind of antibody from millions of possible alternatives. T-receptor diversity is thought to be generated in much the same way.

Mechanisms of genetic variability Evolution depends on genetic variation. The recent discoveries of introns, exons, pseudogenes, repetitive DNA, and the excisions of genetic material within antibody genes suggest that the eucaryotic genome may itself provide variation upon which evolution can act.

The structure of the antibody genes, and genes for many other proteins as well, suggest that some regions of such genes arose from duplications and rearrangements of earlier gene sequences. The duplication of exons (or genes) can occur through breakage and fusion, or from unequal crossing over. Alternatively, viral reverse transcriptase may catalyze transcription of the host's mRNA into cDNA and the cDNA can be inserted into the host's chromosome as a duplication.

Duplications may also be caused by *transposons*.

These are mobile genetic elements (composed of a gene or genes and/or a control element) that can move from their site to another target site on the same chromosome. The move may involve simple excision from the chromosome or the transposon may move only after being copied, so a complete copy of itself remains at its original location. The transposon then recognizes a target sequence and inserts itself into the host chromosome.

Eucaryotic gene evolution may depend in part on gene duplication followed by changes in base sequences that give rise to functionally different products. It may also be possible for individual exons from different genes to be brought together to produce new genes (a process called exon recombination). If the product is partially functional, the new gene could survive and be acted on by natural selection.

QUESTIONS

Testing recall

For each of the items or characteristics listed below, write T *if it is generally associated with* T *lymphocytes,* B *if it is generally associated with* B *lymphocytes, and* B, T *if it is associated with both.*

1 cell-mediated immunity (p. 766)

2 humoral immunity (p. 766)

3 plasma cells (p. 767)s

4 circulating antibodies (p. 767)

5 memory cells (p. 767)

6 mature in the thymus (p. 766)

7 mature in the bone marrow (p. 766)

8 generally stimulated by bacteria and toxins free in the plasma (p. 766)

9 particularly effective against cancer cells and virus-infected cells (p. 766)

10 secretes interleukins and lymphokines (p. 769)

11 membrane-mounted receptors specific for two antigens (p. 768)

Testing knowledge and understanding

Choose the one best answer.

12 Which organ or tissue is *not* an integral part of the immune system?

  a pancreas   *d* spleen
  b lymph nodes   *e* bone marrow
  c thymus   (p. 765)

13 Which one of the following statements about immunity is *false*?

  a The vertebrate body apparently has all the genetic information necessary to make antibodies before it encounters stimulating antigens.
  b Each individual B lymphocyte can make several different kinds of antibodies.
  c The T lymphocytes function primarily in cell-mediated immune responses.
  d When an antigen enters a host organism, it combines with cells bearing surface antibodies complementary to it, and initiates proliferation of these cells.
  e The plasma cells formed during an immune response produce the circulating antibodies.   (pp. 766–71)

14 The memory cells of the immune system

  a are nonspecific; each reacts to a variety of antigens.
  b produce circulating antibodies.
  c are produced by the thymus.
  d are responsible for an accelerated response upon second exposure to an antigen.
  e remain in circulation for a short time and are then destroyed.   (p. 767)

15 A clone of activated B lymphocytes may give rise to all of the following *except*

  a plasma cells.
  b T cells.
  c circulating antibodies.
  d memory cells.   (p. 767)

16 Which one of the following is *true* of T lymphocytes?

  a responsible for the humoral immune response
  b produce circulating antibodies
  c primarily effective against bacteria and free viruses
  d secrete the chemical mediators lymphokines and interleukins
  e activate the complement system to cause lysis of the cell   (pp. 766–68)

17 Which of the following statements about immunity is *false*?

a An individual has all the genes required to make antibodies before it is exposed to antigens.
b Gene rearrangements to produce specific antibodies are induced by exposure of the lymphocyte to an antigen.
c Each lymphocyte can make only one type of antibody.
d When an antigen enters a host organism it combines with cells bearing surface antibodies complementary to it, and initiates proliferation of these cells.
e T lymphocytes function primarily in cell-mediated immune responses.
(pp. 766–71)

18 The receptors of the T lymphocytes differ from the membrane-mounted antibodies on B lymphocytes in that

a the two sides of the receptor molecule recognize different antigens.
b each receptor molecule can recognize many different antigens.
c each receptor molecule is specific for a single antigen.
d upon exposure to an antigen the receptor molecules are released into circulation to fight the infection. (p. 770)

19 Which of the following statements best describes the effect of an antigen on the body's immune system when it is first encountered?

a The antigen molecules cause developing lymphocytes to rearrange their genes.
b The antigen molecules are used by the lymphocytes as "models" in order to make correctly shaped antibodies.
c The antigen molecules cause those lymphocytes that carry appropriately shaped antibody molecules to divide.
d Antigen molecules increase the number of mutations occurring in antibody genes.
e Antigen molecules catalyze the formation of disulfide bonds between light and heavy chains. (p. 770)

20 Which one of the following is *true* concerning the immune system's ability to distinguish "self" and "not-self"?

a The ability to distinguish between "self" and "not-self" is acquired late in childhood.
b The ability to distinguish between "self" and "not-self" can be permanently reduced by brief treatment with immunosuppressive drugs.
c The ability to distinguish between "self" and "not-self" is permanent and never breaks down.

d The ability to recognize "not-self" is largely determined by the presence of different MHC antigens.
e Two of the above are correct. (p. 773)

21 Which statement about antibodies is *false*?

a Each antibody combines with a specific antigen.
b Two sites on each antibody molecule can bind antigen.
c Antibodies are globulin proteins.
d Antibodies are found in vertebrate animals.
e Human blood contains high levels of circulating antibodies against all possible antigens. (pp. 766, 775)

22 An antibody molecule is made up of four polypeptide chains joined by disulfide bonds. Which one of the following statements concerning these polypeptide chains is *correct*?

a All four chains are of equal length.
b Only the two light chains have variable regions.
c Only the two heavy chains have variable regions.
d Only the heavy chains have constant regions.
e All four chains have variable regions. (p. 775)

23 Which one of the following statements concerning antibodies is *correct*?

a Each antibody has two chains.
b Antibodies are usually proteins but sometimes they are polysaccharides.
c Antigens are bound by the variable regions of the antibody molecules.
d The heavy, but not the light, chains of the antibody molecule have variable and constant regions.
e The tail region is important in antigen specificity. (pp. 775–76)

24 The number of genes that code for an antibody molecule is

a one.   d four.
b two.   e more than four.
c three.   (p. 776)

25 The enormous variety of antibodies produced by an individual is due to

a synthesis of antibodies upon exposure to an antigen using the antigen as a template for antibody formation.
b specific rearrangement of genes in developing lymphocytes induced by antigens in the nucleus.
c excision of genetic material during fatal development and exon removal during mRNA processing.

d induction of cell division by cross-linking of the antibody molecules.

e mutations within the genome.

(pp. 776–78)

26 Genes and/or exons can be duplicated by all of the following *except*

a breakage and fusion during meiosis.
b reverse transcription.
c transposition.
d unequal crossing over during meiosis.
e mRNA processing in the nucleus.

(pp. 779–81)

27 Sections of DNA containing one or several genes and a control element that can move from one part of the genome to another are called

a transposons.   *d* plasmids.
b exons.   *e* promoters.
c introns.   (p. 781)

28 Eucaryotic gene evolution may involve

a gene duplications followed by mutations in the gene sequences.
b exon duplication and recombination.
c plasmid introduction and transformation.
d *a* and *b*.   (pp. 783–84)

For further thought

1 The Sabin poliomyelitis vaccine consists of three strains of weakened live viruses.

a Explain the process by which your body develops immunity to the polio viruses after you receive an oral dose of this vaccine.

b Why are babies usually given two or three separate doses of the vaccine rather than one dose?

c Children are given a "booster" dose of the vaccine at the age of 1½ years and again before they start school. What is the purpose of the boosters? (pp. 471–74)

2 It has been suggested that a number of human degenerative diseases, such as rheumatoid arthritis and multiple sclerosis, may be the result of slow-virus infections. In a slow-virus infection the virus can remain in a cell and replicate, while allowing the infected cell to continue to function. However, the slow virus does alter the protein composition of the cell surface. Explain how the altered cell surface may be responsible for much of the damage found in degenerative diseases.

ANSWERS

Testing recall

1	T	5	B, T	9	T
2	B	6	T	10	T
3	B	7	B	11	T
4	B	8	B		

Testing knowledge and understanding

12	*a*	17	*b*	21	*e*	25	*c*
13	*b*	18	*a*	22	*e*	26	*e*
14	*d*	19	*c*	23	*c*	27	*a*
15	*b*	20	*d*	24	*c*	28	*d*
16	*d*						

Chapter 31

ORGANIZATION OF DEVELOPMENT IN ANIMALS AND PLANTS

A GENERAL GUIDE TO THE READING

This chapter examines the principal events in the development of multicellular animals and plants. It first focuses on animal development, then moves to plant development, and includes a comparison of the two patterns of embryonic development. The next chapter discusses differentiation, with reference to both animals and plants. As you read Chapter 31 in your text, you will want to concentrate on the following topics.

1   Fertilization. You may be surprised to learn from the section on this topic (pp. 788–90) that development begins with penetration of the sperm rather than with fertilization itself.

2   Early cleavage and morphogenetic stages. Many new terms are presented in the discussion of this topic (pp. 790–96). You will find it helpful to take the time to learn them before reading further, as these terms are used frequently. Careful attention to Figures

31.3–31.9 (pp. 790–95) will also prove helpful, particularly as you study the material on gastrulation and neurulation, the crucially important processes that establish the shape and form of the embryo.

3   Development of the human embryo. The fascinating account on page 797 illustrated by Figure 31.11 (p. 796) details some of the principal events in human embryology.

4   Larval development and metamorphosis. The text (pp. 800–04) describes a pattern of development characteristic of many animals. You will want to understand the difference between gradual and complete metamorphosis.

5   The seed and its germination. You will want to learn the embryonic stages of development in a plant, the structure of the seed, and how the seed germinates (pp. 809–10). The differences between embryonic development in plants and in animals are summarized nicely on page 809.

6 Growth and differentiation of the plant body. Pay close attention to the section on this topic (pp. 810–17); the patterns of growth and differentiation in plants are quite different from those in animals.

KEY CONCEPTS

1 Development in sexually reproducing organisms begins with the penetration of the sperm into the ovum and ends with the organism's death. (pp. 788–806)

2 Development proceeds step by step and is essentially irreversible. (pp. 790–96)

3 In animals, the developmental processes of cell division, cell growth, cell differentiation, and morphogenetic movements convert the fertilized egg into the mature organism. (pp. 790–96)

4 The postembryonic organism is not a static entity; it continues to change, and hence to develop, until death brings the developmental process to an end. (pp. 798–806)

5 In plants, as in animals, the developmental processes of cell division, cell growth, and cell differentiation are important. However, morphogenetic movements do not occur in plants; instead, morphogenesis is achieved by differential patterns of cell division and growth. (pp. 806–17)

6 In plants a few cells remain forever embryonic; thus new organs can be formed and growth will continue throughout the plant's life. (pp. 810–17)

OBJECTIVES

After studying this chapter and reflecting on it, you should be able to carry out the following objectives.

1 Describe the events triggered by the penetration of an animal egg by the sperm, and discuss the process of fertilization. Explain what happens to prevent more than one sperm from fertilizing the egg. (pp. 788–90)

2 Using diagrams like those in Figures 31.3A–E (p. 790), 31.6A–D (p. 792), 31.8 (p. 794), and 31.9A–C (p. 795), describe the principal events occurring during the cleavage stages of animal embryos. Indicate whether the embryos increase in size during cleavage, and

explain how the amount of yolk in the egg affects the cleavage pattern. (pp. 790–92, 793–94)

3 Using diagrams like those in Figures 31.3E–G (p. 790), 31.6D–F (p. 792), and 31.9C–D (p. 795), describe gastrulation in amphioxus and indicate how the process differs in amphioxus, in frogs, and in birds. (pp. 790, 792, 795)

4 Discuss the process of neurulation in amphioxus and in frogs, and name the structures that the neural tube gives rise to (Figures 31.5 and 31.6, p. 792, may be helpful). (pp. 794–95)

5 Using diagrams like those in Figures 31.3 (p. 790), 31.5 (p. 792), and 31.6 (p. 792), point out the morula, gastrula, blastula, ectoderm, endoderm, mesoderm, archenteron, blastopore, and neural folds. (pp. 790–95)

6 Name the three primary cell layers, and indicate which primary cell layer gives rise to each of the following adult structures or tissues:

fingernails	hair
brain	lining of digestive tract
notochord	nerve cord
lungs	muscle
anus	skin (epithelial portion)
connective tissue	bone
blood	liver (p. 794)

7 Describe the role of growth in the postembryonic development of an animal. (pp. 799–800)

8 Define metamorphosis, and explain the adaptive significance of the larval stage. Distinguish between complete and gradual metamorphosis in insects. (pp. 800–4)

9 Describe the process of aging, indicating which kinds of cells tend to age and which do not. List three factors that appear to contribute to aging. (pp. 804–6)

10 Using diagrams like those in Figure 31.21 (p. 807), describe the early embryology of an angiosperm plant. In doing so, explain the function of the cotyledons, epicotyl, hypocotyl, radicle, and apical meristems; then specify what each of the following tissues gives rise to in the mature plant: the protoderm, the provascular tissue, and the ground tissue. (pp. 806–8)

11 Using a diagram such as Figure 31.22 (p. 808), identify the different parts of a seed: the cotyledon, endosperm, radicle, seed coat, epicotyl, and hypocotyl. Give two differences between the seed of a dicot and that of a

monocot. Then, using Figures 31.23 (p. 808), 31.24 (p. 809), and 31.25 (p. 810), describe patterns of seed germination. (pp. 806–10)

12 Compare embryonic development in a plant and in an animal with respect to cleavage and cell growth, morphogenesis, and organogenesis. (p. 809)

13 Using a diagram or photograph as given in Figures 31.26 (p. 811) and 31.34 (p. 815), point out the areas of cell division, cell enlargement, and cell maturation. Briefly explain the role auxin is thought to play in inducing cell-wall plasticity and cell enlargement. (pp. 810–14)

14 Describe the process by which leaves are produced along the growing young stem. Using a drawing such as Figure 31.35 (p. 816), point out a terminal bud, axillary bud, node, internode, bundle scar, and bud-scale scar. Specify what the axillary bud gives rise to. (pp. 814–16)

15 Using diagrams like those in Figure 31.36 (p. 817), explain how differentiation proceeds in a young root. Be sure to mention the role of the cambium. Explain how branch roots are formed. (pp. 810–12)

KEY TERMS

The following terms are important in this chapter; you should become familiar with them.

fertilization (p. 788)
pronucleus (p. 789)
cleavage (p. 790)
morphogenesis (p. 790)
blastocoel (p. 790)
blastula (p. 790)
gastrula (p. 791)
gastrulation (p. 791)
animal hemisphere (p. 791)
vegetal hemisphere (p. 791)
blastopore (p. 793)
archenteron (p. 793)
ectoderm (p. 793)
endoderm (p. 793)
mesoderm (p. 793)
notochord (p. 793)
yolk (p. 793)
blastodisc (p. 794)
epiblast (p. 794)
hypoblast (p. 794)
neurulation (p. 794)
larval stage (p. 800)
metamorphosis (p. 801)

pupa (p. 801)
imaginal discs (p. 801)
complete metamorphosis (p. 801)
gradual metamorphosis (p. 804)
aging (p. 804)
embryo (p. 806)
endosperm (p. 806)
seed (p. 806)
seed coat (p. 806)
suspensor (p. 807)
protoderm (p. 807)
provascular tissue (p. 807)
ground tissue (p. 807)
cotyledon (p. 807)
hypocotyl (p. 807)
apical meristem of the shoot (p. 808)
apical meristem of the root (p. 808)
epicotyl (p. 808)
radicle (p. 808)
germination (p. 809)
lateral root (p. 814)
node (p. 814)
internode (p. 814)
bud (p. 814)
axillary bud (p. 816)
differentiation (p. 816)
cork cambium (p. 816)

SUMMARY

Development of a multicellular animal The process of development in a sexually reproducing multicellular animal begins with the penetration of the sperm into the ovum. The membrane of the sperm fuses with that of the egg and the sperm nucleus moves into the egg. Vesicles in the outer region of the egg then release their contents, changing the structure of the egg membrane, preventing penetration by other sperm. Penetration also initiates the completion of oogenic meiosis. *Fertilization* occurs when the two gamete nuclei fuse.

The zygote then undergoes a rapid series of mitotic divisions called *cleavage*. The cytoplasm of the one large cell is partitioned into many new smaller cells. As cleavage continues, the cells become arranged in a hollow ball called a *blastula*. Two regions are often evident: the *animal hemisphere*, made up of small dark cells, and the *vegetal hemisphere*, made up of larger yolky cells.

Next begins a series of complex movements that are important in establishing the shape and pattern of the developing organism (*morphogenesis*). The blastula is converted into an embryo called the *gastrula*. Cleavage and gastrulation are greatly influenced by the amount of yolk in the egg. Gastrulation first produces an embryo with two layers, an outer *ectoderm* and an inner *endoderm*. A third layer, the *mesoderm*, forms between them. The ectoderm gives rise to the outermost layers of the body, the nervous system, and the sense organs; the endoderm to the lining of the digestive tract and associated structures; and the mesoderm to the supportive tissue—muscles and connective tissues.

The morphogenetic movements of gastrulation and neurulation give form and shape to the embryo, and bring masses of cells into proper position for their later differentiation into the principal tissues of the adult body. The developmental processes of cell division, cell growth, cell differentiation, and morphogenetic movement convert the gastrula into a young animal ready for birth.

The predominant factor in postembryonic development in most animals is growth in size. Growth does not occur at the same rate and at the same time in all parts of the body.

Many aquatic animals and certain groups of terrestrial insects go through *larval* stages that bear little resemblance to the adult. The complex series of developmental stages that convert an immature animal into an adult is called *metamorphosis*. Some insects undergo *complete metamorphosis*, which begins with a wormlike larval stage quite unlike the adult. After a period of growth the larva enters the *pupal stage*, during which it is extensively reorganized to form the adult. Other insects undergo *gradual metamorphosis*; the young, which already resemble the adult, go through a series of molts that make them more and more like the adult.

Aging is another aspect of development; it is a complex series of developmental changes that lead to the deterioration of the mature organism and ultimately to its death. In both plants and animals the aging process seems to be correlated with the degree of cellular specialization; cells that remain relatively unspecialized and continue to divide do not age as rapidly as cells that have lost the capacity to divide. Aging of the whole multicellular organism results from the deterioration and death of irreplaceable cells and tissues.

Aging in animals seems to be related to such factors as the replacement of diseased or injured tissue by connective tissue, the increased burden on remaining cells, altered hormonal balance, the deposition of waste material in older cells, and changes in the intercellular connective tissue.

Development of an angiosperm plant The egg cell of an angiosperm plant is fertilized in the ovary of the maternal plant by a sperm nucleus from a pollen grain. The first cell division gives rise to two unequal cells, a terminal cell and a basal cell. The terminal cell will give rise to the embryo proper. Three types of tissues begin to differentiate: a surface layer of *protoderm*, which will form epidermal tissue; an inner core of *provascular* tissue, which will form the cambium and vascular tissues; and a middle layer of *ground tissue*, which will form the cortex. Next, two embryo leaves, the *cotyledons*, arise. The part of the embryonic axis below the cotyledons is the *hypocotyl*; it will form the first part of the stem. Small clumps of tissue at each end of the embryonic axis remain undifferentiated and become the *apical meristems* of the shoot and root. The embryo, together with the food-storage tissue, the endosperm, becomes enclosed in a seed coat. The resulting structure is the seed.

Embryonic development of the angiosperm differs from that of most animals in that cell division and cell growth occur together, and morphogenesis is governed by patterns of cell growth and division rather than cell migration. Also, some plant cells remain forever embryonic (meristematic) and organogenesis continues throughout the life of the plant.

When the seed germinates, the hypocotyl emerges and turns downward; the *radicle* at its lower end forms the root. The *epicotyl* then begins to elongate; it forms most of the shoot. Growth of the shoot or root involves the production of new cells by the apical meristem, and then the elongation and differentiation of these cells.

Cell elongation is controlled by hormones, particularly auxin. Auxin appears to activate membrane receptors that transport hydrogen ions from the cytoplasm into the wall, breaking hydrogen bonds between the polysaccharide fibrils. The hydrostatic pressure of the cell contents causes stretching of the loosened walls and the cell elongates, but without synthesis of new cytoplasm. Auxin also seems to influence the insertion of new wall material.

In stems, certain areas of the apical meristem give rise to *nodes*, swellings where leaf primordia arise. The length of stem between two successive nodes is an *internode*. Most increase to stem length results from elongation of the cells in the young internodes. At the tip of the stem is a *bud*, which consists of the apical meristem and unelongated internodes enclosed within leaf primordia. When the bud opens, the internodes elongate and mitosis within the leaf primordia produces the leaves. Small areas of meristematic tissue (the *axillary buds*) arise between the leaf and internode.

The cells produced by the apical meristems will differentiate to form the various primary tissues of the plant. Increase in circumference of a root or stem depends upon the formation of secondary tissues derived from the lateral meristems.

KEY DIAGRAM

The following coloring exercise is designed to help you learn the early embryology of amphioxus, frog, and chick. The diagrams on the next page show selected diagrams from the blastula through the neurula stage. Color the titles and the appropriate structures in the order given, using a different color for each. (Hint: In most embryology books ectoderm is colored blue, endoderm yellow, and mesoderm red. You may want to follow this convention when you do your coloring.) We suggest you color in each required structure, if present, on all the diagrams so that you will see the relationships among them. For example, the first item to be colored in is ecto-derm, so color in all the ectoderm in each of the diagrams, then go on to endoderm, etc. You will find that some of the structures will be found on only one or two of the diagrams, others will be on almost every one. (Reference: Figures 31.3, p. 790; 31.5, p. 792; 31.6, p. 792; and 31.9, p. 795)

Ask yourself

What primary cell layers are the notochord and spinal cord derived from?

QUESTIONS

Testing recall

Below are listed some important terms in animal embryology. Match each statement with the proper term.

a   animal pole
b   archenteron
c   blastocoel
d   blastopore
e   blastula
f   gastrula
g   ectoderm

h   endoderm
i   mesoderm
j   morphogenesis
k   neurula
l   vegetal pole
m   zygote

1   stage in which neural tube forms (p. 794)

2   cavity of the blastula (p. 790)

3   opening into the cavity of the digestive tract (p. 793)

4   embryo as a hollow ball (p. 790)

5   process establishing shape and pattern (p. 801)

6   fertilized egg (p. 788)

7   a two-layered, later three-layered, animal embryonic stage (p. 791)

8   cavity of the digestive tract (p. 793)

9   part of the embryo where the invagination of gastrulation usually occurs (p. 791)

10   tissue that gives rise to the hair, skin, and fingernails (p. 793)

11   tissue that gives rise to the lining of the digestive tract and pancreas (p. 793)

12   tissue that gives rise to the brain and spinal cord (p. 793)

13   tissue that gives rise to muscles and bones (p. 793)

14   Which of the following are correctly matched? (pp. 806–17)

a   hypocotyl – radicle
b   endosperm – stored food
c   morphogenesis in plants – cell movement
d   cotyledons – embryonic leaves
e   seed dormancy – lowered gibberellin concentration
f   epicotyl – primary root
g   meristematic cells – differentiated cells
h   ground tissue – cortex
i   terminal cell – plant embryo
j   apical meristem – secondary tissues
k   shoot growth – cell division and cell elongation
l   auxin – increased plasticity of cell wall
m   node – leaf primordia
n   vascular cambium – elongation of the stem
o   seed – embryo, endosperm, seed coat

Testing knowledge and understanding

Choose the one best answer.

15   An egg cell differs from a sperm cell in that the egg cell

a   has cytoplasm.
b   has mitochondria.
c   contains a haploid set of genes.
d   is a product of meiosis.
e   has much greater energy reserves.

(p. 788)

amphioxus

frog

PRIMARY CELL LAYERS
(and their derivatives):

ECTODERM

ENDODERM

MESODERM

CAVITIES (color lightly):

BLASTOCOEL

ARCHENTERON

COELOM

NERVOUS AND
SUPPORTIVE STRUCTURES:

SPINAL CORD

NOTOCHORD

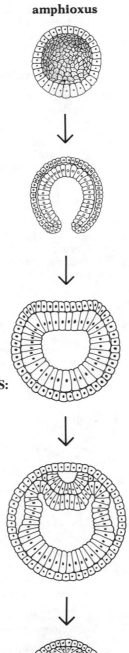

chick

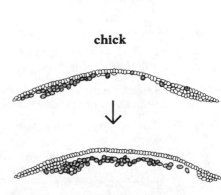

16 The fertilization process can be regarded as completed when

    a a sperm comes into contact with an egg.
    b a fertilization membrane is formed.
    c the sperm penetrates the cell membrane.
    d the sperm enters the egg cytoplasm.
    e the sperm nucleus fuses with the egg nucleus.   (p. 789)

17 The process of development begins when

    a the sperm nucleus fuses with the egg nucleus.
    b the sperm penetrates the egg.
    c egg oogenesis is completed.
    d cleavage begins.
    e the cell enters the M stage.   (p. 789)

18 In early cleavage, with each successive division the

    a number of chromosomes per cell is reduced by half.
    b number of chromosomes is doubled.
    c volume of cytoplasm per cell is reduced.
    d cytoplasmic contents are equally distributed.
    e extensive growth in size of the embryo begins.   (p. 790)

19 Which one of the following represents a correct sequence in the developmental process?

    a zygote → gastrula → neurula → blastula
    b fertilization → cleavage → blastula → gastrula
    c fertilization → gastrula → blastula → neurula
    d zygote → neurula → blastocoel → gastrula
    e fertilization → zygote → blastula → neurula → gastrula   (pp. 788–93)

20 In the early-cleavage embryo of the chick

    a the blastocoel forms by invagination.
    b all cells have an equal amount of yolk.
    c the heart has already begun to beat.
    d there is no cleavage through the yolk.
    e cell divisions are restricted to the egg white.   (pp. 794–95)

21 In most embryos, the archenteron

    a usually disappears during gastrulation.
    b is filled with mesoderm.
    c is formed by invagination.
    d is present in the blastula.
    e eventually forms the neural tube.   (p. 792)

22 Which one of the following statements is *false*?

    a In a vertebrate embryo, the blastula is formed by mitotic cell divisions without significant cell growth.
    b The embryonic archenteron becomes the cavity of the digestive tract in the adult.

    c Cleavage is restricted to the cytoplasmic disc of the ovum in birds.
    d The details of the gastrulation process vary considerably from species to species, depending on the amount of yolk in the ovum.
    e The embryonic blastocoel opens to the outside via the blastopore.   (pp. 788–94)

23 Which one of the following is *false* concerning the process of morphogenesis?

    a It refers to the production of form and pattern.
    b It involves cell movements.
    c It plays a major role in the development process.
    d It is synonymous with differentiation.
    e It is exhibited in the formation of the neural tube.   (p. 791)

Questions 24–29 refer to the following diagram of an amphioxus embryo during gastrulation.

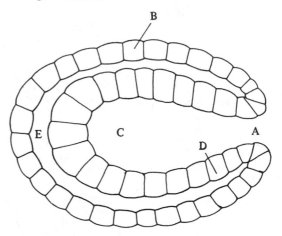

24 The area marked A

    a is called the blastopore.
    b forms during neurulation.
    c opens into the blastocoel.
    d forms during cleavage.
    e two of the above   (p. 793)

25 From which cell layer (B or D) will the nervous system develop? (p. 794)

26 From which cell layer will the lining of the digestive tract develop? (p. 794)

27 Which letter marks the blastocoel?

28 Which letter marks the archenteron?

29 All of the following are shown in the diagram *except*

    a ectoderm   *d* archenteron
    b endoderm   *e* two of the above
    c mesoderm   (p. 793)

30 Which one of the following is an *incorrect* match?

  a neural tube–neurula
  b three germ layers–gastrula
  c archenteron–gastrula
  d blastopore–blastula
  e hollow ball–blastula   (pp. 790–94)

31 Which one of the following sets of structures is derived primarily from mesoderm?

  a eye, brain, spinal cord
  b epidermis, hair, fingernails
  c blood, skeleton, muscle
  d lining of gut, lining of lungs, liver
  e epidermis, nervous system, skeleton
                          (p. 793)

32 Many insects undergo a complete metamorphosis during development. If the following four stages were placed in order, which one would be third?

  a pupa   *c* larva
  b egg   *d* adult   (p. 801)

33 Groups of cells present in the larva which give rise to the adult are called

  a blastomeres.
  b imaginal discs.
  c blastodiscs.
  d morphogenetic cells.
  e protoderm.   (p. 801)

34 As one grows older

  a unspecialized cells age quickly.
  b cell division in all tissues ceases.
  c somatic mutations may accumulate, leading to errors in the DNA.
  d enzymes that repair breaks in DNA become more efficient.   (pp. 804–5)

Questions 35–37 refer to the following diagram of a plant embryo.

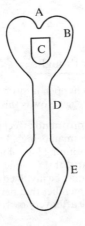

35 From what part will the cotyledons be produced? (p. 807)

36 From what part will the apical meristem arise? (p. 807)

37 Which part will give rise to xylem and phloem? (p. 807)

38 Which one of the following areas is closest to the tip of a plant root?

  a axillary bud
  b zone of cell elongation
  c apical meristem
  d zone of cell differentiation
  e zone of cell maturation   (p. 807)

39 The primary function of the cotyledons is to

  a give rise to the epicotyl.
  b define the first node of the embryo.
  c absorb nutrients from the endosperm.
  d protect the hypocotyl as it grows out of the seed coat.
  e none of the above   (p. 807)

40 Provascular tissue

  a is found only in the root.
  b gives rise to highly modified epidermal cells that include leaf hairs and root hairs.
  c differentiates into the primary xylem and phloem.
  d is found between the protoderm and the ground tissue.
  e none of the above   (p. 807)

41 Which of the following seed structures is *not* part of the embryo?

  a cotyledons
  b endosperm
  c shoot apical meristem
  d radicle
  e hypocotyl   (pp. 807–808)

42 Which of the following statements about plant development is *incorrect*?

  a All the organs of the mature plant are present in the embryo.
  b The first divisions of the zygote are accompanied by cell growth.
  c Plant cells do not migrate during morphogenesis.
  d Some plant cells continue to divide as the plant grows.
  e The root–shoot axis is established in the embryo before seed germination. (p. 809)

43 Which one of the following statements about a bean seed and its germination is *false*?

  a In the dormant state the seed consumes oxygen, but at a very low rate.

b As germination begins, much water is absorbed and oxygen consumption rises.

c The part of the embryonic axis called the epicotyl grows down and gives rise to the plant root.

d The two cotyledons contain stored food and provide the principal source of nourishment upon which the early growth of the embryo depends.

e As the young seedling develops, its increase in length is due to mitotic activity in the apical meristems of the root tips and terminal buds, followed by cell elongation in regions immediately behind the meristems. (p. 809)

44 Which one of the following is typical of growth of animals but not of plants?

a Cell growth is by an increase in the vacuole.

b Cell growth is by an increase in cytoplasm.

c Cell division and cell growth typically occur together.

d Growth continues throughout the life of the organism.

e Growth is by cell elongation. (pp. 796, 809)

45 Which one of the following statements is *false*?

a Lateral roots usually arise from the pericycle.

b Much of the size increase in plant cells involves absorbing more water into the cell vacuole.

c Provascular tissue is derived from apical meristems.

d The developed plant embryo contains rudiments of all the organs of the mature plant.

e Auxin promotes cell elongation. (pp. 809–14)

46 Which one of the following processes is important in animal morphogenesis but not in plant morphogenesis?

a cell division   *d* cell differentiation
b cell migration   *e* cell death
c cell growth   (p. 809)

Questions 47–50 refer to the following diagram of a root.

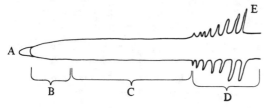

47 Which is the region of elongation? (p. 811)

48 Which is the region of cell division? (p. 811)

49 Which is the region where mature xylem and phloem are found? (p. 811)

50 Which is the region that produces the root-cap cells? (p. 811)

Use the following diagram to answer the next four questions.

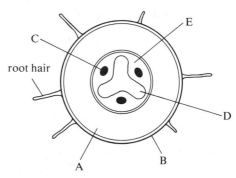

51 Which tissue carries sugar from the shoot to the root? (pp. 816–17)

52 Which tissue carries water from the root to the shoot? (pp. 816–17)

53 From which tissue will a lateral root be produced? (p. 814)

54 From which zone is this root cross section taken?

a zone of cell division
b root cap
c zone of cell maturation
d zone of elongation
e apical meristem (p. 811)

55 Which one of the following meristematic regions is ordinarily *least* important in increasing the overall size (height and circumference) of a tree over a period of 20 years?

a apical buds   *c* vascular cambium
b root tips   *d* cork cambium
                                (p. 816)

For further thought

1 What would happen if more than one sperm fertilized an egg? How is this prevented?

2 Aging has a profound effect on cerebral function in human beings. From midlife on there is a continuing loss of neurons from the brain and parts of the spinal cord. Suggest factors that may be involved in this aging process. (Remember that neurons normally do not divide after very early childhood.)

ANSWERS

Testing recall

1	*k*	5	*j*	8	*b*	11	*h*
2	*c*	6	*m*	9	*l*	12	*g*
3	*d*	7	*f*	10	*g*	13	*i*
4	*e*						

14 *a, b, d, e, h, i, k, l, m, o*

Testing knowledge and understanding

15	*e*	26	D	36	A	46	*b*
16	*e*	27	E	37	C	47	C
17	*b*	28	C	38	*c*	48	B
18	*c*	29	*c*	39	*c*	49	D
19	*b*	30	*d*	40	*c*	50	B
20	*d*	31	*c*	41	*b*	51	E
21	*c*	32	*a*	42	*a*	52	D
22	*e*	33	*b*	43	*c*	53	E
23	*d*	34	*c*	44	*b*	54	*c*
24	*a*	35	B	45	*d*	55	*c*
25	B						

MECHANISMS OF DEVELOPMENT: DIFFERENTIATION AND PATTERN FORMATION

A GENERAL GUIDE TO THE READING

This chapter, the last in Part III "The Perpetuation of Life," focuses on the two basic processes involved in development: the differentiation of cells and the spatial arrangement of those tissues to form the various morphologies. Many important concepts are presented in this chapter, for it describes the processes by which cells with identical genetic content become specialized to perform different functions. As you read Chapter 32 in the text you will want to concentrate on the following topics.

1   The effects of cleavage on differentiation. The concept of determinate and indeterminate cleavage is introduced on pages 819–23. It will be referred to again in Chapter 42.

2   Tissue interactions in development. In the discussion of this topic (pp. 823–31), read with special care the material on organizers and inducers, since embryonic induction plays a vitally important role in development.

The examples in Figures 32.8–32.10 (pp. 824–25) will help you understand induction.

3   The gradualness of differentiation. The developmental landscape model in Figure 32.20 (p. 832) is a graphic and easily understood model of the developmental process, as discussed on pages 831–32.

4   Reversibility of differentiation. The question of the reversibility of differentiation has long been of interest to embryologists. Some of their research is summarized on pages 834–37; note particularly the information on nuclear transplants.

5   The organization of neural development. The development of the nervous system provides an excellent example of many aspects of development. By learning how neurons migrate to their proper place and form axons and synapses (pp. 837–40), you will gain some understanding of how development as a whole may proceed.

KEY CONCEPTS

1 All the cells of a single organism arise by repeated division of the fertilized egg and are genetically identical. Which genes are active, and hence which potentialities are expressed, is determined in part by the nonuniform distribution of cytoplasmic substances in dividing cells. (pp. 819–21)

2 In both plants and animals, there is evidence of the presence of morphogenetic fields: as a particular tissue or organ develops it releases substances that inhibit formation of the same kind of tissue or organ in the immediate area. (p. 827)

3 Pattern formation is an important aspect of development; not only must cells differentiate properly but they must also have positional information available to them so that they will be arranged in proper order. (pp. 827–30)

4 As cells and tissues become more differentiated, the chemicals they secrete alter the environment of nearby cells and thereby profoundly affect the activity of their genes. (pp. 831–34)

5 Differentiation is a matter of progressive determination; development is gradually restricted to one of the many initially possible pathways.

6 In at least some organisms, differentiation appears to be reversible. (pp. 834–37)

7 The highly integrated functional network characterizing the nervous system is accomplished by the organized migration of neurons to their proper location and the subsequent growth of axons and formation of synapses on appropriate target cells. (pp. 837–40)

OBJECTIVES

After studying this chapter and reflecting on it, you should be able to carry out the following objectives.

1 Using drawings like those in Figures 32.2–32.4 (pp. 819–21), explain how the polarity of an egg cell and the location of the plane of cleavage influence development. Distinguish between mosaic eggs and regulative eggs and between determinate and indeterminate cleavage. (pp. 819–21)

2 Explain what is meant by embryonic induction, using as an example the role of the dorsal lip of the blastopore in a salamander (as shown in Fig. 32.8, p. 824) or the role of

optic vesicles in the induction of lenses in a frog or mammalian eye (see Figures 32.9, p. 824, and 32.10, p. 825). (pp. 823–26)

3 Distinguish between instructive and permissive inducers, and classify plant and animal hormones according to their inductive role. (p. 825)

4 Explain, using an example, the concept of morphogenetic fields. (p. 827)

5 Cite evidence to indicate that cells have positional information to ensure proper pattern formation. (pp. 827–30)

6 Discuss the role of the physical environment in guiding development.

7 Define differentiation. Then, use C. H. Waddington's developmental landscape model (Figure 32.20, p. 822) to explain the gradualness of differentiation. Give experimental evidence that supports this model. (pp. 831–32)

8 Cite experiments with plants and with animals that indicate that at least for some multicellular organisms, differentiation of the nucleus is reversible. (pp. 834–35)

9 Describe the process by which neurons become organized into an integrated, functional network, mentioning the roles of migration, growth cones, synapse formation, and cell death in your answer. (pp. 837–40)

KEY TERMS

The following terms are important in this chapter; you should become familiar with them.

gray crescent (p. 819)
totipotent (p. 819)
chimera (p. 819)
regulative eggs (p. 820)
mosaic eggs (p. 820)
determinate cleavage (p. 821)
indeterminate cleavage (p. 821)
dorsal lip of the blastopore (p. 823)
chordamesoderm (p. 823)
instructive inducers (p. 825)
permissive inducers (p. 825)
morphogenetic fields (p. 827)
pattern formation (p. 827)
growth cone (p. 839)

SUMMARY

Because all cells in an organism have the same genetic potential, differentiation and pattern formation depend on the differential activity of various genes, on chemical and electrical signals from other cells, and on information derived from various environmental factors.

Polarity of cells and the plane of cleavage Although the genetic content of all embryonic cells is identical, their cytoplasm is not. The cytoplasm of the zygote is rarely homogeneous. Therefore, the daughter cells produced by cleavage of the egg cell may not share equally in all cytoplasmic materials, depending on the orientation of the first plane of cleavage. If the cleavage partitions critical cytoplasmic constituents (i.e. ones that help activate or repress specific genes) equally, the cleavage is *indeterminate*; the new cells are totipotent—that is, they have the full developmental potential of the zygote. These eggs are called *regulative eggs*. If, however, the first cleavage partitions critical cytoplasmic constituents unequally, the cleavage is *determinate*; when the daughter cells are separated they do not have equivalent development potentialities. Eggs with marked cytoplasmic differences are called *mosaic eggs*. Indeterminate cleavage is characteristic of most echinoderms and vertebrates; determinate cleavage is characteristic of annelids and molluscs. Even in the species where the first few cleavages are indeterminate, a determinate cleavage eventually occurs.

The mechanisms establishing the polarity of the egg cytoplasm are largely unknown, but in some organisms the point of sperm penetration or various asymmetric environmental conditions are known to play a role. Polarity continues to be an important aspect of organization throughout the life of the organism.

Tissue interactions in development Developing cells are influenced by neighboring cells, by diffusible chemicals they secrete, by hormones, and by many parameters of their physical environment (e.g. temperature, light, humidity, gravity, pressure). Certain *organizer* areas of the embryo (e.g. the *gray crescent* and *dorsal lip* of the blastopore of a frog embryo) have an especially strong inductive function; they control the differentiation of other tissues. Some cell-to-cell interactions may be a direct effect of cell contact, but most developmental interactions are probably chemically mediated. Many chemicals can act as intercellular inducers in embryonic development. Some inducers are *instructive* and restrict developmental potential; others are *permissive* and facilitate the already determined potentialities.

In plants, hormones influence nearly all phases of development and may have both instructive and permissive roles. In vertebrates, hormones play a mainly permissive role.

Some chemicals influencing development are inhibitory. As a particular structure develops, it may release substances that inhibit the formation of the same kind of structure in the immediate area (called a *morphogenetic field*).

Cells apparently have some sort of positional information available to them that ensures proper pattern formation. In addition, a number of physical factors from the environment may affect the activity of a developing cell.

The course of differentiation If all the cells of an organism have the same genetic potential, other factors determine which potentialities are expressed. The first restrictions on the potential result from qualitative differences established during the early cleavages, which may distribute gene products already synthesized into different cells, and may give the nuclei of the different cells different cytoplasmic environments and thus bring about the activation of different genes. As development proceeds, the extracellular environment of the cells becomes less uniform. Moreover, as cells become more differentiated, they exert an increasing influence on all other cells in their vicinity via chemicals they secrete, and they are, in turn, influenced by other cells. The environment of each cell is constantly changing as development proceeds, and the changes in the environment profoundly affect the activity of genes.

As development proceeds, the individual cells become more and more committed to one particular course of differentiation. Differentiation is thus a matter of progressive determination; development is gradually restricted to one of the many initially possible pathways. Determination refers to a cell's fate, whereas differentiation refers to a cell's specialized chemistry and morphology.

As cells become more differentiated, the activity of the cells' genetic material changes. In plants, the changes undergone by cell nuclei during differentiation appear to be entirely reversible. In animals, the situation is less clear. Nuclear transplant experiments in frogs indicate that in some cases the differentiated nuclei had apparently lost their former developmental potential; but other experiments produced the opposite result.

Some animals retain the ability to regenerate

lost parts at maturity, but most lose this capacity. Chemical and/or electrical factors may play a role in inducing regeneration.

The organization of neural development Nerve cells are born, migrate to their proper places, send axons to specific target locations, and form a highly integrated functional network. A newly formed presumptive neuron will move from where it was formed to where it is supposed to be. Such movement is well oriented and appears to involve diffusing chemicals, cell-surface chemicals, and tactile clues. Once the neuron has reached its proper location, it sends axons to specific target cells. A *growth cone* forms and sends out filopodia that react to specific chemical and tactile information which guide axon growth.

In most animal nervous systems many more cells are produced than are used; those not needed atrophy and die.

QUESTIONS

Testing recall

1 Which of the following play a direct role in animal cell differentiation?

   a pattern of cleavage (p. 821)
   b cell migration (p. 809)
   c polarity of the egg (pp. 820–21)
   d cell division (p. 809)
   e induction by chemicals (p. 825)
   f hormones (p. 827)
   g morphogenetic fields (p. 827)
   h physical factors—light, temperature, gravity, pH (pp. 830–31)
   i cell elongation (p. 809)
   j pattern formation (p. 827)

Testing knowledge and understanding

Choose the one best answer.

2 In silkworms, silk gland cells are devoted to the synthesis of large quantities of the protein known as silk fibroin. These same cells do not make blood-specific proteins. One would expect that silk gland cells have

   a only silk fibroin genes.
   b the genes for both blood protein and silk fibroin.
   c silk fibroin genes and some other genes, but not blood protein genes.
   d blood protein genes but not silk fibroin genes.
   e fewer genes than the zygote. (p. 818)

3 The gray crescent region is very important in the early development of a frog embryo. If the first cleavage division divides this region into two and these two cells are separated,

   a both cells will die.
   b one cell will produce a normal tadpole, the other will produce a mass of unspecialized cells.
   c both cells will produce an abnormal tadpole.
   d both cells will produce a normal tadpole.
   e both cells will produce a mass of mesoderm tissue. (p. 819)

4 Which one of the following is *least* likely to be a significant factor in determining the differentiation pattern of embryonic cells?

   a cytoplasmic distribution during early cleavage stages
   b distance from the surface of the embryo
   c inducer substances secreted by neighboring cells
   d the pH and light intensity to which the cells are exposed
   e gene recombination during the cleavage stages (pp. 819–31)

5 The dorsal lip of the frog egg blastopore gives rise to mesodermal tissue, which normally underlies the neural tube. What would happen if the dorsal lip were removed prior to invagination?

   a A normal frog would develop.
   b A frog without lips would develop.
   c No neural tube would develop.
   d The gut tract underlying the neural tube would not develop.
   e Two neural tubes would develop. (pp. 823–24)

6 During development of the eye, the optic vesicles form as lateral outpockets from the brain. Early during this process the cells of the optic vesicle do not appear to be specialized but when they are cut from the brain and transplanted to another site they continue to develop as eyes. At this early stage the cells of the optic vesicles are

   a determined and fully differentiated.
   b fully differentiated but not determined.
   c determined but not fully differentiated.
   d neither determined nor fully differentiated. (p. 825)

7 Chemicals that restrict the developmental potential of the target cell and thus help determine the course of differentiation are called

   a instructive inducers.
   b permissive inducers.

c imaginal discs.
d morphogenetic fields.
e determinate inducers.   (p. 825)

8 Which one of the following inductive influences is inhibitory?

a morphogenetic fields
b dorsal lip of the blastopore
c low concentrations of auxin in plant tissues
d male sex hormone in the developing male gonad
e cells of the optic vesicle on overlying epidermal tissue   (p. 827)

9 The nuclei of highly differentiated human cells

a have lost certain genes.
b have specialized functions as a result of selective mutation.
c transcribe messenger RNA from all of the genes.
d are in a state of specific gene repression.
e can easily undergo a process of "dedifferentiation."   (p. 834)

10 If a *Xenopus* egg is enucleated, and a nucleus is transplanted to it from a more differentiated *Xenopus* cell, what is the most advanced developmental stage from which the nucleus can come if it is to be capable of directing full development of the egg into a normal adult?

a blastula   d tadpole
b gastrula   e adult
c neurula   (p. 835)

11 The results of transplantation of frog nuclei into eggs allow for which conclusion?

a The nuclei of differentiated cells generally lack some important genes.
b As a cell becomes more differentiated, its nucleus is less likely to support development of an egg into a tadpole after nuclear transplantation.
c The differentiated state is not reversible.
d The differentiated state is always reversible.   (p. 835)

12 Which one of the following statements is *false*?

a Visible morphological changes sometimes occur in the regions of a chromosome presumed to be most active.
b Enucleated frog eggs into which nuclei from adult cells have been transplanted usually (but not always) fail to develop normally.
c Under some conditions, differentiated plant cells can regain the capacity to divide mitotically.
d Radioactive tracer studies have shown that many genes are usually lost from the chromosomes of differentiated mammalian cells.
e Inducer chemicals probably play an important role in determining which genes will be active at any given time.
(pp. 753, 824–26, 834)

13 Which substance might potentially be the most effective experimental chemical to apply to the stump of a rat leg accidentally severed, in an attempt to cause regeneration of the lost tissue?

a *Xenopus* egg extract
b extract from the lost leg
c rat brain extract
d sea urchin blastula extract
e rat embryo endoderm cell extract
(pp. 836–37)

14 Before nerve cells join together in a network, many must move from their place of origin to their final location. All of the following are probably involved in this movement *except*

a gradients of diffusing chemicals
b growth cones of the axons
c cell-surface chemicals
d tactile clues   (pp. 837–39)

For further thought
_____

1 Explain the various factors in embryonic development that cause cells to have different characteristics despite their identical genetic content.

2 Normally, the lens of a frog's eye begins to differentiate from the outer layer (ectoderm) of the embryo when part of the brain touches the ectoderm. If the brain is transplanted to the tail of a frog embryo, a lens will form where the brain touches the tail ectoderm. If a piece of plastic is placed between the growing brain and the ectoderm, the lens will not develop. What do these experiments tell you about the stimulus for the differentiation of the lens?

ANSWERS

Testing recall

1 *a, c, e, f, g, h, j*

Testing knowledge and understanding

2 *b*	6 *c*	9 *d*	12 *d*
3 *d*	7 *a*	10 *e*	13 *c*
4 *e*	8 *a*	11 *b*	14 *b*
5 *c*			

Chapter 33

EVOLUTION: ADAPTATION

A GENERAL GUIDE TO THE READING

This chapter and the next discuss in detail a process that was first introduced in Chapter 1—the process of evolution. You may therefore find it helpful to begin your study of evolution by reviewing pages 12–20, with particular attention to the summary on page 20. The subject matter in both chapters is easily understood and most interesting. As you read this chapter in your text, you will want to concentrate on the following topics.

1   Sources of genetic variation. Pay careful attention to the discussion of this topic (pp. 846–48). Make sure you understand that evolution can act only through variation that is inherited. Also, learn the difference between the theory of evolution by natural selection and the theory of evolution by the inheritance of acquired characteristics.

2   The gene pool. The essential point on pages 848–49 is that evolution involves the gene pool of a *population*. A population can

evolve; an individual cannot. The term gene pool refers to the total of all the genes in a population.

3   The Hardy-Weinberg Law. This important law (pp. 849–55) sets up the conditions under which evolution would not occur. The fact that these conditions cannot all be met provides an indirect demonstration that all populations are evolving. Study carefully the four conditions for equilibrium. Notice too that the law can be used not only to predict the allelic and genotypic frequencies in successive generations in a population that is not evolving but also to calculate allelic frequencies when only phenotypic frequencies can be measured directly. Details on how this is done are presented in the box on page 852, and a study guide for population genetics is included in this chapter of the *Study Guide*.

4   The role of natural selection. Although the text concentrates on natural selection as being the most important force in bringing

about evolution, it is important for you to
realize that evolution does not depend
exclusively on any one mechanism. For
perspective, read carefully the two para-
graphs at the top of page 855 before reading
pages 855–64. There will be more discussion
of this topic in Chapter 34. There are a wealth
of examples provided in this section that will
help clarify the concept of natural selection
for you.

5 Adaptation. Begin your study of this topic by
learning the exact meanings of "adaptation"
and "fitness" in evolutionary biology; the
definitions of these terms on page 864 are
probably different from those you have
learned previously. The text (pp. 864–81)
provides a host of examples of different
adaptations. As you read about them, try to
see how each increases fitness.

KEY CONCEPTS

1 Modern evolutionary theory is based on two
concepts: that the genetically determined
characteristics of living things change with
time, and that this change is directed by
natural selection. (p. 845)

2 The evolutionary raw material is genetic
variation in a population. Natural selection
can act on genetic variation only when it is
expressed as phenotypic variation. (pp.
846–48)

3 Evolutionary change means change in allelic
frequencies (and hence in genotypic ratios) in
populations of organisms, not in individuals.
(pp. 848–55)

4 By natural selection, we mean nonrandom
reproduction, or, more specifically, reproduction that is to some degree correlated with
genotype. (p. 853)

5 Evolutionary change is not automatic; it
occurs only when something disturbs the
genetic equilibrium. Mutation pressure and
selection pressure are always disturbing this
equilibrium, and migration and genetic drift
may also do so. (pp. 851–55)

6 Many characteristics are both advantageous
and disadvantageous; the evolutionary fate
of such characteristics depends on the algebraic sum of the separate selection pressures.
(pp. 862–63)

7 Adaptations are genetically determined
characteristics that enhance an organism's

chances of perpetuating its genes in future
generations. (p. 863)

8 An adaptation need not be one hundred
percent effective to give the individuals that
possess it a significantly greater chance of
surviving to reproduce. (pp. 864–66)

OBJECTIVES

*After studying this chapter and reflecting on it,
you should be able to carry out the following
objectives.*

1 List the processes that can lead to variation in
the genetic material, and explain why changes
that affect only somatic cells, such as variations produced by practice, education, diet,
or medical treatment, cannot bring about
evolution. Contrast Lamarck's theory of
evolution by the inheritance of acquired
characteristics with Darwin and Wallace's
theory of evolution by natural selection.
(p. 848)

2 Explain the concept of the gene pool. Given
the frequency of two alleles, calculate the
ratios of the genotypes produced by them,
using a Punnett square or the algebraic
formula given on page 854. Your instructor
may pose additional problems for you to
solve. (pp. 848–52)

3 State the Hardy-Weinberg Law, and discuss
its four conditions for the maintenance of
genetic equilibrium, describing the forces
that prevent these conditions from being
fulfilled. In doing so, use the following
terms: mutational equilibrium, genetic drift,
mutation pressure, gene flow, selection
pressure, and random reproduction. (pp.
850–55)

4 Explain how natural selection on a phenotype
in one generation can affect the genotype of
the next generation (p. 855, bottom), and
calculate how changes in allelic frequencies in
the parental generation can alter allelic
frequencies in the offspring. (pp. 852–56)

5 Using diagrams like those in Figure 33.10 (p.
861), contrast directional selection, stabilizing
selection, and disruptive selection. Indicate
which type of selection is involved in the
experiment shown in Figure 33.9 (p. 860).
(pp. 859–62)

6 Contrast the roles of selection and mutation
in directing evolutionary change. Indicate
other factors that may be important in
evolution. (pp. 853–55)

7 Using an example, explain how a characteristic can have both positive and negative effects, and indicate what determines whether or not a trait will increase or decrease in the population. Include a discussion of heterozygote superiority, showing how this condition affects the fate of certain alleles. (pp. 862–63)

8 Define adaptation and fitness in their evolutionary sense, and explain, using examples, how such phenomena as flower structure, defensive secretions, cryptic appearance (including polymorphism), warning coloration, and mimicry are adaptive. Distinguish between Batesian and Müllerian mimicry. (pp. 864–77)

9 List the three types of symbiotic interactions, and show how they differ from one another. Distinguish between external and internal parasites, and discuss two adaptations of internal parasites. Explain why most well-adapted parasites do not generally kill or seriously harm their host species. (pp. 877–81)

KEY TERMS

The following terms are important in this chapter; you should become familiar with them.

evolution by natural selection (p. 848)
evolution by the inheritance of acquired
   characteristics (p. 848)
gene pool (p. 848)
allelic frequency (p. 849)
genetic equilibrium (p. 849)
genotype frequencies (p. 849)
Hardy-Weinberg Law (p. 850)
genetic drift (p. 851)
neutral selection (p. 851)
mutation pressure (p. 853)
gene flow (p. 853)
selection pressure (p. 854)
preadapted (p. 857)
polygenic characters (p. 857)
directional selection (p. 859)
disruptive selection (p. 861)
stabilizing selection (p. 861)
discontinuous phenotypes (p. 863)
balanced polymorphism (p. 863)
heterozygote superiority (p. 864)
adaptation (p. 864)
fitness (p. 864)
coevolution (p. 867)
cryptic appearance (p. 870)
warning coloration (p. 874)
aposematic (p. 874)

Batesian mimicry (p. 875)
Müllerian mimicry (p. 876)
commensalism (p. 877)
mutualism (p. 877)
parasitism (p. 877)
parasitoids (p. 879)
degeneracy (p. 880)
specialization (p. 880)

SUMMARY

Fundamental to the modern theory of evolution are two concepts—that the characteristics of living things change with time and that this change is directed by natural selection. Evolution is the change in the genetic makeup of a population in successive generations. There are always variations among members of a population; if there is selection against certain variants and for other variants, the overall makeup of the population may change with time.

Natural selection can act on genetic variation only when it is expressed phenotypically. Nongenetic variations and variations produced by somatic mutations are not evolutionary raw material. Somatic mutations do not affect the genotype of the germ cells. Lamarck's hypothesis of acquired characteristics is not tenable.

Population genetics is based on the concept of the *gene pool*, the sum total of all the genes possessed by all the individuals in the population. The frequencies of the various alleles of a given gene are used to characterize the gene pool. Evolution is a change in the allelic frequencies within gene pools.

Evolutionary change is not automatic; it occurs only when something disturbs the genetic equilibrium. According to the Hardy-Weinberg Law, both allelic frequencies and genotypic ratios remain constant from generation to generation in sexually reproducing populations if the following conditions for stability are met: large population (so there will be no *genetic drift*), no mutation (or else mutation equilibrium), no migration (i.e. no *gene flow* into or out of the population), and totally random reproduction. In reality, many populations are large, and some populations exist without migration, but the conditions of no mutation and random reproduction are never met in any population. Mutations are always happening and mutational equilibrium is rare; hence *mutation pressure* can cause slow shifts in allelic frequencies. Reproduction is never totally random; no aspect of reproduction is completely devoid of correlation with genotype. Nonrandom reproduction, or natural selection, is the universal rule.

Thus the Hardy-Weinberg Law describes the conditions under which there would be no evolution—and since these conditions cannot be met in nature, it follows that evolution is always occurring. Evolutionary change is a fundamental characteristic of the life of all populations.

The role of natural selection All populations are subject to *selection pressure*, which disturbs the genetic equilibrium. Even very slight selection pressures can lead to major changes in allelic frequencies over time. Changing environmental conditions can exert *directional selection*, which causes the population to evolve along a particular functional line.

Natural selection can be creative. Even in the absence of new mutation, selection can produce new phenotypes by combining old genes in new ways. Mutation, by contrast, is not usually a major directing force in evolution; its principal evolutionary role is to provide new variations upon which future selections can act.

Sometimes a population is subject to two or more opposing directional selection pressures. Such *disruptive selection* may divide the population into distinct groups.

Natural selection also plays a conservative role; it acts to preserve favorable gene combinations by eliminating less adaptive new combinations created by recombination and mutation. This sort of selection has been termed *stabilizing selection*.

Many characteristics have both advantageous and disadvantageous effects, and a single allele may influence multiple characteristics (*pleiotropy*). The evolutionary fate of such characteristics or alleles depends on whether the sum of all the various positive selection pressures acting on them is greater or less than the sum of the negative selection pressures.

Sometimes the effects of a given allele are more advantageous in the heterozygous than in the homozygous condition. In such cases, *heterozygote superiority* may lead to balanced polymorphic variation because it favors the retention of both alleles in the population. *Polymorphism*, the occurrence in a population of two or more distinct forms of a genetically determined character, can be advantageous because the different forms may exploit more completely the subdivisions of a variable environment.

Adaptations Adaptations are genetically controlled characteristics that enhance an organism's chances of perpetuating its genes in succeeding generations. Adaptations can be structural, physiological, or behavioral; genetically simple or complex; highly specific or general.

The flowering plants depend on external agents (e.g. wind, birds, insects) to carry pollen from the male to the female parts of the plant. The flowers are adapted in shape, structure, color, and odor to their particular pollinating agent. The plants and their pollinators have evolved together, each becoming more finely tuned to the other's peculiarities. Such evolutionary interaction is called *coevolution*.

Defensive secretions, ambiguous body orientation, and *cryptic* (concealing) appearances are all adaptations that help animals escape predation. Kettlewell's experiments on the light and dark forms of pepper moths (*Biston betularia*) showed clearly that those moths that most closely resemble their background have the best chance of escaping predation.

Some animals that are disagreeable to predators (because of sting, bad taste, or smell) have evolved warning (*aposematic*) coloration. They benefit by being gaudily colored and conspicuous because predators can easily learn to recognize and avoid them after one or two unpleasant encounters. The avoidance of aposematic insects may not depend solely on learning; this avoidance response may evolve in predators.

Another protective adaptation is *mimicry*: members of different species resemble (mimic) one another. There are two types of mimicry. In *Batesian mimicry*, an unprotected species resembles a distasteful species; the mimicry is based on deception. *Müllerian mimicry* involves the evolution of similar appearances by two or more distasteful species. This type of mimicry is advantageous to both because predators learn more easily to avoid both prey species.

Symbiosis Many organisms have evolved adaptations for *symbiosis*—for living together. There are three types of symbiosis: commensalism, mutualism, and parasitism. However, there are no sharp boundaries between them; they grade into one another.

Commensalism is a relationship in which one species benefits while the other neither benefits nor is harmed. The advantages the commensal species receives from its host frequently include shelter, support, transport, and/or food. Often it is difficult to determine whether a relationship is commensalism or *mutualism*, in which both species benefit.

Parasitism is a symbiotic relationship in which one species benefits while the other species is harmed. External parasites live on the outer surface of their host, while internal parasites live inside the host's body. Internal parasitism is usually marked by more extreme specializations than external parasitism; these include structural degeneracy (evolutionary loss of structures), resistant body walls, a complex life cycle, and large reproductive potential. Internal parasites usually tend to evolve towards greater specificity, both towards their host and towards the part of the host's body they inhabit. Over time the host and the parasite undergo coevolution, eventually reaching a dynamic balance in which both can survive without serious damage. Most long-established host-parasite relationships are balanced ones.

STUDY GUIDE FOR POPULATION-GENETICS PROBLEMS

The best way to gain an understanding of population genetics is to work with it. The following information is intended to get you started on problems; additional problems are provided at the end of this section. Before you begin, review page 852 in your text.

1   Allelic frequencies. We can express the frequency of alleles in a population numerically. Example 1: Suppose we have 100 alleles of a particular gene in a population, with 70 of these being allele A and 30 being allele a:

$$\text{Frequency } A = \frac{70}{100} = 0.7, \text{ or } 70\%$$

$$\text{Frequency } a = \frac{30}{100} = 0.3, \text{ or } 30\%$$

Example 2: Calculating allelic frequencies in alleles showing intermediate inheritance: In cats, tail length is determined by a single pair of alleles. Long-tailed cats have the genotype I^L/I^L, short-tailed I^L/I^t and no-tailed I^t/I^t. In a given population, 49 cats had long tails, 42 had short tails, and 9 had no tails. What are the frequencies of the I^L and I^t alleles?

Solution
The total number of individuals is $49 + 42 + 9 = 100$. Since each cat has two alleles, the total number of alleles in the population is $2 \times 100 = 200$.

The number of I^L alleles = $(2 \times 49) + 42$

$$= \frac{140}{200} = 0.7$$

The number of I^t alleles = $(2 \times 9) + 42$

$$= \frac{60}{200} = 0.3$$

Note that the frequency of the two alleles adds up to one.

2   The expression $p + q = 1$. In population genetics the symbol p represents the frequency of one allele; q, the frequency of the other. Thus in the preceding example, $p = 0.7$ and $q = 0.3$. We can then use the expression $p + q = 1$ to represent the relationship between the two alleles; since all individuals in this population have either allele A or allele a, the two frequencies must add up to 1. When we know the frequency of the alleles we can calculate the frequency of the various genotypes they will produce:

	0.7 A	0.3 a
0.7 A	0.49 A/A	0.21 A/a
0.3 a	0.21 A/a	0.09 a/a

The Punnett square tells us that 49 percent of the population will be A/A; 21 plus 21, or 42 percent, will be A/a; and 9 percent will be a/a.

3   The Hardy-Weinberg equilibrium. The same results can be obtained algebraically by means of the formula

$$p^2 + 2pq + q^2 = 1$$

Here, p^2 is the frequency in the offspring of the homozygous dominant genotype, $2pq$ is the frequency of the heterozygous genotype, and q^2 is the frequency of the homozygous recessive genotype.

$$(0.7)^2 + 2(0.7)(0.3) + (0.3)(0.3) = 1$$
$$0.49 + \quad 0.42 \quad + \quad 0.09 \quad = 1$$

As shown in your text (p. 851), the allelic frequencies in the offspring generation are the same as they were in the parent generation. The population is therefore said to be in genetic equilibrium.

4   Solving problems. We can use the two equations presented above to solve problems.

1   In a rabbit population 10 percent of the genes for coat color are for albino (b), while 90 percent are for black (B). What percentage of the rabbits are heterozygous if the Hardy-Weinberg assumptions hold true?

Solution
Frequency of $B = 90\% = 0.9 = p$
Frequency of $b = 10\% = 0.1 = q$
$p^2 + 2pq + q^2 = 1$
Heterozygotes = $2pq = 2(0.9)(0.1) = 0.18$
18% of the rabbits are heterozygous.

2   The frequency of the recessive allele b in a population is 0.2. Then a sudden catastrophe exerts a selective pressure,

reducing the frequency to 0.16 in a single generation. What will be the frequencies of all genotypes (*BB, Bb, bb*) in the next generation?

Solution

Frequency of $b = 0.16 = q$

$p + q \quad = 1$

$p + 0.16 = 1$

$\quad p = 0.84$

$$p^2 \quad + \quad 2pq \quad + \quad q^2 \quad = 1$$
$$(0.84)^2 + 2(.84)(.16) + (.16)^2 = 1$$

Genotypic fre- $p^2 = 0.70 = B/B$
quencies in the next $2pq = 0.27 = B/b$
generation (rounded) $q^2 = 0.03 = b/b$

3 Allele *A*, for unattached earlobes, is dominant over allele *a*, for attached earlobes. If 36 percent of a population has attached earlobes, what are the frequencies of allele *A* and allele *a* in this population?

Solution

$p^2 + 2pq + q^2 = 1$

Frequency of attached
    earlobes (a/a) = $0.36 = q^2$

$$\sqrt{0.36} = q$$

$\quad 0.6 = q =$ frequency of allele *a*

$p + q \quad = 1$

$p + 0.6 = 1$

$\quad p = 0.4 =$ frequency of allele *A*

4 Allele *T*, for the ability to taste a chemical called PTC, is dominant over allele *t*, for the absence of this ability. In a population of 1,000 individuals, 750 are tasters and 250 are nontasters. What are the frequencies of allele *T* and allele *t* in this population?

Solution

$p^2 + 2pq + q^2 = 1$

Frequency of non-
  tasters (t/t) $= \dfrac{250}{1000} = q^2$

$\quad 0.25 = q^2$

$$\sqrt{0.25} = q$$

$\quad 0.5 = q =$ frequency of allele *t*

$p + q \quad = 1$

$p + 0.5 = 1$

$\quad p = 0.5 =$ frequency of allele *T*

Population-genetics problems

1 The frequency of two alleles in a gene pool is 0.1 (*A*) and 0.9 (*a*). What is the percentage in the population of heterozygous individuals? of homozygous recessives? Assume the population is in Hardy-Weinberg equilibrium.

2 Allele *B* for white wool, is dominant over allele *b*, for black wool. In a sample of 900 sheep, 891 are white and 9 are black. Estimate the allelic frequencies in this sample, assuming the population is in equilibrium.

3 In a population that is in Hardy-Weinberg equilibrium, the frequency of the recessive homozygote genotype of a certain trait is 0.09. What is the percentage of individuals homozygous for the dominant allele?

4 In a population that is in Hardy-Weinberg equilibrium, 36 percent of the individuals are recessive homozygotes for a certain trait. For the same trait, what is the percentage in this population of homozygous dominant individuals? of heterozygous individuals?

5 Allele *T*, for the ability to taste a particular chemical, is dominant over allele *t*, for the inability to taste it. At Cornell University, out of 400 surveyed students, 64 were found to be nontasters. What is the percentage of heterozygous students? Assume the population is in equilibrium.

6 In humans, Rh-positive individuals have the Rh antigen on their red blood cells, while Rh-negative individuals do not. Assume that the Rh-positive phenotype is produced by a dominant gene *Rh*, and the Rh-negative phenotype is produced by its recessive allele *rh*. In a population that is in Hardy-Weinberg equilibrium, if 84 percent of the individuals are Rh-positive, what are the frequencies of the *Rh* allele and the *rh* allele at this locus?

7 In corn, yellow kernel color is governed by a dominant allele; white, by its recessive allele. A random sample of 1,000 kernels from a population that is in equilibrium reveals that 910 are yellow and 90 are white. What are the frequencies of the yellow and white alleles in this population? What is the percentage of heterozygotes in this population?

8 A rare disease due to a recessive allele which is lethal when homozygous occurs with a frequency of one in a million. How many individuals in a town of 14,000 can be expected to carry this allele?

Answers to population-genetics problems

1 18 percent heterozygous individuals, 81 percent homozygous recessives
2 0.9 white, 0.1 black
3 49 percent
4 16 percent homozygous dominant individuals, 48 percent heterozygous
5 48 percent
6 *Rh* 0.6, *rh* 0.4
7 yellow 0.7, white 0.3, 42 percent heterozygotes
8 about 28 individuals

QUESTIONS

Testing recall

Determine whether each of the following statements is true or false. If false, correct the statement.

1 For evolution to occur in a population, reproduction must be totally random. (pp. 850–51)

2 Genetic drift has more evolutionary impact on small populations than on large ones. (p. 851)

3 Somatic mutation does not provide raw material for evolution. (p. 848)

4 Many genes are simultaneously subject to both positive and negative selection pressures. (pp. 862–63)

5 Polymorphism is the occurrence in a population of two or more distinct forms of a genetically determined character. (pp. 863–64)

6 Cryptically colored insects are usually distasteful to predators. (pp. 871–74)

7 Industrial pollution caused the English pepper moth (*Biston betularia*) to absorb coal dust and become dark. (pp. 873–74)

8 Two unrelated sympatric beetle species look very much alike. Both produce a noxious chemical that makes them taste bad to potential avian predators. This is probably an example of Müllerian mimicry. (p. 876)

9 In a large population with random mating, mutational equilibrium, and no migration, a disadvantageous allele will gradually disappear. (pp. 850–51)

10 Selection of heterozygotes can result in an increase in the frequency of lethal recessive alleles. (pp. 862–63)

The following important terms are sometimes confused by students. Distinguish between the terms within each group.

11 adapted – preadapted (pp. 856–57)

12 natural selection – neutral selection (pp. 851, 853–54)

13 directional selection – disruptive selection – stabilizing selection (pp. 857–62)

14 Batesian mimicry – Müllerian mimicry (pp. 875–76)

15 cryptic appearance – warning coloration (pp. 870–75)

16 symbiosis – mutualism – commensalism – parasitism (pp. 877–79)

17 parasite – predator – parasitoid (p. 879)

Testing knowledge and understanding

Choose the one best answer.

Questions 18–21 refer to the following situation.

Two pirates and three Polynesian beauties settled on a previously uninhabited tropical island. All five of these settlers had brown eyes, but only one man carried the recessive allele for blue eyes (his genotype was *B/b*).

18 What are the frequencies of allele *B* and allele *b* in the population of this island?

  a B = 0.8; b = 0.2
  b B = 0.9; b = 0.1
  c B = 0.1; b = 0.9
  d B = 0.2; b = 0.8
  e none of the above   (p. 849)

19 If you assume a Hardy-Weinberg equilibrium for the eye-color alleles (admittedly very improbable), about how many people would you expect to have blue eyes when the island population reaches 20,000?

  a 0   *d* 1000
  b 20   *e* 2000
  c 200   (p. 852)

20 No new settlers arrive on the island, and no one leaves this tropical paradise. If the number of blue-eyed individuals differs greatly from the answer to the previous question, this could be due to a combination of which factors?

  a gene flow and genetic drift
  b gene flow and natural selection
  c genetic drift and natural selection
  d gene flow and nonrandom mating
                               (pp. 854–55)

21 Twenty percent of this island population of 20,000 has type AB blood, but the population our original two pirates had come from had only 1 percent type AB individuals. This is most likely an example of

    a genetic drift.
    b natural selection.
    c selective mutation.
    d gene flow.
    e nonrandom mating.   (p. 854)

22 In a population that is in Hardy-Weinberg equilibrium, there are two alleles for a certain gene, A and a. The frequency of A is 80 percent. What percentage of the population will be heterozygous for this gene?

    a 4 percent   *d* 64 percent
    b 16 percent   *e* 100 percent
    c 32 percent   (p. 852)

23 According to the Hardy-Weinberg Law, which of the following is *not* a condition for genetic equilibrium?

    a Mutations must not occur or must be at mutational equilibrium.
    b There must be no immigration or emigration.
    c Reproduction must be nonrandom.
    d The population must be large.   (p. 851)

24 A large population of a certain species of freshwater fish lives in South America. No close relatives of this species are known. Suppose you could somehow cause all mutations to cease in this population and prevent all immigration into the population and emigration from it. Which one of the following statements *best* expresses the probable future of the population?

    a All evolution will promptly cease because without mutation there will be no raw material for evolution.
    b The population will begin to deteriorate after four or five generations because of excessive inbreeding that will result from the absence of immigration and emigration.
    c The population will continue to evolve for a long time as selection acts on the variability produced by recombination.
    d Major evolutionary changes will continue in the population because of genetic drift.
    e Though the population will cease to evolve, it may survive for a long time if the environment remains constant.   (pp. 850–52)

25 In the peppered moth the allele for light color is recessive and the allele for dark color is dominant. In a population that is in Hardy-Weinberg equilibrium, if there are 640 light

moths and 360 dark moths, how many of the moths are heterozygous?

    a 96 percent   *d* 32 percent
    b 64 percent   *e* 8 percent
    c 48 percent   (p. 852)

26 Natural selection can best be defined as

    a survival of the best-adapted individuals.
    b nonrandom reproduction.
    c differential population-growth rates.
    d enhanced survival of those individuals best adapted to attract mates.
    e the elimination of the weak by the strong.   (pp. 853–54)

27 The theory of natural selection postulates that

    a in each generation, all the individuals well adapted for their environment live longer and produce more progeny than the less well adapted individuals.
    b the deaths of individuals occur completely at random with respect to the physical environment.
    c the deaths of individuals occur completely at random with respect to their genotypes.
    d to at least a small extent the survival and reproductive success of individuals depend upon the extent to which they are genetically adapted to their environment.
    e most deaths of individual organisms occur soon after fertilization, as a result of hereditary deficiencies.   (pp. 853–54)

28 A city was intensively sprayed with DDT in 1953 in an effort to control houseflies. The number of flies was immediately greatly reduced. Each year thereafter the city was again sprayed with DDT, but the flies gradually increased in numbers until ten years later they were almost as abundant as they were when the control program began. Which one of the following is the most likely explanation?

    a Flies from other areas moved in and replaced the ones killed by the DDT.
    b The few flies that were affected by DDT but survived developed antibodies to DDT, which they passed on to their descendants.
    c The DDT caused new mutations to occur in the surviving flies, and this resulted in resistance to DDT.
    d The DDT killed susceptible flies but the few that were naturally resistant lived and reproduced, and their offspring repopulated the area.   (pp. 856–57)

29 Careful measurements have shown that duck eggs of average size are more likely than smaller or larger eggs to produce viable ducklings. This is an example of

a disruptive selection.
b stabilizing selection.
c directional selection.
d polymorphism.
e convergence. (pp. 857–59)

30 Flowers pollinated by hummingbirds are often red and usually odorless; humming-birds see red easily but have a poor sense of smell. This is an example of

a mimicry.
b polymorphism.
c coevolution.
d stabilizing selection.
e convergence. (p. 867)

31 Of the following methods of pollination, which one would result in the least selection pressure for the evolution of colored or odoriferous flowers?

a pollination by night-flying bats
b pollination by butterflies
c pollination by hummingbirds
d pollination by wind
e pollination by flies and bees (p. 867)

32 In England the peppered moth has two color forms. The prevalence in an area of one or the other form is determined by the amount of soot polluting the area. Both forms of the moth exhibit

a cryptic appearance.
b Batesian mimicry.
c Müllerian mimicry.
d warning coloration.
e coevolution. (pp. 873–74)

33 The edible viceroy butterfly looks like the distasteful monarch butterfly and hence is protected from predation. This is an example of

a cryptic coloration.
b cryptic appearance.
c Batesian mimicry.
d coevolution.
e Müllerian mimicry. (pp. 875–76)

34 Many mosses live on the trunks and branches of trees. They flourish but do not seem to harm the trees in any way, since they are occupying only the bark surface, which is dead tissue. The interaction between these two organisms could *best* be described as

a symbiosis.   d parasitism.
b commensalism. e predation.
c mutualism.   (pp. 877–78)

35 *Pseudomyrmex* ants live in an intimate association with the *Acacia* plant, a Central American plant with very large thorns. The queen ant bores a hole in one of the large thorns, hollows it out, and lays her eggs. As the colony grows, more thorns are excavated and inhabited. The ants continuously patrol the plant and remove herbivorous insects and competing vegetation. The relationship between the ants and the *Acacia* could *best* be described as

a symbiosis.   d predation.
b mutualism.   e commensalism.
c parasitism.   (pp. 877–78)

36 The relationship between a farmer and his flock of egg-producing hens could *best* be described as

a predation.   d mutualism.
b parasitism.   e competition.
c commensalism.   (pp. 877–78)

For further thought

[Answers to these questions are included at the end of the answer section of this chapter.]

1 In a certain African tribe, 4 percent of the population are born with sickle-cell anemia. What is the percentage in the tribe of individuals who enjoy the selective advantage of the sickle-cell gene, being more resistant to malaria than those homozygous for "normal" hemoglobin?

2 Approximately one child in 10,000 is born with PKU (phenylketonuria). What is the frequency of this allele in the population? What is the frequency of the normal allele? What is the percentage in the population of carriers of this trait?

3 In white human beings, hair straightness or curliness is thought to be governed by a single pair of alleles showing intermediate inheritance. Individuals with straight hair are homozygous for the I^s allele while those with curly hair are homozygous for the I^c allele. Individuals with wavy hair are heterozygous. In a population of 1,000 individuals, 245 were found to have straight hair, 393 had curly hair, and 362 had wavy hair. Give the allelic frequencies of the I^s and I^c alleles. Is this population in Hardy-Weinberg equilibrium? Explain your answer and perform a chi-square test (Table on page 665 of the text) to support your answer.
(Hint: If the population is in equilibrium, then the distribution of genotypes in the population must fit $p^2 + 2pq + q^2 = 1$. First, calculate the allelic frequencies of the I^s and

I^c genes, and, using the formula, calculate the expected distribution. Does the observed = the expected? Perform a chi-square test to see if the data fit the equilibrium formula.)

4   In poultry, the autosomal gene F^B produces black feather color and another allele, F^W, produces white feathers. The heterozygous condition produces blue andalusian. A white hen is mated to a black rooster and the F_2 was found to contain 78 black, 206 blue, and 116 white chickens. What are the frequencies of the black and white alleles? Is this population in Hardy-Weinberg equilibrium? Explain, using a chi-square test.

5   Artificial selection can alter the distribution of phenotypes in a population. Suppose the distribution of phenotypes within a population looks like this:

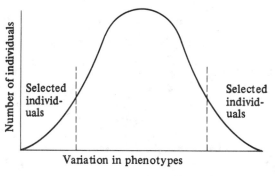

Disruptive selection is then practiced—only those organisms at both extremes of the distribution are chosen for breeding; those in the central portion are rejected. Draw a curve showing the expected phenotypic distribution of the F_1.

6   The introduction of DDT to control mosquitoes in the 1940s brought rapid control over the mosquitoes carrying the malarial pathogen. Today, resistance to DDT has developed in many mosquito populations and is reducing the effectiveness of malarial-control programs. Explain how it was possible for the mosquitoes to evolve DDT resistance in such a short time.

7   Children born with Tay-Sachs disease die within two or three years of birth. The disease is inherited as an autosomal recessive. Considering the strong selection pressure against individuals with the disease, explain why the allele persists in the population.

8   The fossil record shows that the few surviving species of horseshoe crabs have existed unchanged for the past 200 million years. How do you account for the fact that there has been essentially no evolution of the species over this time span?

ANSWERS

Testing recall

1   false—nonrandom
2   true
3   true
4   true
5   true
6   false—tasteful
7   false—to exhibit an increase in the frequency of the dark forms
8   true
9   false—no change in frequency
10 true

11 *Adapted*, the more general of the terms, refers to an organism's possession of any genetically controlled characteristic that increases its fitness to survive in the existing environment. *Preadapted* refers to an organism's possession of genes that increase its fitness to survive if the environment changes in certain respects; the organism is adapted to a new environment before it encounters that environment.

12 *Natural selection* refers to nonrandom reproduction; those individuals with certain characteristics have a better chance of surviving and reproducing than individuals with other characteristics, leading to an increase in the frequency of some alleles and a decrease in the frequency of others. *Neutral selection* is an evolutionary change in a small population due to genetic drift (chance); the direction of change is not influenced by the relative adaptiveness of the alleles.

13 *Directional selection* acts against individuals exhibiting one extreme of a character, causing

the population to evolve along a particular line; *disruptive selection*, which occurs when two opposing selection pressures are present, favors individuals exhibiting both extremes of a character, thereby dividing the population into two distinct types; and *stabilizing selection* acts against individuals exhibiting characters that are too different from the mean condition, thus maintaining stability in the population.

14 In *Batesian mimicry*, a palatable species mimics in appearance some warningly colored unpalatable species; in *Müllerian mimicry* two or more unpalatable species mimic each other in appearance.

15 Animals with a *cryptic appearance* tend to blend into their surroundings so as to be undetectable; those with *warning coloration* have evolved colors and patterns that contrast boldly with their environment. The latter are often distasteful to predators.

16 *Symbiosis* means "living together." *Mutualism* (in which both species benefit from the association), *commensalism* (in which one species benefits and the other neither benefits nor is harmed), and *parasitism* (in which one species benefits at the expense of the other) are types of symbiosis.

17 A *parasite* passes much of its life on or in the body of the living host from which it obtains its food, whereas a *predator* eats its prey very quickly and goes its way. *Parasites* generally do not kill their hosts; *parasitoids* do.

Testing knowledge and understanding

18	b	23	c	28	d	33	c
19	c	24	c	29	b	34	b
20	c	25	d	30	c	35	b
21	a	26	b	31	d	36	d
22	c	27	d	32	a		

For further thought

1 16 percent are heterozygous and resistant to malaria

2 198 are carriers

3 $I^s = 0.43$; $I^c = 0.57$. Population is not in equilibrium. $\chi^2 = 56.9$ which at 2 degrees of freedom is highly significant. The deviation from the expected genotypes is due to something other than chance. (Expected distribution = 180 straight hair, 330 curly, and 490 wavy.)

4 $F^B = 0.45$; $F^W = 0.55$. Population is in equilibrium. Observed distribution is close to the expected $\chi^2 = 0.64$ which is not significant at 2 degrees of freedom; a deviation this large is due to chance.

Chapter 34

EVOLUTION: SPECIATION AND PHYLOGENY

A GENERAL GUIDE TO THE READING

This chapter applies the evolutionary concepts presented in Chapter 33 to the process of speciation—the process by which an ancestral species may split, giving rise to two or more different descendant species. This process has been exceedingly frequent, and is believed responsible for the immense diversity of life today. This chapter is a very important one, and as you read it you will want to concentrate on the following.

1 Species and speciation. You will want first to learn the definition of "species" (p. 883) and then to concentrate on how speciation occurs (pp. 887–94). This material is crucially important; when you finish the chapter, you should have a thorough understanding of divergent speciation.

2 Sympatric speciation. You will want to understand what is meant by this term, and how sympatric speciation differs from divergent speciation.

3 Adaptive radiation. Darwin's finches are used to illustrate the process of adaptive radiation and to provide a model by which the evolution of the enormous diversity of species living today can be understood.

4 Competition. The relative importance of competition and natural catastrophies to the evolution of populations is described on pages 903–5. This is an interesting and provocative section.

5 Gradualism vs. punctuated equilibrium. The current controversy over the mechanism of evolution is described on pages 905–7. Figure 32.22 (p. 905) provides a good example of the two hypotheses in action. When you read this section note that the so-called "instantaneous" speciation is hardly that—speciation may take thousands of years.

6 Phylogeny. You will want to understand some of the problems faced by the systematist (pp. 908–15) and to learn the classification

hierarchy (p. 918). A help in remembering the order of phylogenetic categories is the mnemonic sentence "*K*indly *p*rofessors *c*annot *o*ften *f*ail *g*ood *s*tudents," in which the first letter of each word is the first letter of a category.

KEY CONCEPTS

1 In the modern view, a species is a genetically distinctive group of natural populations that share a common gene pool and that are reproductively isolated from all other such groups. (pp. 883–87)

2 Divergent speciation usually begins when two sets of populations are separated geographically by external barriers; as the two population systems evolve independently, they accumulate differences that lead in time to the development of intrinsic isolating mechanisms. (pp. 887–94)

3 Sympatric speciation does not involve geographical isolation. Species that arise by polyploidy are genetically distinctive and are reproductively isolated from the parent species. Habitat preference, host-specificity, and certain behavioral isolating mechanisms can also produce reproductive isolation leading to speciation. (pp. 894–97)

4 Divergent evolution—the evolutionary splitting of lineages into many separate lineages—has occurred very frequently; it has produced the immense variety of living things. (pp. 897–903)

5 Although competition is believed to be the predominant force leading to speciation over time, the catastrophic effects of rare crises may be important in the evolution of certain populations. (pp. 903–5)

6 Two opposing hypotheses of evolutionary change have been proposed: gradualism, which states that speciation occurs by the gradual accumulation of small changes over long periods of time, and punctuated equilibrium, which states that most evolution is marked by long periods of equilibrium followed by a crisis or some other event that leads to rapid changes in the population, resulting in speciation. (pp. 905–7)

7 The phylogeny of a group of organisms is the evolutionary history of that group. Biologists attempt to reconstruct evolutionary history by using a systematic approach to discover the relationship among species and then fitting these relationships into an orderly scheme. (pp. 910–20)

OBJECTIVES

After studying this chapter and reflecting on it, you should be able to carry out the following objectives.

1 Discuss intraspecific geographic variation; in doing so, distinguish among the following: population, deme, race or subspecies, cline. (pp. 882–86)

2 Give a definition of a species and, using a diagram such as Figure 34.7 (p. 890), explain the geographic-isolation model of divergent speciation. In doing so, be sure to take into account the roles of mutation, recombination, natural selection, and the gene pool; and distinguish between extrinsic and intrinsic isolating mechanisms. Then briefly describe five intrinsic isolating mechanisms. (pp. 887–94)

3 Describe speciation by polyploidy and differentiate between autopolyploidy and allopolyploidy. Indicate why plant breeders frequently take advantage of this process. (pp. 894–95)

4 Differentiate between sympatric speciation and divergent speciation. (pp. 888–89, 894–97)

5 Explain what is meant by adaptive radiation, and discuss the evidence for this phenomenon, using the Galápagos finches as an example. (pp. 897–903)

6 Contrast the role of competition and natural catastrophes on the evolution of natural populations, explaining what is meant by an evolutionary bottleneck. Indicate which of the above is believed to be the predominant force leading to speciation over time. (pp. 903–5)

7 Compare the hypothesis of punctuated equilibrium with that of gradualism, and give an example supporting each hypothesis. Finally, using Figure 34.22 (p. 905), explain how the giraffe evolved from the pre-okapi, using the two different hypotheses. (pp. 905–7)

8 Describe the sources and types of information used by systematists to determine phylogenetic relationships. Include information on phenetics, cladistics, and molecular taxonomy. (pp. 910–20)

9 Explain why the modern definition of a species is difficult to apply to sexual organisms, fossil organisms, populations at an intermediate stage of divergence, and allopatric populations. (pp. 907–9)

10 Distinguish among divergent, parallel, and convergent evolution, and give examples of

each. Indicate why convergence poses a problem for the systematist. Explain the difference between homologous and analogous similarities. (p. 912)

11 List in order the categories of the classification hierarchy used today. Then explain how a species is named. (pp. 918–20)

KEY TERMS

The following terms are important in this chapter; you should become familiar with them.

population (p. 882)
deme (p. 882)
species (p. 883)
intraspecific variation (p. 884)
cline (p. 886)
subspecies (p. 886)
races (p. 886)
speciation (p. 887)
founder effect (p. 888)
intrinsic reproductive isolation (p. 889)
allopatric (p. 890)
sympatric (p. 890)
ecogeographic isolation (p. 890)
habitat isolation (p. 890)
seasonal isolation (p. 890)
behavioral isolation (p. 891)
sibling species (p. 891)
mechanical isolation (p. 891)
gametic isolation (p. 892)
developmental isolation (p. 893)
hybrid inviability (p. 893)
hybrid sterility (p. 893)
selective hybrid elimination (p. 893)
character displacement (p. 894)
polyploidy (p. 894)
sympatric speciation (p. 894)
autopolyploidy (p. 894)
allopolyploidy (p. 895)
nonchromosomal sympatric speciation (p. 896)
evolutionary bottleneck (p. 904)
punctuated equilibrium (p. 905)
stasis (p. 905)
gradualism (p. 905)
mosaic evolution (p. 906)
convergence (p. 912)
homologous (p. 912)
analogous (p. 912)
phenetics (p. 914)
cladistics (p. 914)
shared derived characters (p. 914)
molecular taxonomy (p. 915)

SUMMARY

Species and speciation A *population* of sexually reproducing organisms is a group of individuals that share a common gene pool. A *deme* is a small local population; demes are usually temporary units of population that intergrade with other such units.

The existence of discrete clusters of living things that can be called species has long been recognized, but the concept of what a species is has changed many times. In the modern view, a *species* is a genetically distinctive group of natural populations (demes) that share a common gene pool and that are *reproductively isolated* from all other such groups.

The vast majority of plant and animal species show geographic variation, most of which probably reflects differences in selection pressures resulting from local environmental conditions. Gradual variation correlated with geography is called a *cline*. An abrupt shift in a character in a geographically variable species may occur; the populations involved may be designated as *races* or *subspecies*. These are groups of natural populations within a species that differ genetically and that are partly isolated from each other reproductively because they have different ranges.

Divergent speciation is the process by which one ancestral species gives rise to two or more descendant species, which grow increasingly unlike as they evolve. The initiating factor is usually geographic separation; if a population is divided by some physical or ecological barrier the separated (*allopatric*) populations will no longer be able to exchange genes. In time, the populations will evolve in different directions because they start out with different gene frequencies, they experience different mutations, and they are exposed to different environmental selection pressures. Eventually, the populations may become genetically so different that they develop *intrinsic isolating mechanisms*—biological characteristics that prevent effective interbreeding should they again become *sympatric*.

Many intrinsic isolating mechanisms act by preventing effective mating. The two populations may be so specialized ecologically that they cannot become sympatric, or they may occupy different habitats; they may breed at different times, be behaviorally isolated, or be physically unable to mate. Even if mating occurs, fertilization may not take place or the embryo may not survive. If hybrids are born, they may be inviable or sterile.

The members of two closely related populations may be able to breed and produce fertile offspring. If the hybrids are as well adapted as the parents, the populations will fuse back together. If the hybrids are less well adapted than the parents, there will be strong selection pressure favoring forms of intrinsic isolation that prevent wrong matings, and the two populations will diverge more rapidly (*character displacement*) until mating between them is impossible.

Among plants there is another process of speciation—*sympatric speciation*, which does not involve geographic isolation. One type is *polyploidy*. Polyploidy is the occurrence in cell nuclei of more than two complete sets of chromosomes. In autopolyploidy there is a sudden multiplication of the number of chromosomes, usually as a result of nondisjunction in meiosis. In allopolyploidy there is a multiplication of the number of chromosomes in a hybrid between two species.

Sympatric speciation can also occur without polyploidy or other major chromosomal rearrangement. Reproductive isolation can be achieved by means other than geographic isolation; such factors as habitat preference, host-specificity, and sexual imprinting may create the reproductive isolation necessary for speciation.

Divergent evolution—the evolutionary splitting of species into many separate descendant species—results in an *adaptive radiation*. The rate of evolutionary divergence is not always constant; when conditions change rapidly and organisms have new evolutionary opportunities available to them, they may undergo a rapid evolutionary burst. Adaptive radiation like that of Darwin's finches on the Galápagos Islands helps account for the tremendous diversity among living things on the earth today.

Competition, leading to *character displacement*, is thought to be the predominant force in the speciation process. According to Gause's competitive exclusion principle, no two species occupying the same niche can long coexist. However, the assumption that competition is the major mode of speciation is being challenged by the view that chance crises may have been important in the evolution of certain populations. Crises can be caused by any environmental factor that severely affects a population so as to cause major changes in allelic frequencies (called an *evolutionary bottleneck*). As the population passes through the evolutionary bottleneck caused by extraordinary environmental conditions, one character or another may gain ascendancy and rapid evolutionary change results.

Two opposing hypotheses of evolutionary change have been proposed: *gradualism*, which states that speciation occurs by the gradual accumulation of small changes over long periods of time, and *punctuated equilibrium*, which states that most evolution is marked by long periods of equilibrium (*statis*) followed by a crisis or some other event that leads to rapid change in the population, resulting in speciation.

The modern definition of a species is hard to apply to asexual organisms, fossil organisms, populations in intermediate stages of evolution, and allopatric populations.

The concept of phylogeny One of the tasks of systematic biology is to discover the relationships among species and to trace the ancestors from which they are descended. There are four major approaches—classical evolutionary taxonomy, phenetics, cladistics, and molecular taxonomy.

Classical taxonomy uses information on the morphology of the adult and embryo, and combines this with information from the fossil record and from life histories to formulate phylogenetic hypotheses. Comparative physiology, comparative behavior, and comparative ecology have also supplied valuable information. However, reconstructing the evolutionary history, or *phylogeny*, of any group of organisms entails considerable speculation. *Phenetics* (numerical taxonomy) and *cladistics* (which focuses on *shared derived characters*) are new approaches that attempt to use more objective methods in classification. More recently *molecular taxonomy* has been used to compare the DNA and proteins of different species and to classify species on the degree of similarity and differences.

One problem in interpreting the similarities among organisms is *convergence*. Organisms that are not closely related may come to resemble each other because they occupy similar habitats and adopt similar environmental roles. Systematists must try to determine whether the similarities are *homologous* (inherited from a common ancestor) or merely *analogous* (similar in function and often in superficial structure) before speculating about the degree of relationship between the organisms.

It is also necessary in reconstructing phylogeny to consider development in time. Is a characteristic *primitive* (older and more like the ancestral condition) or *advanced* (less like the ancestral condition)? These are often confused with the terms *specialized* (adapted to a narrow way of life) and *generalized* (broadly adapted). Generalized characters are likely to be more primitive, and specialized ones more advanced.

Classification The classification system used today attempts to encode information about the organism's evolutionary history. It uses a hierarchy of categories in which each category (taxon) is a collective unit containing one or more groups from the next-lower level. The principal categories are: kingdom, phylum, class, order, family, genus, and species. In the modern system of nomenclature each species is given a name consisting of two Latin words; the first names the genus to which the species belongs, the second the individual species designation.

QUESTIONS

Testing recall

Determine whether each of the following statements is true or false; if false, correct the statement.

1 By definition, no hybridization between two closely related species is possible. (pp. 883, 893)

2 When two or more closely related species are sympatric, differences in their courtship displays often play an important role in preventing hybridization between them. (p. 891)

3 Character displacement is more likely to be observed in two related bird species from different islands than in two such species from the same island. (p. 894)

4 Two species can never produce viable hybrids. (p. 893)

5 Two races (or subspecies) of the same species cannot be sympatric for long without fusing. (p. 890)

6 Speciation by polyploidy is much more common in animals than in plants. (p. 894)

7 The catastrophic effects of severe environmental crises may be an important force leading to speciation in certain populations. (p. 905)

8 The hypothesis of adaptive radiation states that most evolution is marked by long periods of stasis followed by short bursts of rapid change. (p. 905)

9 The wings of birds and the wings of butterflies are examples of convergently evolved structures. (p. 912)

10 The modern system of scientific naming of plants and animals dates from the work of the Swedish naturalist Carolus Linnaeus. (p. 919)

The following important terms are sometimes confused by students. Distinguish between the terms within each of the following groups.

11 species − subspecies

12 genetic drift − founder effect

13 divergent speciation − sympatric speciation

14 intrinsic isolating mechanism − extrinsic isolating mechanism

15 sympatry − allopatry

16 autopolyploidy − allopolyploidy

17 homologous − analogous

18 convergence − divergence

Testing knowledge and understanding

Choose the one best answer.

19 Which of the following pairs represents two species?
a German shepherd, poodle
b honeybee worker, queen honeybee
c monarch butterfly, viceroy butterfly
d housefly larva, housefly adult
e Eskimo, Peruvian Indian   (pp. 883, 875)

20 In North America there is a progressive decrease in the size of song sparrows from the north to the south of their range. This is an example of
a a deme.
b interspecific variation.
c the founder effect.
d behavioral isolation.
e a cline.   (pp. 882−86)

21 Population A occurs only in Africa. The closely related population B occurs only in South America. Only a few minor structural differences between A and B can be found. Population C, which is closely related to both A and B and is sympatric with B, differs more noticeably from B in several characters. In an attempt to determine whether A and B belong to the same or to different species, individuals from the two populations are mated in the laboratory; they produce viable offspring. Populations A and B must be regarded as
a members of the same species and subspecies.

b members of the same species but of different subspecies.
c members of different species.
d uncertainly related, because insufficient evidence is available to settle the matter.
(p. 883)

22 Gradual variation of a characteristic correlated with geography is an example of

a a cline.
b polymorphism.
c phenotypic variation.
d genetic drift.
e gene flow. (p. 886)

23 The first step in speciation in animals is ordinarily

a geographical separation of populations.
b evolution of hybrid sterility.
c evolution of intrinsic isolating mechanisms.
d character displacement. (p. 888)

24 Which one of the following isolating mechanisms is generally considered most important in *initiating* the process of divergent speciation?

a geographic isolation
b behavioral isolation
c developmental isolation
d seasonal isolation
e hybrid sterility (p. 888)

25 You carefully study populations of two very similar meadow mice, one from the Northeast and one from Texas. You want to know whether the populations belong to the same species or to two different species. You could *most* confidently decide this if you could

a show that the ranges of the two mice overlap without hybridization occurring.
b bring the two types of mice into the laboratory and show that they can cross and produce viable offspring.
c demonstrate that the natural ranges of the two types of mice are entirely allopatric.
d show that there are statistically significant structural differences between the two types.
e show that the northeastern mice live in wetter habitats than the Texan mice.
(pp. 890–93)

26 A biologist studies a large field in which hundreds of plants, all belonging to a single species, are growing. Plants of this species occur nowhere else in the state, and no other species of the same genus are known to science. The biologist demonstrates that all the plants in the field are probably interfertile.

Five years later the biologist again visits the field and notices a few plants that look slightly different from the others. Upon conducting cross-pollination experiments, he discovers that these unusual plants can cross with one another but not with the normal plants; they appear to be members of a second species closely related to the original one. The most likely explanation is that

a a new mutant gene arose in some members of the original population, causing it to split into two species.
b genetic drift in this relatively small population led to speciation.
c habitat differences in two different parts of the field led to divergent adaptation that resulted in speciation.
d the unusual plants are polyploids.
e the original species first evolved polymorphism, and then each form became a separate species. (pp. 894–95)

27 One plant species has a diploid number of 8. A second plant species has a diploid number of 10. These two plant species hybridize and a single fertile hybrid is produced. This new species produces gametes which probably have how many chromosomes?

a 4   d 18
b 5   e 36
c 9   (pp. 894–95)

28 All of the following are intrinsic isolating mechanisms except one which is an extrinsic isolating mechanism. Which one is the extrinsic isolating mechanism?

a developmental isolation
b geographical isolation
c habitat isolation
d behavioral isolation
e mechanical isolation (pp. 888–93)

29 All of the following are examples of intrinsic reproductive isolation *except*

a two populations separated by a mountain range
b two populations with different habitats that keep potential mates from each population separate
c populations that breed during different seasons of the year
d organisms that can mate but fail to produce viable zygotes
e two populations with very different courtship displays (pp. 890–93)

30 New plant species may arise in one generation as a result of

a seasonal isolation.
b behavioral isolation.

    c   polyploidy.
    d   divergent speciation.
    e   the founder effect.   (p. 894)

31 The evolution of the Galápagos finches from a common ancestral form is thought to have occurred by the mechanism of

    a   convergence.
    b   coevolution.
    c   parallel evolution.
    d   adaptive radiation.
    e   sympatric speciation.   (pp. 897–903)

32 All of the following factors may be involved in sympatric speciation *except*

    a   polyploidy
    b   host-specificity
    c   sexual imprinting
    d   habitat isolation
    e   geological isolation   (pp. 894–97)

Questions 33–34 refer to the following evolutionary tree. Each letter in the tree represents a different species; C, D, E, and F live on different continents.

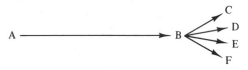

33 The evolution of species C, D, E, and F from B looks like an example of

    a   adaptive radiation.
    b   parallel evolution.
    c   convergent evolution.
    d   polyploid speciation.   (p. 897)

34 Species D and F are

    a   sympatric.
    b   allopatric.   (p. 890)

35 Natural catastrophes that cause major changes in allelic frequencies in a population are referred to as a(n)

    a   evolutionary bottleneck.
    b   geographical isolation.
    c   extrinsic isolating mechanism.
    d   intrinsic isolating mechanism.
    e   adaptive radiation.   (p. 904)

36 The term "punctuated equilibrium" refers to the concept that

    a   there have been long periods in the history of the earth when no new species evolved.
    b   species tend to be static for long periods and then sudden "speciation events" occur.
    c   the number of species on the earth remains at equilibrium.

    d   evolution requires geographic isolation.
    e   small species evolve more rapidly than those with larger body size.   (pp. 904–7)

37 The Australian mole is actually a marsupial rather than a placental mammal like the North American or European mole. The two animals are similar in appearance because

    a   they evolved from a mole-like common ancestor.
    b   there are practically no placental mammals in Australia.
    c   the selection pressures on both were similar.
    d   they have undergone a long period of coevolution.
    e   marsupials and placental mammals are closely related.   (p. 912)

38 Convergent evolution between two species would be most likely to occur as a result of

    a   a series of identical mutations occurring in both species.
    b   hybridization between the two species.
    c   interbreeding by both species with members of a third species.
    d   exposure of both species to similar selection pressures.
    e   genetic drift between the species.   (p. 912)

39 In their similarities, the wing of a butterfly and the wing of a bat exemplify

    a   parallel evolution.
    b   divergent evolution.
    c   convergent evolution.
    d   adaptive radiation.
    e   coevolution.   (p. 912)

40 Which one of the following is a correct hierarchical sequence of taxonomic groups?

    a   class − order − family − genus
    b   order − class − family − genus − species
    c   family − order − species − genus
    d   class − phylum − order − family   (p. 918)

41 The scientific name of the human species is properly written

    a   homo sapiens.   *d*   *Homo Sapiens.*
    b   *homo sapiens.*   *e*   *Homo sapiens.*
    c   Homo Sapiens.   (p. 921)

For further thought

1 The apple maggot, *Rhagoletis pomonella*, parasitizes apple trees. The adult flies generally mate on the fruit tree the maggots parasitize and the female lays her eggs there. About fifteen years ago, some apple maggots were found on cherry trees in an area where cherry orchards and apple orchards were

close to one another. Today the two populations are genetically distinctive and cherry-preferring maggots emerge in early summer, when the cherries ripen, while the apple-preferring maggots emerge in late summer when the apples ripen. Hybridization between the two populations is possible. Can these two populations be considered as two distinct species? What type of speciation is going on here? Justify your answers to both these questions.

2 Discuss the way in which one species may give rise to two or more new species. Be sure to explain the role of mutation, recombination, gene pool, extrinsic isolating mechanisms, and intrinsic isolating mechanisms, giving specific examples. Discuss the role that natural selection plays in this process.

ANSWERS

Testing recall

1 false—hybridization may occur, but the hybrids are often invisible or sterile
2 true
3 false—more likely in species from the same island
4 false—sometimes; the hybrids may be sterile
5 true
6 false—more common in plants
7 true
8 false—punctuated equilibrium
9 true
10 true

11 A *species* is a genetically distinctive group of natural populations that share a common gene pool and are reproductively isolated from all other such groups. *Subspecies* are groups of natural populations within a species that differ genetically and are partly isolated from each other genetically because they have different ranges.

12 *Genetic drift* is evolutionary change that is caused by chance. *Founder effect* is a type of

genetic drift; it occurs when a new population is formed by a small number of individuals whose allelic frequencies differ from those of the parental population.

13 *Divergent speciation* requires geographic isolation for speciation; *sympatric speciation* does not require geographic isolation but relies on some other mechanism such as polyploidy or some intrinsic isolating mechanism to achieve reproductive isolation for speciation.

14 An *extrinsic isolating mechanism* is a physical or geographic barrier that divides a population and prevents interbreeding between the two segments. An *intrinsic isolating mechanism* is any biological characteristic that prevents two populations from interbreeding effectively.

15 *Sympatric* means occupying the same range, whereas *allopatric* means occupying different ranges.

16 *Autopolyploidy* involves a sudden multiplication of the number of chromosomes in a plant, whereas *allopolyploidy* involves a multiplication of the number of chromosomes in a hybrid between two species.

17 When characters in different organisms are similar because they are inherited from a common ancestor, they are *homologous*; when characters in different organisms are similar in function but of different evolutionary origin, they are *analogous*.

18 *Convergence* occurs when organisms that are not closely related become more similar because of independent adaptations to similar environmental situations; *divergence* occurs when organisms become more dissimilar over time.

Testing knowledge and understanding

19	c	25	a	31	d	37	c
20	e	26	d	32	e	38	d
21	d	27	c	33	a	39	c
22	a	28	b	34	b	40	a
23	a	29	a	35	a	41	e
24	a	30	c	36	b		

Chapter 35

ECOLOGY

A GENERAL GUIDE TO THE READING

This chapter and the next examine the interactions among organisms, and their relationships to the chemical, biological, and physical environment. This chapter will explore the interactions between individuals within a population, between species within a community, and between the community and the environment. Ecology differs from other areas of biology in that the emphasis is on these interactions and relationships rather than on the individual organism or part of an organism. Ecologists are fond of saying, "You cannot do just one thing." Elimination of one species can cause a ripple effect through the whole ecosystem. As you read Chapter 35 in your text, try to appreciate the complexity of the interrelationships among organisms. You will want to focus on the following.

1   Basic terminology. Terms used frequently are defined on pages 421–22; you need to learn them before continuing your reading.

2   Population growth. You will want to be familiar with the various types of population growth curves, as shown in Figures 35.4 (p. 925), 35.5 (p. 926), and 35.7 (p. 927).

3   Population regulation. In the section on this topic (pp. 931–40), the differences between r-selected species and K-selected species are important, as are the various types of limitations on population growth. Pay particular attention, too, to the concept of the niche.

4   Dominance, diversity, and community stability. The concept of stability in an ecosystem is well described on pages 940–43.

5   Ecological succession. In reading this material (pp. 943–50), pay particular attention to the causes of succession and the trends in succession. A helpful summary of the trends is presented on pages 948–50.

KEY CONCEPTS

1 Ecology is the study of interactions between organisms and their environment. The various ecosystems are linked by biological, chemical, and physical processes. (pp. 921–22)

2 Many natural populations show an initial period of exponential growth at low densities, followed by a deceleration in growth at higher densities, and by an eventual leveling off as the density approaches the carrying capacity of the environment. Such a population is primarily limited by influences that provide feedback control in that they depend at least partially on the density of the population itself. (pp. 926–27)

3 In some populations density grows exponentially, but then falls precipitously before reaching the carrying capacity of the environment, as a result of density-independent limiting factors. (pp. 924–25, 927–28)

4 In natural populations, growth is subject to a number of limitations; the regulation of population density can occur in different ways in different populations. (pp. 927–40)

5 The niche is the function and position of an organism in the ecosystem. The more the niches of two different species overlap, the more intense the competition between them will be. (pp. 934–36)

6 A species does not exist as an isolated entity; it is always interacting in a variety of ways with other species in the community to which it belongs. (pp. 940–43)

7 The biotic community formed by the various species in an area can be considered a unit of life, with its own structure and functional interrelationships. (pp. 942–43)

8 The trend of most ecological successions is toward a more complex and stable ecosystem, in which less energy is wasted and hence a greater biomass can be supported without further increase in the supply of energy. (pp. 943–50)

9 Most ecological successions eventually reach a climax stage that is more stable than the stages preceding it. The more complex organization of the climax community, and its larger organic structure and more balanced metabolism, enable it to buffer its own physical environment to such an extent that it can perpetuate itself as long as the environment remains essentially the same. (pp. 950–55)

OBJECTIVES

After studying this chapter and reflecting on it, you should be able to carry out the following objectives.

1 Define population, community, ecosystem, and biosphere, indicating how each is related to the others. (pp. 921–22)

2 Give two methods biologists use to estimate population densities and distinguish between uniform, clumped, and random distributions, and indicate the conditions under which each occurs, and which one is the most common. (pp. 922–24)

3 Draw an exponential growth curve, and write the equation for the curve, defining all the terms used in the equation. (pp. 924–25)

4 Draw a logistic curve, and label the carrying capacity, the inflection point, the portion of the curve showing an accelerating rate of population growth, and the portion showing a decelerating rate. Compare this curve with the exponential growth curve. Then explain what is meant by zero population growth, and describe how this condition is reached. (pp. 925–26)

5 Draw an exponential growth curve with sudden crash, and list factors that might cause the crash. Distinguish between density-dependent and density-independent limitations on population growth, and indicate which type of limitation regulates growth in populations showing respectively an exponential growth curve with sudden crash and a logistic growth curve. (pp. 927–28)

6 On a single graph draw type I, type II, and type III survivorship curves; then state which curve is most common in natural populations. (p. 928)

7 Distinguish between an *r*-selected species and a *K*-selected species with respect to body size, life-span, number of offspring, relative time of reproduction (earlier or later in life), type of survivorship curve, type of environment (stable or unstable), and type of growth curve (S-shaped or boom-and-bust). (pp. 931–32)

8 Using examples, discuss the ways in which parasitism, predation, intraspecific competition, emigration, mutualism, and physiological and behavioral mechanisms can act as density-dependent limitations on population growth. Explain, using an example, how destroying the balance between predator and prey in a community can upset the ecology of an area. (pp. 932–40)

9 Carefully define the concept of ecological niche, and explain its significance with respect to the competitive exclusion principle and the principle of limiting similarity. Specify the three possible results of intense interspecific competition. (pp. 934–36)

10 Discuss, using an example, the relationship between species diversity and complexity and community stability. Describe the effect of human intervention in biological communities. (pp. 940–43)

11 Describe the process of ecological succession, indicating why the species in a given area change over time. Distinguish between primary and secondary successions, and give an example of each. Also, summarize the trends seen in many successions, and explain what is meant by a climax community. (pp. 943–51)

KEY TERMS

The following terms are important in this chapter; you should become familiar with them.

physiological ecology (p. 921)
populations (p. 921)
communities (p. 921)
ecosystems (p. 921)
population density (p. 922)
uniform distribution (p. 923)
random distribution (p. 923)
clumped distribution (p. 923)
exponential growth curve (p. 924)
intrinsic rate of increase (p. 925)
logistic growth curve (p. 926)
density-dependent limitation (p. 926)
zero population growth (p. 927)
maximum sustained yield (p. 927)
density-independent limitation (p. 928)
mortality and survivorship (p. 928)
r-selected species (p. 931)
K-selected species (p. 931)
intraspecific competition (p. 933)
interspecific competition (p. 934)
niche (p. 934)
principle of limiting similarity (p. 935)
microhabitats (p. 936)
emigration (p. 937)
primary succession (p. 944)
secondary succession (p. 944)
climax community (p. 950)

SUMMARY

Ecology is the study of the interaction between organisms and their environment. There are three higher levels of organization: *populations*, groups of individuals belonging to the same species; *communities*, units composed of all the populations living in a given area; and *ecosystems*, the sum total of the communities and their physical environment considered together. The various ecosystems are linked to one another by biological, chemical, and physical processes. The entire earth is a true ecosystem; this global ecosystem is called the *biosphere*.

Populations as units of structure and function
The distribution of individuals within an area may be uniform, random, or clumped. *Uniform* and *random* distributions are relatively rare and occur only where environmental conditions are fairly uniform. A uniform distribution results from intense competition or antagonism between individuals; a random distribution occurs when there is no competition, antagonism, or tendency to aggregate. *Clumping* is the most common distribution because environmental conditions are seldom uniform, reproductive patterns favor clumping, and animal behavior patterns often lead to congregation. The optimum density for population growth and survival is often an intermediate one; undercrowding can be as harmful as overcrowding.

All organisms have the potential for explosive growth; under ideal conditions their growth curve would be *exponential*. The equation for an exponential growth curve is $I = rN$ where I is the rate of increase of the population, r is the *intrinsic rate of increase* of the population (average birth rate − average death rate) and N the number of individuals in the population at a given moment. If r is positive, the population will grow at an ever accelerating rate.

The exponential growth of many real populations begins to level off as the density approaches the *carrying capacity* (K) of the environment. Such a growth curve is called an S-shaped or *logistic growth curve* and results from a changing ratio between births and deaths. After the curve has leveled off, births and deaths are in balance and the population has *zero population growth*. This occurs because environmental limitations become increasingly effective in slowing population growth as the population density rises. When the density approaches the carrying capacity, the limitation becomes severe. A *density-dependent limi-*

tation, $(K - N)/K$, is one whose density is determined by the density of the very population it helps limit. The equation for the logistic growth curve is

$$I = r\left(\frac{K - N}{K}\right)N$$

At low population densities, population growth is exponential; the rate of increase reaches its maximum at the inflection point of the curve. When the population exceeds the carrying capacity, I becomes negative, and the population decreases.

The populations of many small short-lived animals, or those living in variable environments, go through a period of exponential growth, followed by a sudden crash (boom-and-bust curve). The crash occurs before the populations reach the carrying capacity; it is due to a *density-independent* limitation such as weather or other physical environmental factors. The operation of such a limitation does not depend on the density of the organisms.

In addition to the birth rate and death rate, the potential life-span, the average life expectancy, and the average age of reproduction are important determinants of the makeup of a population. Determining the mortality rates for the various age groups in the population gives a survivorship curve. A type I curve is one in which all the organisms live to old age and die quickly; a type II curve shows a constant mortality rate at all ages; and a type III curve is typical of populations where the mortality among the young is very high, but those who survive the early stages tend to live for a long time. In nature, high mortality among the young is the rule.

Population regulation   The regulation of population density in organisms with boom-and-bust curves is primarily due to the work of density-independent limitations. The maximum density achieved before the decline is primarily a function of the organisms' high reproductive rates. Since these organisms have evolved high intrinsic rates of increase (i.e. high r), they are called *r-selected species*.

Organisms with S-shaped growth curves, whose population limitation is primarily density-dependent, are called *K-selected species*; the maximum density for population stability is determined largely by the environment's carrying capacity. The fitness of these organisms depends on their ability to exploit the limited environmental resources efficiently.

Both parasitism and predation usually influence the prey (or host) species in a density-dependent manner. In general, the density of predators or parasites fluctuates in direct proportion to the changes in the density of the prey. Predator-prey relationships, like long-established host-parasite relationships, tend to evolve toward a dynamic balance in which predation is a regulatory influence in the life of the prey species, but not a real threat to its survival.

Intraspecific competition is one of the chief density-dependent limiting factors. As population density rises, competition for limited environmental resources becomes increasingly intense and acts as a brake on population growth.

Interspecific competition can also act as a density-dependent limitation. The more the niches of the species overlap, the more intense the interspecific competition. *Niche* refers to the functional role and position of an organism in an ecosystem. Every aspect of an organism's existence helps define that organism's niche. Two species cannot for long simultaneously occupy the same niche in the same place.

The more similar two niches are, the more likely it is that both species will be competing for a limited resource. According to the principle of limiting similarity, there is a limit to the amount of niche overlap compatible with coexistence. Extinction of one of the competing species, range restriction, character displacement, or a combination of the last two, will usually be the outcome of intense interspecific competition.

If two species live in a mutualistic association, the density of each will affect the density of the other. Thus mutualism can also act as a density-dependent limitation. In some animals, crowding induces physiological and behavioral changes that result in increased emigration from the crowded region. Dense populations often experience disease epidemics. Numerous laboratory experiments with mice indicate that crowding induces hormonal changes which reduce the reproductive rate; the importance of this mechanism in nature is unclear.

The biotic community   The species comprising a community interact with each other and with the physical environment. The biotic community they form can be considered a unit of life, with its own characteristic structure and functional interrelationships. Species diversity and complexity of interactions influence community stability. A simple community responds violently to a disturbance but often recovers quickly. The complex community responds less dramatically but may continue to show effects over a longer period. A diverse physical environment seems to favor community stability. Monoculture, pollution, and other human activities have increased community instability by decreasing species diversity and structural complexity.

Ecological succession is an orderly process of community change involving the replacement, over time, of the dominant species within a given area by other species. Succession results in part from the modification of the habitat by the organisms themselves or by physiographic changes, and in part from differences in dispersal and growth patterns among species. The species that predominate in the early stages tend to be rapid growers that are easily dispersed.

Successions in different places and at different times are not identical. The sequence of changes in *primary successions* (those occurring in newly formed habitats) is longer and slower than in *secondary successions* (those occurring in areas where previous communities were destroyed). The trend of most successions is toward a more complex and longer-lasting ecosystem in which less energy is wasted and hence a greater biomass can be supported without further increase in the supply of energy.

If no disruptive factor interferes, most successions reach a stage that is more stable than those that preceded it—the *climax community*. It will persist as long as the climate, physiography, and other environmental factors remain the same. Earlier ecologists supported the monoclimax hypothesis, which held that all succession in a given climatic region will converge to the same climax type. Modern ecologists, however, argue that since each species is distributed according to its own particular biological potentialities, climax has meaning only in relation to the individual site and its environmental conditions.

QUESTIONS

Testing recall

Match each of the descriptive phrases listed below with the term it best defines. Use each answer only once.

a abiotic
b biosphere
c climax community
d community
e competition
f ecological niche
g ecosystem
h K
i mortality rate
j population
k r
l succession

1 global ecosystem (p. 922)
2 all organisms in a given place at a given time (p. 921)
3 functional role of an organism in its community (p. 934)
4 progressive change in the plant and animal life of an area (p. 943)
5 intrinsic growth rate of a population (p. 925)
6 a major determiner of population density (p. 925)
7 carrying capacity of the environment (p. 926)
8 group of individuals belonging to the same species (p. 921)
9 stable stage of succession (p. 950)
10 density-dependent limitation on population growth (p. 933)
11 sum total of the physical features and organisms in a given area (p. 921)

Decide whether the following statements are true or false, and correct the false statements.

12 Clumped spacing of members of a population is more common than uniform or random spacing. (p. 923)
13 A population with more offspring per generation will always have a faster growth rate (r) than a similar population with fewer offspring per generation. (p. 952)
14 The housefly, which has a short life-span and produces a large number of eggs, could be considered a K-selected species. (p. 931)
15 Predation usually acts as a density-independent limiting factor on populations. (p. 932)
16 The term niche designates the part of an organism's habitat that includes physical environmental factors, such as temperature, humidity, and soil pH. (p. 934)
17 Interspecific competition between two closely related species living in the same area may lead to the extinction of one of the competing species. (p. 935)
18 The species composition of the community in a given area changes markedly during the course of ecological succession. (p. 948)
19 Energy utilization is usually less efficient in a climax community than in a pioneer community. (pp. 943–44)

20 All succession in a given large climatic area will eventually converge to the same climax type; there is only one type of climax community for a given region. (pp. 950–51)

Testing knowledge and understanding

Questions 21–25 refer to the following situation.

A New York State farmer stocked his farm pond with 580 bluegill fish fingerlings. Bluegills usually reproduce first as yearlings and regularly thereafter. The farmer recorded the number of fish each year for the next ten years. He obtained the following data:

Year	Number of fish
Stocked	580
1	600
2	750
3	1200
4	1400
5	1460
6	1440
7	1450
8	1460
9	1450
10	1460

21 Plot these data on the graph.

22 What kind of growth curve is seen? (pp. 926–27)

23 Would you consider the bluegill a *K*-selected species or an *r*-selected species? (p. 931)

24 Mark the point on the curve where the rate of increase is greatest. (p. 926)

25 What factors might be involved in slowing the population growth from the fourth year on? (pp. 926–27)

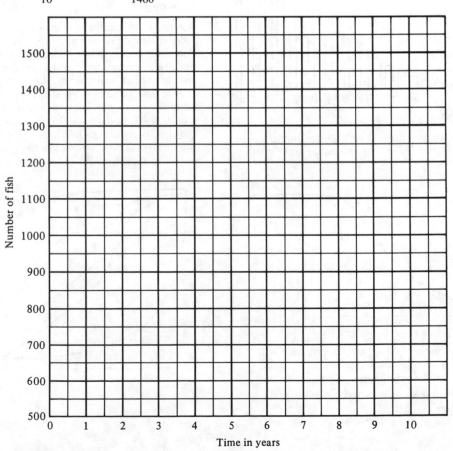

Choose the one best answer.

26 To an ecologist, an "interbreeding group of individuals that occupies a specific geographic area" is a(n)

   a population.   *d* gene pool.
   b species.   *e* ecosystem.
   c community.   (p. 921)

27 The graph below, showing the total number of persons in the United States in each census, 1790–1970, indicates that the population

   a was growing exponentially.
   b had reached carrying capacity in 1970.
   c was at zero population growth in 1970.
   d exhibited an S-shaped growth curve.
   e none of the above   (p. 925)

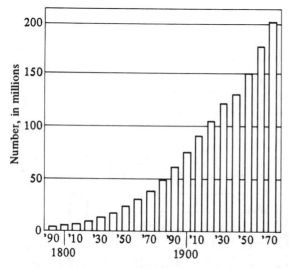

28 In the curve below, at which stage of population growth is the birth rate equal to the death rate? (p. 925)

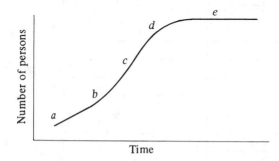

29 Which one of the following populations will show the greatest population increase in the next year, assuming exponential growth?

   a population A, with 200,000 individuals, when $r = .020$

   b population B, with 500,000 individuals, when $r = .040$
   c population C, with 2 million individuals, when $r = .008$
   d population D, with 10 million individuals, when $r = .002$
   e population E, with 30 million individuals, when $r = .001$ (p. 925)

30 Suppose that you study the growth of a population. You find that it increases in size exponentially and then reaches steady state. Which of the following factors is highly *unlikely* to have been a factor in limiting the population?

   a predation   *d* competition
   b parasitism   *e* emigration
   c fire   (pp. 926–27)

31 Which one of the following would be most likely to act as a limiting factor for a population of insects showing a boom-and-bust growth curve?

   a predation
   b competition
   c disease
   d a period of unfavorable weather
   e physiological changes induced by crowding   (pp. 927–28)

32 A certain fish species lives in large lakes. It feeds on small insects and other invertebrates. Spawning occurs in late spring, when males and females congregate in shallow water and engage in brief courtship displays, after which they release gametes in great clouds. The adult fish then swim back to deeper water. Young fish become sexually mature when they are three years old. Which of the following survivorship curves is most likely to apply to this species? (p. 928)

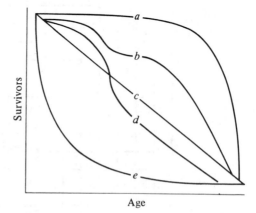

33 Which one of the following characteristics is *not* typical of an *r*- selected species?

  a lives in an unpredictable environment
  b has a high reproductive rate
  c is small in size
  d has a long life-span
  e exhibits little or no parental care of young
  (p. 931)

34 Which one of the following statements is *false*?

  a Theoretically, the ecological niche of any given species is determined by an almost infinitely large number of different factors.
  b If two species are living in a mutualistic association, the density of one species affects the density of the other.
  c The reproductive potential of any species of plant or animal usually equals the actual reproductive rate.
  d Severe interspecific competition often leads to extinction, range restriction, or evolutionary divergence.   (pp. 931–40)

35 Which of the following statements is *correct*?

  a Two species may not live in the same habitat.
  b The more dissimilar the niches of two species, the stronger is their competition.
  c No two species can occupy the same niche in the same habitat.
  d No two species may occupy the same ecosystem.   (p. 934)

36 If two sympatric species occupy very similar niches under natural conditions, one would expect

  a the species to hybridize.
  b intense interspecific competition to occur.
  c extensive interspecific cooperation to occur.
  d the carrying capacity of the environment to be reduced.
  e a mutualistic symbiosis to develop.
  (p. 934)

37 Which ecosystem is the most unstable?

  a Sahara desert
  b African grassland
  c Alpine tundra
  d Montana wheat field
  e Costa Rican tropical rain forest   (p. 943)

38 Which one of the following is *not* a trend in ecological succession?

  a an increase in the number of trophic levels
  b an increase in productivity
  c an increase in community stability
  d a decrease in nonliving organic material
  e an increase in species diversity
  (pp. 948–51)

39 Which one of the following is *least* likely to be true of an ecological succession?

  a The species composition of the community changes continuously during the succession.
  b The total number of species rises initially, then stabilizes.
  c The total biomass in the ecosystem declines after the initial stages.
  d The total amount of nonliving organic matter in the ecosystem increases.
  e Though the amount of new organic matter synthesized by the producers remains approximately the same after the initial stages, the percentage utilized at the various trophic levels rises.
  (pp. 948–51)

40 The climax stage of a biotic succession

  a persists until the environment changes significantly.
  b changes rapidly from time to time, seldom remaining at any stage for more than a decade or so.
  c is the first stage in the reclamation of land from a lake bottom.
  d is a stage in which the dicot plants are always dominant.   (p. 950)

For further thought

1 The population of Brazil in 1978 was estimated to be 113,100,000. If the birth rate was 37.1/1,000 and the death rate was 8.8/1,000, what was *r* for this population? What must have been the approximate population at the end of 1979? Contrast the annual population growth of Brazil with that of the People's Republic of China, where the population in 1978 was estimated to be 950 million, the birth rate was 26.5/1,000, and the death rate was 10.3/1,000. Which country showed a greater population increase in the next five years if the growth rates remained unchanged?

2 Look at the human population growth curve shown in question 27 on the previous page. Can our population continue to grow indefinitely? What are some of the limitations on population growth that may occur as the population density increases?

ANSWERS

Testing recall

1	b	4	l	7	h	10	e
2	d	5	k	8	j	11	g
3	f	6	i	9	c		

12 true
13 false—often
14 false—r-selected
15 false—density-dependent
16 false—niche includes more than environmental factors; it is functional role and position of the organism in its ecosystem
17 true
18 true
19 false—more efficient
20 false—species comprising a given community are products of local environmental conditions —there is no single climax community for a region

Testing knowledge and understanding

21 The curve should be similar to the one in Figure 35.5 (p. 926).
22 logistic
23 K-selected
24 The rate of increase is greatest at the inflection point (Figure 35.5, p. 926).
25 density-dependent factors, especially competition

26	a	30	c	34	c	38	d
27	a	31	d	35	c	39	c
28	e	32	e	36	b	40	a
29	e	33	d	37	d		

Chapter 36

ECOSYSTEMS AND BIOGEOGRAPHY

A GENERAL GUIDE TO THE READING

This chapter, the second on the topic of ecology, considers some of the ways in which the movement of energy and materials binds together the community and the physical environment as a functioning system. First we shall look at the movement of energy and nutrients through ecosystems, and then go on to consider how physical forces create regional variations in climate and thus influence the dispersal of species and the distribution of communities on earth. As you read this chapter you will want to focus on the following.

1   The flow of energy. Learn the terms presented on pages 952–55, and make sure you understand why the flow of energy is always noncyclic.

2   The cycle of materials. At first the four cycles presented on pages 956–62 may look forbiddingly detailed, but careful study will show that they share certain general features. In all of them, materials cycle between living and nonliving systems. You will find the cycles easier to learn if for each you identify the reservoir in the nonliving system, recognize how the chemicals enter and leave the cycle, and understand how they are passed from organism to organism within the living system.

3   Biological magnification. You have doubtless heard for years about the effects of DDT and other persistent chemicals; the discussion in the text shows the reason for the problem. Figure 36.11 (p. 962) is helpful in understanding biological magnification.

4   Biomes. The major biomes of the world are described (pp. 970–79). The accompanying photographs will help you recognize the characteristics of each.

5   Continental drift. Especially important in the section on biogeography (pp. 979–85) is the material on the concept of continental drift, since it helps explain the present distribution of plants and animals.

6 The dispersal of species. This section describes the factors governing species dispersal. The information on island biogeography (pp. 988–90) is particularly important. Figures 36.40, 36.41, and 36.42 (pp. 988–89) are most helpful in understanding this material.

KEY CONCEPTS

1 The movement of energy and materials knits a community together and binds it with the physical environment to form a functioning system. (pp. 952–62)

2 Radiant energy from the sun is the ultimate energy source for life on earth. This energy is captured by the producers and passed on to the consumers. (pp. 952–54)

3 Energy is constantly lost from an ecosystem as it is passed along the links of a food chain. Energy flow is always noncyclic. (p. 954)

4 The water and mineral components of the biosphere are used over and over again; they cycle through an ecosystem, and can be passed around it indefinitely. (pp. 956–62)

5 The sun is the ultimate energy source for life, and the distribution of its energy helps determine the distribution of living things. Because the earth's surface curves away from the path of incident light, areas at different latitudes receive different amounts of sunlight and precipitation. (pp. 969–70)

6 A limited number of major categories of climax formations, called biomes, can be recognized; classification of a region as belonging to a particular biome is determined by which climax community is the most common in the region. (pp. 970–79)

7 The distribution of biomes is a consequence of climate, physiography, and other environmental factors within each geographic area. (pp. 970–79)

8 Understanding the distribution of life today requires combining knowledge of present conditions with evidence from the fossil record and with geological evidence of the past configurations and past climates of the earth's land masses. (pp. 979–85)

9 Since its intrinsic rate of increase always exceeds the carrying capacity of the environment, a population is always under pressure to expand its niche or to extend its range. To spread successfully into a new area, a species must have the physiological potential to survive and reproduce there, an ecological opportunity, and physical access. (pp. 985–93)

10 The size of an island and its distance from a source of colonists affects the number of species established on the island. (p. 988)

11 As the number of species established on an island increases, the rate of immigration of species not already established falls and the rate of extinction of established species rises. When the point is reached at which immigration and extinction are equal, the island has the equilibrium number of species; it cannot support more species unless the immigration rate increases or the extinction rate decreases. (pp. 988–90)

OBJECTIVES

After studying this chapter and reflecting on it, you should be able to carry out the following objectives.

1 Diagram the flow of energy through an ecosystem, including the following: producers, decomposers, primary consumers, secondary consumers, first trophic level, second trophic level, and third trophic level. State the difference between a food chain and a food web. (pp. 952–54)

2 Distinguish between gross primary productivity and net primary productivity. (p. 952)

3 Explain why the distribution of productivity within an ecosystem can always be represented as a pyramid, and why there are seldom more than four or five levels in a food chain. In doing so, specify the percentage of energy present at one trophic level that can usually be passed on to the next, and indicate the reason for the great decrease from level to level in the amount of available energy. (pp. 952–53)

4 Describe the pyramid of biomass and the pyramid of numbers and explain why these pyramids do not apply to all populations. (p. 953)

5 Using a diagram such as Figure 36.5 (p. 956), describe the water cycle, specifying the roles of evaporation, transpiration, and rainfall. (pp. 956–57)

6 Using a diagram such as Figure 36.6 (p. 957), describe the carbon cycle. In doing so,

explain how carbon enters the living system, and how it leaves, indicate the role of microorganisms in the cycle, and identify the reservoir for carbon. (pp. 957–58)

7   Using a diagram such as Figure 36.7 (p. 959), describe the nitrogen cycle. In doing so, discuss nitrogen fixation, nitrification, and denitrification, indicate the role of microorganisms in the cycle, and identify the reservoir for nitrogen. (pp. 959–61)

8   Using a diagram such as Figure 36.9 (p. 960), describe the phosphorus cycle. In doing so, explain how phosphorus enters the living system, indicate the role of microorganisms such as bacteria in the cycle, and identify the reservoir for phosphorus. Then describe the effect of increased phosphorus levels in a freshwater ecosystem, showing how they may accelerate eutrophication. (pp. 961–62)

9   Using a diagram such as Figure 36.11 (p. 962), explain the process of biological magnification. (p. 962)

10 Describe how the structure and properties of soil affect the availability of water, oxygen, and minerals to plants. (pp. 963–67)

11 Explain how the earth's curvature and axis of rotation influence the amount of sunlight reaching a given area, and how this influences the temperature and precipitation in that area. (pp. 969–70)

12 List the major biomes of the world; for each, specify the principal characteristics of the area, some representative plants and animals, and the climatic factors that influence it. Using a diagram such as Figure 36.28 (p. 976), contrast altitudinal biomes with latitudinal biomes. (pp. 970–79)

13 Distinguish between each of the following pairs of terms: benthic division and pelagic division; euphotic zone and aphotic zone; neritic province and oceanic province. (pp. 977–78)

14 Outline the main features of the continental-drift concept, and indicate how it helps explain the present geography of life. (pp. 979–85)

15 Discuss the similarities and the histories of the two island continents, and indicate how their isolation from other continents is reflected in the character of their plants and animals. (pp. 972–73)

16 Name the four main biogeographical regions of the World Continent, and explain why the plants and animals of these regions are more alike than are those of the island continents. Indicate what barriers separate these regions from one another. (pp. 984–85)

17 Discuss the three major conditions that influence a species' success at colonization and describe how the Law of the Minimum and the Law of Tolerance are related to success at colonization and the distribution of species. (pp. 985–87)

18 Using Figure 36.40 (p. 988), state the relationship between the size of an island and its species diversity. (p. 988)

19 Explain the equilibrium model of species diversity (Figures 36.41, p. 988, and 36.42, p. 989) and give reasons why some species become extinct after arriving in a new area. Then, explain the effect of distance of an island from potential colonizers on the island's equilibrium, giving reasons for this effect. (pp. 988–90)

KEY TERMS

The following terms are important in this chapter; you should become familiar with them.

gross primary productivity (p. 952)
net primary productivity (p. 952)
detritus (p. 953)
food chain (p. 953)
food web (p. 953)
producers (p. 954)
primary consumers (p. 954)
secondary consumers (p. 954)
trophic levels (p. 954)
pyramid of productivity (p. 954)
pyramid of biomass (p. 955)
pyramid of numbers (p. 955)
top predators (p. 955)
water cycle (p. 956)
carbon cycle (p. 957)
nitrogen cycle (p. 958)
nitrogen fixation (p. 958)
nodules (p. 959)
nitrification (p. 960)
denitrification (p. 961)
eutrophication (p. 961)
biological magnification (p. 962)
loam (p. 964)
humus (p. 964)
acid rain (p. 965)
biomes (p. 970)
tundra (p. 970)
taiga (p. 973)
deciduous forests (p. 973)
tropical rain forests (p. 974)

SUMMARY

The flow of energy and materials knits a given community together and binds it with the physical environment as a functioning system. Almost all forms of life obtain their high-energy organic nutrients, directly or indirectly, from photosynthesis. The total amount of energy bound into organic matter by photosynthesis is called *gross primary productivity*; *net primary productivity* is the amount left after subtracting the amount the plant uses in respiration. Heterotrophs obtain their energy by consuming green plants, other heterotrophs, or the dead bodies or wastes of other organisms.

The *food chain* is the sequence of organisms, including the *producers* (autotrophic organisms), *primary consumers* (herbivores), *secondary consumers* (herbivore-eating carnivores), and *decomposers*, through which energy and materials may move in a community. In most communities the food chains are completely intertwined to form a *food web*. The successive levels of nourishment in the food chains are called *trophic levels*. The producers constitute the first trophic level, the primary consumers the second, the herbivore-eating carnivores the third, and so on. Since many species eat a varied diet, trophic levels are not hard-and-fast categories.

At each successive trophic level there is loss of energy from the system; only about 10 percent of the energy at one trophic level is available for the next. The distribution of productivity within a community can be represented by a *pyramid of productivity*, with the producers at the base and the last consumer level at the apex. In general, the decrease of energy at each successive trophic level means that less biomass can be supported at each level; thus many communities show a *pyramid of biomass*. Some communities also show a *pyramid of numbers*: there are fewer individual herbivores than plants, and fewer carnivores than herbivores.

The endless cycling of water to earth as rain, back to the atmosphere through evaporation, and back to earth again as rain maintains the freshwater environments and supplies water for life on land. The *water cycle* is also a major factor in modifying temperatures and in transporting chemical nutrients through ecosystems.

Carbon cycles from the inorganic reservoir to living organisms and back again. Carbon dioxide in the atmosphere or in water is converted into organic compounds by photosynthesis; the resulting organic compounds may be released as CO_2 by respiration or be consumed by animals or decomposers. Eventually the carbon will be released as CO_2 and the cycle will begin again. Human activities have increased the CO_2 level in the atmosphere; eventually this may lead to a change in the earth's temperature since the heat radiated from earth is absorbed by atmospheric CO_2 and radiated back to warm the earth in a "greenhouse effect."

Biological *nitrogen fixation* by microorganisms, some of them living symbiotically in root nodules, provides most of the usable nitrogen for the earth's ecosystems. The microorganisms reduce atmospheric N_2 to NH_3, which is often in the form of NH_4^+. *Nitrification* may then occur; different groups of bacteria convert NH_4^+ to NO_2^- and then to NO_3^-, which may be absorbed by roots and converted into organic nitrogen. When the plant dies, nitrogen is returned to the soil as NH_3, which can be recycled. Some bacteria carry out *denitrification*, converting NH_3, NO_2^- or NO_3^- into N_2 gas.

Phosphorus also cycles from the inorganic reservoir to living organisms and back again. Phosphate rock dissolves slowly and becomes available to plants, which pass it to animals. Some is excreted by animals; the rest is released from organic compounds when the organism dies. Huge quantities of phosphates flow into aquatic environments in runoff water. Sewage, detergents, and fertilizer runoff have greatly increased the phosphate level, which accelerates the *eutrophication* (aging) process of lakes.

Modern industry and agriculture have been releasing vast quantities of chemicals into the environment. Some, such as mercury, are harmless when released, but are made toxic by microorganisms. Some chemicals (e.g. DDT) show *biological magnification*. When ingested, these persistent chemicals are retained in the body and tend to become increasingly concentrated as they are passed up the food chain to the top predator.

The properties of soils—their particle sizes, amount of organic material, and pH, among others—determine how rapidly water and minerals move through the soils and how available to plants they will be. The proportions of clay, silt, and sand particles help determine many of these soil properties. Decaying organic material (*humus*) promotes proper drainage and aeration.

A complex equilibrium exists between the ions free in the soil water and those adsorbed on clay and organic particles. Many factors, acidity in particular, can shift this equilibrium; for example, acid rain, a by-product of air pollution, damages the soil. The plants and animals living in or on the soil also profoundly affect soil structure and chemistry. The cutting down of forests, poor farming practices, and irrigation have ruined many soils.

The biosphere The sun is the ultimate energy source for life, and the distribution of its energy helps determine the distribution of living things. Because the earth's surface curves away from the path of incident light, areas at different latitudes receive different amounts of sunlight and precipitation.

Most biologists recognize a number of major climatic regions called *biomes*. The *tundra* is the northernmost biome of North America, Europe, and Asia. The subsoil is permanently frozen. There are many organisms on the tundra but relatively few species. South of the tundra lies the zone dominated by the coniferous forests, the *taiga*. More different species live in the taiga than on the tundra.

The biomes south of the taiga show much variation in rainfall and thus more variation in climax communities. The *deciduous forests* predominate in temperate zones with abundant rainfall and long, warm summers. Tropical areas with abundant rainfall are usually covered by *tropical rain forests*, which include some of the most complex communities on earth; the diversity of species is enormous. Huge areas in both the temperate and tropical regions are covered by *grassland* biomes. These occur in areas of low or uneven rainfall. Places where the rainfall is very low form the *desert* biomes. Deserts are subject to the most extreme temperature fluctuations of any biome type.

A series of different biomes can also be found on the slopes of tall mountains. Climatic conditions change with altitude, and biotic communities change correspondingly.

Aquatic ecosystems also vary with varying physical conditions. Oceanic ecosystems may be classified as the *benthic division* (the ocean bottom with all bottom-dwelling organisms) or the *pelagic division* (the water above the bottom with all the swimming and floating organisms). Another system distinguishes between an upper well-lighted *euphotic zone* and a deeper lightless *aphotic zone*. Still another possible distinction is between the *neritic province* above the continental shelves and the *oceanic province* of the main ocean basin. The most complex oceanic communities occur in the *littoral zone*, the shallow waters along the beach. These subdivisions are not fully analogous to terrestrial biomes since energy and materials flow between the different subdivisions.

The evolution of biogeographic regions To understand the present geography of life, we must combine knowledge of present conditions with evidence from the fossil record and with geological evidence of past configurations of the earth's land masses and their climates. Geological evidence indicates that 225 million years ago all the earth's land masses were combined in a single supercontinent called *Pangaea*. Pangaea broke up into a northern supercontinent called *Laurasia* and a southern one called *Gondwana*. Soon Gondwana broke up; India drifted to the north and the African-South American mass separated from the Antarctic-Australian mass. Later, each of these masses split. The division of Laurasia into North America and Eurasia was one of the last to occur. As the continents moved, their climates changed, altering the distribution of organisms. In addition, fossil evidence indicates that the earth's climate has undergone many changes over time.

The biota (flora and fauna) of the *Australian region* is most unusual; many species, particularly the mammalian fauna, that are common in Australia exist nowhere else. Most of the ecological niches that are filled by placental mammals on other continents are filled by marsupials in Australia. The unusual biota can be explained by the long isolation of Australia from the other continents.

The South American continent, the *Neotropical region*, has also been an island continent through much of its history, but its nearness to North America and its recent connection via the Central American land bridge have given it a more diverse biota.

Europe, Asia, Africa, and North America form the World Continent; their biotas are relatively similar. The World Continent can be divided into the *Nearctic* (North America), *Palaearctic* (Europe, northern Asia), *Oriental* (southern Asia), and *Ethiopian* (Africa south of the Sahara). After the division of Laurasia, North America and Eurasia were connected through much of their history by the Siberian land bridge.

The dispersal of species Since its intrinsic rate of increase always exceeds the carrying capacity of the environment, a population is always under pressure to expand its niche or to extend its range. A species must meet three conditions to spread into a new area: it must possess the *physiological potential* to survive and reproduce in that area; it must have the *ecological opportunity* to become established; and it must have *physical access* to the new area.

Colonization is possible only if the colonizers are already at least minimally preadapted to survive under the new environmental conditions. Climate and weather are important limitations on the distribution of species. Often the extremes of the weather limit the distribution. The same kinds of organisms will not live at all points within a given region since the microenvironmental conditions may vary. Other environmental factors are also important. Liebig's *Law of the Minimum* states that the growth of a plant will be limited by whichever required factor is most deficient in the environment. Shelford's *Law of Tolerance* states that the distribution of a species will be limited by that environmental factor for which the organism has the narrowest range of adaptability.

The colonizing species must have the ecological opportunity to become established in the new area; it must encounter little competition at first and find an available niche. Studies of island biogeography show that the greater the area, the greater the species diversity. Generally, the number of species doubles for every tenfold increase in area. The distance of the island from a source of colonists also affects the number of species on the island. The equilibrium number of species on a near island is greater than the equilibrium number for a distant island.

An organism must also have physical access to a new area. Organisms may be dispersed from one place to another by active locomotion or by passive transport (e.g. by air, water currents, birds, or mammals). The southeast Pacific island of Krakatoa constitutes a natural experiment in colonization of new territory and illustrates different dispersal mechanisms. Studies of island biogeography indicate that the more distant an island is from a major source of new colonists, the lower the species diversity at its equilibrium state. The geological or ecological barriers that separate two regions will be more effective barriers to some species than to others.

QUESTIONS

Testing recall

Determine whether each of the following statements is true or false. If it is false, correct it.

1 Food chains almost always begin with some sort of unicellular autotrophic organism. (p. 953)

2 The pyramid of productivity holds true for all populations. (p. 954)

3 The pyramid of numbers holds true for all populations. (p. 955)

4 In a normal biological community, there will be less usable energy at the carnivore trophic level than at the herbivore trophic level. (p. 954)

5 In most biological communities there are more top predators than primary consumers. (p. 955)

6 Most flowering plants obtain their nitrogen in the form of nitrate. (p. 960)

7 Carnivores are more likely to have high concentrations of DDT in their tissues than comparable herbivores in the same ecosystem. (p. 962)

8 The subsoil of the taiga is permanently frozen. (pp. 970–73)

9 A temperate deciduous forest usually has fewer plant species than a taiga forest. (p. 974)

10   No producers are found in the benthic division in the aphotic zone of the ocean. (p. 977)

11   Once an island has the equilibrium number of species(s), it cannot support more species unless the immigration rate increases or the extinction rate decreases. (p. 988)

12   Islands far from the mainland tend to have fewer species than islands of similar size located closer to the mainland. (p. 988)

13   Italy is in the Nearctic region. (p. 984)

Testing knowledge and understanding

Choose the one best answer.

14   The diagram below shows a particular food web. Each letter represents a different species. Arrows indicate the flow of energy and materials. Which of the following would have the greatest total biomass?

a F	*d* K + M
b J + G	*e* H
c K	(pp. 953–54)

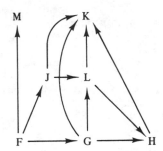

15   Which one of the species is probably carnivorous?

a J	*c* L
b M	*d* G (p. 954)

16   Which species is a decomposer?

a F	*d* K
b G	*e* L
c H	(p. 954)

The arrows in questions 17–22 indicate the direction of energy flow. For each question, evaluate the validity of the direction of the arrow. If the arrow points the correct way, the answer is a; *if the direction is incorrect, the answer is* b.

17   autotroph → heterotroph (p. 954)

18   carnivore → herbivore (pp. 954–55)

19   dead carnivore → decomposer (p. 954)

20   respiration → photosynthesis (p. 955)

21   primary consumer → secondary consumer (p. 954)

22   producer → decomposer (p. 954)

23   The diagram below shows the flow of materials between trophic levels. Which arrow is incorrect? (pp. 954, 957)

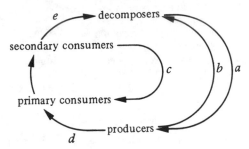

24   When we eat carrot sticks, we are acting as

   a   producers.
   b   primary consumers.
   c   secondary consumers.
   d   tertiary consumers.   (p. 954)

25   In most food chains,

   a   there are fewer individuals at the top predator level than at the second trophic level.
   b   there is less usable energy at the herbivore level than at the carnivore level.
   c   there are few individuals at the decomposer level.
   d   there is less usable energy at the autotrophic level than at the carnivore level.
                                                   (p. 954)

26   If the total weight of producers in an ecosystem was 100 tons, what would you expect the total weight of secondary consumers to be?

a 0.1 ton	*d* 100 tons
b 1 ton	*e* 1,000 tons
c 10 tons	(p. 954)

27   Assume that producers in an ecosystem total 1,000 grams of dry mass per square meter and there is an efficiency of 10 percent in each transfer of dry mass from one trophic level to the next. How many total grams of dry mass will there be in tertiary consumers?

a 100	*d* 0.1
b 10	*e* 700
c 1	(p. 954)

28   Of the five elements that are the most important constituents of living things, the only one that requires the action of microorganisms to enter the living system is

a carbon.	*d* nitrogen.
b hydrogen.	*e* phosphorus.
c oxygen.	(pp. 958–60)

29 Though free nitrogen (N₂) is abundant in the atmosphere, living organisms are unable to use it until it is "fixed" into certain nitrogen compounds. Which one of the following statements best describes nitrogen fixation?

a Only flowering plants are able to fix nitrogen.
b Nitrogen is fixed by both free-living and symbiotic bacteria.
c Only symbiotic bacteria, not free-living bacteria, fix nitrogen.
d Most nitrogen is fixed by lightning during thunderstorms.
e Animals, but not plants, are able to fix nitrogen. (p. 958)

30 Biological magnification is

a the increase in size of organisms at higher trophic levels.
b the increase in number of organisms at higher trophic levels.
c the concentration of stable, nonexcretable chemicals in organisms at higher trophic levels.
d the concentration of nitrates and phosphates in a polluted lake.
e the capture of small organisms by larger organisms at higher trophic levels. (p. 962)

31 DDT has been more of a problem in predatory birds such as the bald eagle than in seed-eating birds. This is because

a seed-eating birds are smaller.
b predatory birds eat more than seed-eating birds.
c DDT is absorbed by animals, but not plants.
d biological magnification concentrates DDT in higher trophic levels.
e seed-eating birds are able to metabolize DDT and excrete its by-products. (p. 962)

32 Which of the following is the major pollutant from electric utilities and an iron smelting operation that acidifies rain?

a SO_2 d HCO_3
b NO_2 e N_2
c H_2CO_3 (p. 965)

33 Large parts of the Soviet Union are covered in coniferous forests. These regions are cold in winter and warm in summer. This biome is known as

a tundra.   d deciduous forest.
b taiga.   e desert.
c grassland.   (p. 973)

34 The step in the nitrogen cycle that provides the nitrogen source used by carnivores is the formation of

a ammonia (NH₃).
b nitrate ions (NO₃⁻).
c nitrate ions (NO₂⁻).
d atmospheric nitrogen (N₂).
e amino acids. (p. 959)

35 Water, oxygen, and minerals are most available to plants growing in soils that are called

a sands.   c clays.
b loams.   d silts.
              (pp. 963–65)

36 The climax formation (biome) of much of the state of Kansas is

a tundra.
b grassland.
c tropical rain forest.
d taiga.
e deciduous forest. (p. 974)

37 Which one of the following statements is false?

a Deserts often display extreme daily variations in temperature.
b Epiphytes are very common in tropical rain forests.
c The subsoil of the tundra is permanently frozen.
d The dominant trees of the taiga are birch and maple.
e Vast numbers of waterfowl nest on the tundra. (pp. 970–75)

38 Vermont is in

a the temperate deciduous biome of the Nearctic region.
b the taiga biome of the Nearctic region.
c the temperate deciduous biome of the Palaearctic region.
d the taiga biome of the Palaearctic region.
e the temperate deciduous biome of the Neotropical region. (pp. 973, 984)

39 If you were climbing a high mountain in Central America, in what order would you cross (1) deciduous forest, (2) taiga, (3) tundra, (4) tropical rain forest?

a 4, 3, 2, 1   d 3, 4, 1, 2
b 4, 3, 1, 2   e 4, 1, 2, 3
c 3, 4, 2, 1   (p. 976)

40 The present distribution of plant and animal life can best be explained by assuming that the earth's land masses have drifted from one place to another. According to this concept of continental drift,

a   South America and Australia were connected through much of geologic time.
b   North America and Europe were connected throughout much of their geologic history.
c   India split away from Asia and drifted southward.
d   the entire Oriental region originally belonged to the same supercontinent as South America and Africa.
e   the separation of Europe from Asia occurred early in geologic history.

(pp. 979–82)

41 When the supercontinent Laurasia broke up, it gave rise to all but one of the following. Which is the exception?

a   North America   d   India
b   Greenland   e   Asia
c   Europe   (p. 980)

42 If the living representatives and all known fossils of a very ancient group of organisms were found only in Africa and Australia, on which of the following continents would you *most* expect to find additional fossils of this group?

a   Antarctica   d   Asia
b   North America   e   Europe
c   Greenland   (pp. 980–81)

43 Which biogeographic regions were island continents?

a   Oriental and Ethiopian
b   Nearctic and Palaearctic
c   Nearctic and Neotropical
d   Australian and Neotropical
e   Australian and Oriental   (p. 982)

For further thought

1   Suppose that the fox in the food web diagramed in Figure 36.1 (p. 953) in your text feeds only on rabbits, squirrels, mice, and seed-eating birds, all of which feed on plant material. How many square meters of plant material are then required to support the fox if the net primary productivity of the plant material is 8,000 Kcal/m^2/year? Assume that a fox's daily caloric requirement is 800 Kcal, and that only 10 percent of the energy at one trophic level can be passed on to the next. If the fox in Figure 36.1 were to feed only on insect-eating birds, how many square meters of plant material would be required?

2   Trace the route that a molecule of CO_2 might follow as it cycled through the ecosystem. (Include at least four organisms in the pathway.)

3   Calcium ions are required nutrients for both plants and animals. The reservoir for calcium is in rocks. Design a calcium cycle, including at least three organisms.

4   As the world's human population continues to grow, the problem of feeding people becomes increasingly difficult. What are some of the environmental consequences of attempting to feed so many people?

ANSWERS

Testing recall

1   false—autotroph may be unicellular or multicellular
2   true
3   false—many
4   true
5   false—fewer
6   true
7   true
8   false—tundra
9   false—more
10 true
11 true
12 true
13 false-Palaearctic

Testing knowledge and understanding

14	a	22	a	30	c	37	d
15	c	23	c	31	d	38	a
16	d	24	b	32	a	39	e
17	a	25	a	33	b	40	b
18	b	26	b	34	e	41	d
19	a	27	c	35	b	42	a
20	b	28	d	36	b	43	d
21	a	29	b				

THE ORIGIN AND EARLY EVOLUTION OF LIFE

A GENERAL GUIDE TO THE READING

This chapter is the first in Part V, "The Genesis and Diversity of Organisms"; it describes the theory of the origin of life widely held by most biologists, and the evidence that supports it. As you read Chapter 37 in your text, you will want to concentrate on the following topics.

1   Formation of the earth and its atmosphere. Carefully read the description of proposed models of the early atmosphere (pp. 999–1000), and compare them with earth's present atmosphere.

2   Formation of small organic molecules. In going over the material on this topic (pp. 1000–2), you will want to pay special attention to Stanley Miller's experiment and its significance; Figure 37.2 (p. 1000) is helpful in this regard.

3   Formation of polymers. You need to understand how polymers are formed from building-block molecules (p. 1002). A review of Figures 3.8 (p. 53), 3.12 (p. 55), and 3.17 (p. 59) will refresh your memory about the condensation reactions that result in the formation of polymers.

4   Formation of molecular aggregates and primitive cells. As the discussion of this topic indicates (pp. 1003–6), coacervate droplets and proteinoid microspheres are suggestive of what prebionts may have been like. Look ahead to Figure 38.11 (p. 1033) and notice the similarities between the bacteria shown there and the proteinoid microspheres shown in Figure 37.5 (p. 1004).

5   Evolution of autotrophy. The evolution of photosynthetic pathways, particularly noncyclic photophosphorylation, was an immensely important event in the evolution of life (pp. 1008–9). Be sure you understand what the oxygen revolution was and why it was significant.

6 The origin of eucaryotic cells. Lynn Margulis's theory of the origin of eucaryotic cells (pp. 1014–16) is fascinating. The material on the endosymbiont hypothesis was first introduced in Chapter 5; see page 136 for a good review.

7 The kingdoms of life. The description of the five-kingdom classification scheme (pp. 1019–20), and the diagrams of it in Figure 37.17A, B (p. 1021), are good preparation for Chapters 38–43, in which the kingdoms are considered one by one.

KEY CONCEPTS

1 Life arose spontaneously from nonliving matter under the conditions prevailing on the early earth; from these beginnings all present life on earth has descended. (pp. 997–98)

2 All the events now hypothesized in the origin of life and all the known characteristics of life seem to fall well within the general laws of the universe; no supernatural event was necessary to the origin of life on earth. (p. 998)

3 There was no abrupt transition from "nonliving" prebionts to "living" cells; the attributes associated with life were acquired gradually. (pp. 1000–6)

4 Living organisms, once they arose, changed their environment and so destroyed the conditions that made possible the origin of life. (pp. 1008–9)

5 The first cells were probably procaryotic; eucaryotic cells may have originated from a symbiotic union of ancient procaryotic cells of several types. (pp. 1013–17)

6 The classification of living organisms into kingdoms is to some degree arbitrary; each classification system has its advantages and disadvantages. (pp. 1017–21)

OBJECTIVES

After studying this chapter and reflecting on it, you should be able to carry out the following objectives.

1 Describe the formation of the earth according to the most widely held hypothesis. Describe two models for the composition of the early atmosphere; then discuss the differences between the two models with respect to the gases thought to be present in the early atmo-sphere, and the differences between these proposed atmospheres and the atmosphere today. (pp. 997–1000)

2 Explain how small organic molecules are thought to have formed in the early oceans, and specify what sources of energy were available to cause molecules to react together to form other compounds. (pp. 1000–2)

3 Describe Stanley Miller's experiment, and explain how its results support Oparin's hypothesis of the origin of life. (pp. 1000–1)

4 Describe one method by which polymers may have been formed from the building-block molecules in the ancient oceans. (pp. 1002–3)

5 Contrast Oparin's coacervate droplets and Fox's proteinoid microspheres, and specify ways in which they resemble living cells. Indicate whether these prebionts have a genetic control system. (pp. 1003–5)

6 Explain why the earliest organisms were most likely heterotrophic procaryotes, and specify the role competition is thought to have played in the evolution of various chemical pathways in the early cells. (pp. 1007–8)

7 Discuss reasons why the evolution of photosynthesis was necessary for life to continue, and indicate which pathway—cyclic or noncyclic phosphorylation—is believed to have evolved first. Explain what is meant by the term oxygen revolution. Give the number of ATP synthesized from one mole of glucose under anaerobic conditions versus the number of ATP that can be synthesized under aerobic conditions. (pp. 1005–9, 187–89)

8 Discuss possible reasons why the evolution of photosynthesis destroyed the conditions that made possible the origin of life. (p. 1009)

9 Evaluate the possibility of life on other planets. (p. 1010)

10 Explain why fossils from the Precambrian are so scarce and why there was such an enormous increase in life forms in the Cambrian. (pp. 1011–13)

11 Discuss the evidence for the endosymbiotic model for the origin of the eucaryotic cell. (pp. 1013–17)

12 Give the rationale for the classification of living things into five kingdoms, and into seven kingdoms. Mention the two alternative ways of considering the Protista (see Figure 37.17A, B, p. 1021). (pp. 1017–20)

13 Name the five kingdoms of organisms used in the classification scheme in your text, and list the distinguishing characteristics of each. (pp. 1018–20)

KEY TERMS

The following terms are important in this chapter; you should become familiar with them.

first atmosphere (p. 999)
second atmosphere (p. 1000)
reducing atmosphere (p. 1000)
abiotic synthesis (p. 1001)
polymer (p. 1002)
coacervate droplet (p. 1003)
proteinoid microsphere (p. 1004)
prebiont (p. 1006)
oxygen revolution (p. 1009)
Precambrian (p. 1011)
Cambrian (p. 1013)
urcaryotes (p. 1014)
endosymbiont hypothesis (p. 1014)

SUMMARY

Scientists today believe that life could and did arise spontaneously from nonliving matter under the conditions that prevailed on the early earth, and that all present earthly life has descended from these beginnings. The basis for the current theory of the origin of life was stated by A. I. Oparin in 1936.

The solar system was probably formed between 4.5 and 5 billion years ago from a cloud of cosmic dust and gas. As the earth condensed, a stratification took place; heavier materials moved toward the center and lighter substances were concentrated at the surface. Eventually the lighter gases of the earth's first atmosphere escaped into space. The intense heat in the interior of the earth drove out various gases through volcanic action; these formed a second atmosphere.

The atmosphere of the early earth was quite different from today's oxidizing atmosphere. Two principal models of the early atmosphere have been proposed. The first states that the early atmosphere contained much H_2 and was a reducing atmosphere. Thus nitrogen was present as ammonia, oxygen as water vapor, and carbon as methane. The second model assumes that the atmosphere was made up primarily of gases that occur in present-day volcano outgassings: H_2O, N_2, H_2S, CO, CO_2, H_2. In either model, the early atmosphere contained almost no O_2.

As the earth's crust cooled, the water vapor condensed into rain and began to form the oceans, in which gases from the atmosphere and salts and minerals from the land dissolved. Ultraviolet radiation, lightning, and heat could have provided the energy for chance bonding of these substances to form the organic building-block molecules. Experimental evidence shows that such abiotic synthesis of organic compounds can occur. These organic compounds could have accumulated slowly in the seas over millions of years since they would not have been destroyed by oxidation or decay.

Some investigators feel that the organic material in the oceans became sufficiently abundant for chance polymerizations to occur; others suggest that various concentrating mechanisms were necessary for polymerization of the building-block molecules.

Oparin speculated that *coacervate droplets* then formed; these are clusters of macromolecules surrounded by an orderly shell of water molecules. Such droplets have a definite internal structure and can absorb substances selectively. Fox proposes, instead, the formation of *proteinoid microspheres*, which exhibit many properties of living cells. Vast numbers of such prebiological systems (*prebionts*) may have arisen in the seas. Some may have contained favorable combinations of materials and grown in size; new droplets could have formed upon fragmentation. Somehow, the nucleotide sequences in nucleic acids came to code for the sequence of amino acids in protein, and transcription and translation evolved. The genetic control system made possible more accurate duplication and more precise control over the chemical reactions taking place within the droplets. Those droplets with particularly favorable characteristics developed into the first cells.

Alternatively, some biologists suggest that the first "living" things were self-replicating nucleic acids that slowly surrounded themselves with cytoplasm and a membrane.

The earliest organisms were heterotrophs that obtained energy from the nutrients available in the early ocean. As the nutrients disappeared, competition between the organisms must have increased. Natural selection favored any new mutation that enhanced an organism's ability to obtain or process food. Over time, various biochemical pathways evolved that enabled organisms to utilize different nutrients. The first form of metabolism using ATP was probably fermentation.

As the free nutrients were used up, some organisms evolved the ability to use another energy source—the sun. Cyclic photophosphorylation probably evolved first, then noncyclic photophosphorylation. From this time onward, life on earth depended on the activity of photosynthetic autotrophs. The oxygen released by algal photosynthesis converted the atmosphere to an oxidizing one. With free O_2 available, organisms could evolve efficient aerobic respiration. The oxygen also gave rise to a layer of ozone in the upper atmosphere that shielded the earth's surface from intense ultraviolet radiation, permitting organisms to move to land.

Since life could have evolved spontaneously from nonliving matter on earth, it is also possible that it could have arisen elsewhere in the universe.

Evolution of the eucaryotic cell The oldest cellular fossils are of bacteria and are about 3.5 billion years old. Most authorities think the cyanobacteria evolved from bacteria about 2.3 billion years ago. Few fossils of higher forms of life are found from the Precambrian, but they are abundant in the Cambrian.

The first fossils of eucaryotic cells are about 1.5 billion years old, but there is no direct evidence of their evolution. Several investigators think that the chloroplasts and mitochondria may be modern descendants of ancient procaryotic cells that became obligate endosymbionts of other cells (called *urcaryotes*) and have evolved in concert with their hosts ever since. A growing list of characteristics indicates that mitochondria and chloroplasts have many features in common with free-living procaryotic organisms. If this model is correct, the chloroplasts are probably derived from cyanobacteria, and the mitochondria from aerobic bacteria. There is little evidence on the origin of the nuclear membrane and the endoplasmic reticulum.

The kingdoms of life The evolutionary relationships between the major groups of organisms are poorly known. Despite this ignorance, science continues to attempt to assign all living things to a few large categories called kingdoms or divisions. Whatever criteria are chosen, it is impossible to make a clean separation between the groups; these are artificial human categories imposed on nature.

The classification system used in your text recognizes five kingdoms: the Monera, Protista, Plantae, Fungi, and Animalia. The kingdom Monera includes the procaryotic organisms. The Protista consist of the unicellular eucaryotic organisms. The protists may be regarded as consisting of three separate groups, which represent the precursors of plants, animals, and fungi, or they can be placed in a separate line that diverged before the other three eucaryotic groups evolved. Three major lineages reached the higher multicellular level. Each exploits a different mode of nutrition—photosynthetic autotrophism is used by plants (kingdom Plantae), absorptive heterotrophism by fungi (kingdom Fungi), and ingestive heterotrophism by animals (kingdom Animalia).

QUESTIONS

Testing recall

1 Arrange the following in the order of their appearance in the evolution of life (pp. 998–1013)

A competition
B formation of the earth
C prebionts
D reducing atmosphere
E oxidizing atmosphere
F primitive procaryotic cell
G abiotic synthesis of building-block molecules
H complex biochemical pathways
I O_2 revolution
J great rains
K polymerization
L eucaryotic cell
M photosynthesis

Testing knowledge and understanding

2 How old is the earth?

a 4.5 million years d 15 billion years
b 15 million years e 4.5 trillion years
c 4.5 billion years (p. 999)

3 Life would probably never have arisen on earth if the early atmosphere had contained

a O_2. d NH_3.
b H_2. e CH_4.
c N_2. (pp. 1000, 1009)

4 In 1953 Stanley Miller conducted a famous experiment in which he simulated conditions of the early earth in an enclosed apparatus. His experiment demonstrated that

a life originated in the early oceans.
b proteins and lipids will self-assemble into membranes.
c amino acids could have been formed from the gases of the early atmosphere.
d amino acids are necessary for life.
e lightning was the source of energy for the synthesis of the first organic molecules.
(p. 1000)

5 In the Miller experiment, gases that were probably present in the early atmosphere were circulated through an apparatus past

electrical discharges from tungsten electrodes. Which of the following were synthesized?

a   proteins
b   long chain nucleic acids
c   amino acids and lactic acid
d   cellulose
e   all of the above   (p. 1001)

6   The earth's primitive atmosphere is best described as

a   oxidizing.
b   nonoxidizing, or reducing.
c   predominantly composed of CH_3, NH_3, H_2O, and O_2.
d   the same in composition as interstellar gas.
e   none of the above   (p. 1000)

7   Which one of the following has been proved experimentally?

a   Life on earth originated by spontaneous generation.
b   Life on earth originated by special creation.
c   Complex organic molecules can be produced abiotically from methane, ammonia, and water vapor.
d   Spontaneous generation cannot occur today.
e   Eucaryotes evolved from procaryotes.
                                        (p. 1000)

8   Which one of the following statements is *false*?

a   Coacervate droplets have a strong tendency toward formation of definite internal structure.
b   The chemical reactions in a coacervate droplet depend in part on the physicochemical organization of the droplet itself.
c   There is a definite interface between a coacervate droplet and the liquid in which it floats.
d   Coacervate droplets have a marked tendency to adsorb and incorporate various substances from the surrounding medium.
e   The regular spatial arrangement of the molecules within a coacervate droplet probably reduces the catalytic activity of proteins in the droplet.   (pp. 1004–5)

9   The first *cellular* organisms on earth were probably most like today's

a   eucaryotic cells.   d   green algae.
b   viruses.   e   Protozoa.
c   bacteria.   (p. 1007)

10 Sidney W. Fox, of the University of Miami, has suggested that the first cells were "proteinoid microspheres." Fox's microspheres are

a   cells with approximately the same degree of organization as bacteria.
b   cells with many membrane-bound organelles, the membranes of which contain protein, but not lipid.
c   much like present day viroids.
d   spherical in shape and look like primitive cells but do not exhibit osmotic properties.
e   tiny spheres which form spontaneously when hot aqueous solutions of polypeptides are cooled.   (p. 1014)

11 Which one of the following statements is *false* concerning Fox's proteinoid microspheres?

a   They have a definite internal structure.
b   Chemical reactions can occur within the droplets.
c   They can adsorb and incorporate various substances from the medium.
d   They will shrink in hypotonic media and swell in hypertonic media.
e   They show an internal movement similar to cytoplasmic streaming.   (p. 1004)

12 Competition probably became important at an early stage in the origin of life because

a   nucleic acids cannot duplicate themselves.
b   available organic materials were being consumed faster than new materials were being synthesized.
c   cells arose early in the sequence as a result of the colloidal characteristic of proteins.
d   the early proteins probably acted as catalysts and thus accelerated the various chemical reactions that were taking place.
e   the ability to carry on photosynthesis had become widespread.   (p. 1007)

13 Which of the following events led most directly to the decrease in the amount of ultraviolet radiation reaching the surface of the earth?

a   evolution of cyclic photophosphorylation
b   evolution of noncyclic photophosphorylation
c   development of a reducing atmosphere
d   evolution of heterotrophy
e   evolution of fermentation   (p. 1009)

14 Early in the evolution of life on earth, metabolic breakdown of energy-rich compounds must have been only by fermentation, because aerobic respiration

a   occurs only in animals, but the earliest organisms were plants.
b   could not occur before the evolution of noncyclic photophosphorylation.
c   could not occur after the so-called oxygen revolution.
d   could not occur before the first vascular plants evolved.

e can only oxidize carbohydrates, but the early nutrients were mostly of other types. (p. 1009)

15 Mode of nutrition is a very important characteristic of organisms. Which evolved first?

a parasitism
b herbivorous feeding
c carnivorous feeding
d heterotrophy
e autotrophy (p. 1008)

16 As the first cellular organism evolved, they began to develop complex biochemical pathways; this development is thought to be due to the

a depletion of abiotically produced nutrients.
b presence of carbon dioxide in the atmosphere.
c prior evolution of autotrophy.
d abundance of abiotically produced nutrients.
e increase in oxygen levels in the atmosphere. (p. 1009)

17 Which of the following events in the history of the earth happened first?

a evolution of eucaryotes
b accumulation of oxygen in the atmosphere
c evolution of Homo sapiens
d evolution of photosynthetic flowering plants
e evolution of procaryotes (p. 1011)

18 Calculations have been made on the probability of life existing on other planets. These calculations suggest that

a conditions needed for the evolution of life are so special that it probably only exists on our planet.
b conditions needed for the evolution of life probably exist on a few planets and there is a small, but real, chance that life exists elsewhere.
c the chances that life has evolved on other planets is about 50/50.
d life probably has evolved on several other planets.
e life has evolved on thousands of other planets at the very least. (p. 1011)

19 Life first appeared on earth approximately

a 6 billion years ago.
b 3.5 billion years ago.
c 2.3 billion years ago.
d 500 million years ago.
e 3.5 million years ago. (p. 1011)

20 The oldest fossils of cyanobacteria are about _____ billion years old.

a 4.5   *d* 1.5
b 3.5   *e* 1.0
c 2.3   (p. 1011)

21 Which one of the following statements is *false* according to current scientific thinking on the origin of life?

a The first autotrophs were probably eucaryotic cells.
b Given great expanses of time, raw materials, and energy, the basic building-block organic compounds could be synthesized in the absence of living cells.
c The first living organisms must have depended on glycolysis and fermentation to obtain usable energy.
d As living organisms evolved, they changed the environment so much that the conditions that made possible the origin of life no longer exist.
e Coacervate droplets have some attributes of living cells; they are orderly arrangements of molecules with a membrane that can selectively absorb substances from the surrounding medium. (pp. 1000–18)

22 According to the Margulis theory of the origin of the eucaryotic cells,

a mitochondria evolved from chloroplasts.
b fungi, plants, and protists obtained their mitochondria from animals.
c present-day procaryotes evolved from primitive eucaryotes.
d mitochondria are remnants of ancient procaryotes that invaded primitive urcaryotic cells.
e mitochondria are remnants of ancient eucaryotes that invaded very primitive, evolving eucaryotic cells. (pp. 1014–16)

23 It has been proposed that mitochondria and chloroplasts (and flagella) are modern descendants of primitive forms of procaryotic cells that took up residence within primitive cells and evolved independently there. Which one of the following is *not* evidence of this view?

a Like procaryotic cells, mitochondria and chloroplasts have DNA that is not closely associated with protein.
b The ribosomes in procaryotic cells, mitochondria, and chloroplasts are very similar chemically, and differ from the cytoplasmic ribosomes of eucaryotes.
c Certain antibiotics inhibit the ribosomes in the mitochondria and chloroplasts but do not affect cytoplasmic ribosomes.

d Mitochondria and chloroplasts are known to have their own genes and ribosomes and to conduct protein synthesis.

e Mitochondria and chloroplasts can be cultured outside cells.   (p. 1016)

24 At present, we are not sure how chloroplasts and mitochondria evolved. However, the most widely accepted theory states that

a they are descendants of ancient procaryotic cells that became permanent residents inside other cells.

b they arose when membranes surrounded existing structures inside cells.

c they were first produced by budding from the plasma membrane and later DNA from the nucleus migrated to the organelles.

d mitochondria evolved from chloroplasts.

e chloroplasts evolved from mitochondria.
                                      (pp. 1014–17)

25 According to the five-kingdom system used in your textbook, the kingdom Protista includes

a most unicellular eucaryotes.

b the procaryotic organisms.

c all unicellular organisms.

d unicellular eucaryotes and the algae.

e unicellular eucaryotes and the fungi.
                                      (pp. 1018–19)

26 The kingdom Monera includes which of the groups listed in question 25? (p. 1018)

27 Which one of the following is an incorrect match?

a Fungi–absorptive heterotrophs

b Animalia–ingestive autotrophs

c Plantae–autotrophs

d Monera–bacteria and cyanobacteria

e Protista–primarily unicellular or colonial eucaryotic groups   (pp. 1018–21)

For further thought

1 Another hypothesis for the origin of life on earth suggests that life did not originate here; it was brought here by some extraterrestrial object such as a meteor. The earth is constantly bombarded by meteorites. Analyses indicate that some meteorites contain molecules (some hydrocarbons) characteristic of living systems. Comment upon this hypothesis.

2 Suppose you were given the following organisms to classify:

a A unicellular organism that has chlorophyll and carries on photosynthesis. It lacks a cell wall and is highly motile.

b A large multinucleated amoeboid heterotrophic organism that carries on phagocytosis. Its reproduction is plantlike; it forms groups of thin-walled spore cases similar to those of fungi.

In which kingdom would you classify each if you were using the five-kingdom system? How would you classify them in the seven-kingdom system? Give the advantages and disadvantages of each system.

ANSWERS

Testing recall

1 B, D, J, G, K, C, F, A, H, M, I, E, L or
   B, D, J, G, K, C, A, H, F, M, I, E, L

Testing knowledge and understanding

2	*c*	9	*c*	16	*a*	22	*d*
3	*a*	10	*e*	17	*e*	23	*e*
4	*c*	11	*d*	18	*e*	24	*a*
5	*c*	12	*b*	19	*b*	25	*a*
6	*b*	13	*b*	20	*c*	26	*b*
7	*c*	14	*b*	21	*a*	27	*b*
8	*e*	15	*d*				

VIRUSES AND MONERA

A GENERAL GUIDE TO THE READING

This chapter focuses on the viruses, which occupy the border between life and nonlife, and on the primitive cellular organisms belonging to the kingdom Monera. As you read Chapter 38 in your text, you will want to concentrate on the following topics.

1   The structure of viruses. In studying this topic (pp. 1022–29), you will find it helpful to review Figure 26.2 (p. 696), showing the structure of the bacteriophage, and to look carefully at Figure 38.1A, B, C (p. 1023) showing the structure of plant and animal viruses.

2   The reproduction of viruses. As the discussion on pages 1024–26 indicates, the reproduction of plant and animal viruses is in some ways quite different from that of the bacteriophage. You will want to review the life cycle of the bacteriophage (Figs. 28.7, p. 734, and 38.2, p. 1025). Make sure, too, that you understand what a provirus is; and since viral

reproduction can be quite complicated, particularly for an RNA virus, study carefully the material on RNA viruses (p. 1026). Some of this material will be a review for you; the reproduction of RNA viruses was discussed in some detail in Chapter 28 (see pages 735–37). Figure 28.8 (p. 736) may be helpful.

3   Bacterial anatomy. To prepare for your study of this topic, review the section on procaryotic cells in Chapter 5 (pp. 134–35). Though bacterial cells (pp. 1032–34) are relatively simple in structure, there is much to learn about them. Figures 38.9–38.13 (pp. 1032–33) will help you appreciate their various forms.

4   Bacterial reproduction. A main point about bacterial reproduction (pp. 1035–36) is that bacteria divide by binary fission rather than by mitosis or meiosis.

5   Bacterial photosynthesis. Try to appreciate the fact that while many of the monerans are photosynthetic, the manner in which photosynthesis is carried out may be quite different

from the reactions we learned back in Chapter 8 (pp. 1038–40). Only the Cyanobacteria and Prochlorophyta possess chlorophyll *a*, carry out noncyclic photophosphorylation, and release O_2.

6   Cyanobacteria. You will want to compare the Cyanobacteria (pp. 1039–40) with the other bacteria and try to understand why the Cyanobacteria are such a successful group of organisms.

7   Prochlorophyta. The Prochlorophyta (pp. 1039–40) may be an important group from an evolutionary standpoint because their pigment system is similar to that of higher plants.

8   Bacteria as agents of disease. The text (pp. 1040–42) describes how bacteria cause disease symptoms and how the body defends itself against bacterial pathogens.

9   Beneficial bacteria. Try to remember that most bacteria are beneficial (pp. 1042–43); for example, recall the vital role that they play in the biogeochemical cycles.

KEY CONCEPTS

1   Viruses are on the borderline between living and nonliving; they lack the metabolic machinery to make ATP and proteins, and they cannot reproduce themselves in the absence of a host, yet they have nucleic acid genes that encode information for their reproduction. (pp. 1023–24)

2   The cells of Archaebacteria and Eubacteria are procaryotic; they lack a nuclear membrane and most other membranous organelles and thus are fundamentally different from all other living organisms. (pp. 1029–32)

3   The Monera are an extraordinarily successful group of organisms; their extreme metabolic versatility and enormous reproductive potential have enabled them to survive in a wide variety of habitats. (pp. 1036–40)

4   The Cyanobacteria are procaryotic cells that produce O_2 as a by-product of their photosynthesis; they may have initiated the oxygen revolution some 2.3 billion years ago. (pp. 1039–40)

5   Many bacteria are beneficial; all other organisms depend directly or indirectly on the activities of the bacteria. (pp. 1042–43)

OBJECTIVES

After studying this chapter and reflecting on it, you should be able to carry out the following objectives.

1   Explain how scientists showed in 1935 that viruses are distinct from bacteria. (p. 1023)

2   Using diagrams and micrographs as given in Figures 38.2 (p. 1023) and 38.4 (p. 1026), describe the basic structure of a virion, and contrast the structure of a bacteriophage virus with that of a plant or animal virus. Indicate whether DNA is the genetic material in all viruses. (pp. 1023–26)

3   Compare the reproductive cycle of a bacteriophage, as it was shown in Figures 28.7 (p. 734) and 38.2 (p. 1025), with that of a virus that infects the cells of higher organisms, pointing out both similarities and differences. Explain how a provirus differs from a vegetative virus. (pp. 1024–26)

4   Describe the replication of an RNA virus, and explain how this process differs from the replication of a virus with DNA genes. Figure 28.8 (p. 736) may be helpful. Explain how replication utilizing RNA replicases differs from replication utilizing reverse transcriptase. Indicate whether eucaryotic cells normally possess RNA replicase or reverse transcriptase and if not, explain where the enzyme comes from. (p. 1026)

5   Using a diagram such as Figure 38.2 (p. 1025), explain how replicated bacteriophage particles normally leave the host cells. Contrast this process with the process by which some plant and animal viruses are released from their host cells, as shown in Figures 38.3 (p. 1026) and 38.5 (p. 1027). Indicate whether virions that infect plant or animal cells ever cause lysis of their host cells. (pp. 1025–26)

6   Compare the viroids and the newly discovered prions with viruses with respect to structure, mode of reproduction, and infectivity. (pp. 1023–26, 1028)

7   Give three hypotheses for the origin of viruses. (pp. 1028–29)

8   Explain why most viral infections do not respond to treatment with antibiotics. Describe the role of interferon in the recovery from viral infection. (p. 1029)

9   List the two sections of the Monera, and give two ways they differ from one another. (pp. 1030–31)

10 For each of the divisions shown in Figure 38.7 (p. 1030), state whether the members of the division are primarily heterotrophic, chemosynthetic, or photosynthetic. (pp. 1032–33)

11 Explain how bacteria differ from viruses. Then list five differences between procaryotic and eucaryotic cells (Table 5.1, p. 135, may be helpful). (p. 1032)

12 List the three different bacterial shapes and explain what the prefices diplo-, staphlo-, and strepto- mean when applied to bacteria. (p. 1032)

13 Contrast the cell walls of bacteria with those of plants and fungi, and explain the significance of the cell wall in the life of bacteria. Explain what a capsule is and what role it plays. (pp. 1033–34)

14 Describe the structure and function of a bacterial endospore. (p. 1034)

15 Give two ways the bacterial flagella differs from a eucaryotic flagella in structure and function. (p. 1035)

16 Using diagrams like those in Figure 23.5 (p. 617), describe the reproduction of a bacterial cell, showing how it differs from that of a eucaryotic cell. Explain how genetic recombination can occur in some bacteria. (pp. 1035–36)

17 Differentiate among aerobes, obligate anaerobes, and facultative anaerobes. (pp. 1036–37)

18 Describe the three nutritive modes found in bacteria, and indicate which mode is the most common. For each division of photosynthetic bacteria, indicate the type(s) of pigments used, whether the process is anaerobic or aerobic, and whether molecular oxygen is produced as a by-product. Indicate the possible evolutionary significance of the Prochlorophyta. (pp. 1036–40)

19 Explain how the oxygen-generating Cyanobacteria are adapted to fix atmospheric nitrogen even though the enzyme that carries out this fixation functions only under anaerobic conditions. (p. 1040)

20 Discuss the role of bacteria as agents of disease and indicate how Koch's postulates can be used to prove that a particular microorganism is the cause of a particular disease. Describe how pathogenic bacteria may harm the body of their host; then specify three kinds of defense by the body against the attacks of such pathogens. (pp. 1040–42)

21 List five ways in which bacteria are beneficial. (p. 1043)

KEY TERMS

The following terms are important in this chapter; you should become familiar with them.

nucleic acid core (p. 1023)
capsid (p. 1023)
envelope (p. 1023)
lytic cycle (p. 1024)
lysogenic cycle (p. 1024)
provirus (p. 1024)
RNA replicase (p. 1025)
reverse transcriptase (p. 1026)
retrovirus (p. 1026)
extrusion (p. 1026)
viroid (p. 1026)
prion (p. 1027)
interferon (p. 1029)
Archaebacteria (p. 1030)
Eubacteria (p. 1031)
coccus (p. 1032)
bacillus (p. 1032)
spirillum (p. 1032)
murein (p. 1033)
capsule (p. 1034)
endospore (p. 1034)
binary fission (p. 1036)
obligate anaerobe (p. 1036)
facultative anaerobe (p. 1037)
phycocyanin (p. 1039)
phycoerythrin (p. 1039)
heterocyst (p. 1040)
Koch's postulates (p. 1041)
toxin (p. 1041)
vaccine (p. 1042)
antiserum (p. 1042)
colostrum (p. 1042)

SUMMARY

Viruses In the late nineteenth century the first evidence of infectious agents so small that they could pass through fine porcelain filters was found. In 1935 the tobacco mosaic virus was crystallized, and the filterable viruses were recognized as different from bacteria.

Viruses are not cells; though some viruses possess a few enzymes, they lack the metabolic machinery for energy generation, and they never have the ribosomes required for protein synthesis. The free virus particle, or *virion*, consists of a nucleic acid core covered with a protein coat (*capsid*). The nucleic acid can be DNA or RNA, double-stranded or single-stranded.

Viruses cannot reproduce themselves; the host cell manufactures new viruses using the genetic instructions provided by the virus. When bacteriophages reproduce, only the nucleic acid enters the host cell; in most other cases the entire virion enters. Once inside the cell, the nucleic acid provides the genetic information for the synthesis of new viral nucleic acid and proteins.

If the nucleic acid is DNA, it acts as a template for the synthesis of both mRNA and new viral DNA. If the nucleic acid is RNA, replication usually takes place in the cytoplasm; an enzyme, *RNA replicase*, uses the RNA as a template to synthesize new RNA. In some RNA viruses, the retroviruses, the RNA is transcribed instead into DNA in a reaction catalyzed by *reverse transcriptase*. The newly formed DNA then becomes integrated as a provirus into the host's chromosomes.

Once new viral components have been synthesized by the host cell, they are assembled into new virions. Some virions are released by lysis of the host cell, others by an extrusion process in which the virus becomes enveloped in a piece of cell membrane.

Viruses can exist in three states: as free infectious virions, as viral nucleic acid directing replication in a host cell, and as provirus.

Recently two new types of infectious agents have been discovered. *Viroids* are small circular pieces of RNA that lack a protein coat. They cause visible disease in certain higher plants. *Prions* are self-replicating infectious proteins that have been implicated in a few degenerative diseases.

Three hypotheses of viral origin have been proposed. The first suggests that viruses are organisms that have reached the extreme of evolutionary specialization for parasitism; they have lost all cellular components except the nucleus. The second hypothesis suggests that modern viruses represent a primitive "nearly living" prebiont stage in the origin of life. The third hypothesis suggests that viruses are fragments of genetic material derived from other organisms.

Viruses cause many diseases in various organisms, and have been associated with some kinds of cancer. Most viral infections do not respond to antibiotic treatment. A protein called *interferon* promotes recovery from viral infections. Produced by the host's cells in response to an invading virus, it is released from infected cells and interacts with uninfected cells, giving them a resistance to infection.

Monera The kingdom Monera has two sections, the Archaebacteria, or "ancient" bacteria, and the "true" bacteria or Eubacteria. The *Archaebacteria* differ from the *Eubacteria* in their cell-wall composition, their membrane lipids, and their translation mechanism. The Archaebacteria typically occupy certain inhospitable environments.

The section Eubacteria has six divisions. The purple bacteria, green bacteria, two divisions of heterotrophic bacteria, and two divisions of aerobic photosynthetic bacteria—the Cyanobacteria and the Prochlorophyta.

The bacteria are cellular; they have the metabolic machinery to generate ATP and to reproduce themselves; and they have ribosomes for protein synthesis. Bacteria occur in three basic shapes: spherical (*cocci*), rod-shaped (*bacillus*), or corkscrew (*spirilla*). Some bacteria remain together after cell division and may form clusters, chains, or other characteristic groupings.

The cell walls of the Eubacteria differ from those of plants and fungi; they are made of murein, a polymer of polysaccharide chains cross-linked by amino acids. Penicillin is toxic to growing bacteria because it inhibits murein formation. Since most bacteria live in a hypotonic environment, they would burst without a cell wall. A few very small parasitic bacteria, the mycoplasmas, do lack a cell wall. Differences in the cell wall chemistry of different bacteria are associated with differences in their staining properties; these can be used for identification. An envelope or capsule may surround the cell wall.

Some bacteria can form special resting cells called *endospores*, which enable them to withstand adverse conditions. When conditions improve, the endospores produce new bacterial cells.

Many Monera are motile. Some have flagella, which are entirely different from the flagella of eucaryotic cells. Other bacteria move with a peculiar gliding motion.

Monera have enormous reproductive potential. The single circular chromosome composed of DNA is found in the *nucleoid* area of the cell; it replicates before division. Bacteria reproduce by *binary fission*, a type of nonmitotic cell division that produces two daughter cells exactly like the parent. Although this process is asexual, many bacteria do have some genetic recombination; new DNA can enter the cell by conjugation, transduction, or transformation.

Most monerans are heterotrophic, being either saprophytic or parasitic. The oxygen requirement of bacteria varies; most are *aerobic* but some are *facultative anaerobes*, and others are *obligate anaerobes*.

Some bacteria are *chemosynthetic autotrophs* and others are *photosynthetic autotrophs*. The chemosynthetic bacteria oxidize certain inorganic compounds and trap the released energy. There are several types of photosynthetic bacteria. The green and purple bacteria, unlike higher plants, lack chlorophyll *a*, do not use water as an electron source, and do not produce O_2. The halophilic Archaebacteria carry out a unique type of photosynthesis that does not use chlorophyll.

The major group of aerobic photosynthetic bacteria are the Cyanobacteria. All Cyanobacteria possess chlorophyll *a*, the pigment found in higher plants, and generate O_2 as a by-product of photosynthesis. The pigments are located in membranous vesicles (thylakoids) but are not contained within chloroplasts.

Many Cyanobacteria can fix atmospheric nitrogen. Certain cells, the *heterocysts*, are specialized for nitrogen fixation. High levels of O_2 inhibit the processes of nitrogen fixation and photosynthesis.

The Prochlorophyta are procaryotic organisms with a pigment system like that of green algae and higher plants; they would seem more likely than the Cyanobacteria to be the progenitor of chloroplasts in eucaryotic cells since they resemble eucaryotic chloroplasts much more than the Cyanobacteria.

While studying pathogenic bacteria, Robert Koch formulated the rules of procedure for proving that a particular microorganism is the cause of a particular disease. *Koch's postulates* are still used today.

Microorganisms cause disease symptoms by interfering with normal function, by destroying cells and tissues, or by· producing poisons called *toxins*.

The human body resists the attacks of pathogenic organisms with the phagocytic action of white blood cells and with the production of antibodies. Some diseases can be prevented by immunization; the patient is injected with a vaccine or an antiserum. *Vaccines* consist of antigenic substances from the pathogen; these produce a long-lasting active immunity by stimulating the production of antibodies in the patient. An *antiserum* contains presynthesized antibodies against a specific antigen, which produce an immediate, short-term, passive immunity.

Beneficial bacteria far outnumber harmful ones. They play a vital role in the cycling of materials in the ecosystem, are important in many food and industrial processes, and are used to produce antibiotics.

QUESTIONS

Testing recall

Fill in the blanks.

A virus consists of an inner core of a single molecule of nucleic acid which may be (1) _____ or _____ . (p. 1024) Surrounding the nucleic acid is a coat of protein called the (2) _____ . (p. 1023) The viruses that attack bacterial cells are called (3) _____ . (p. 1024) They attach by their tail to the bacterial cell wall and the (4) _____ (p. 1025) is injected into the host cell. The energy for this injection comes from the hydrolysis of (5) _____ . (p. 1024) The phage nucleic acid provides the genetic information for the synthesis of new viral (6) _____ and _____ . (p. 1024) The new virions escape by (7) _____ of the cell. (p. 1024)

In plant and animal viruses, the (8) _____ enters the host cell (p. 1025) If the virus is an RNA virus, the RNA molecule can serve as a template for the synthesis of new RNA, provided the enzyme (9) _____ is present to catalyze the reaction. (p. 1025) In the retroviruses, however, the RNA serves as a template for the synthesis of (10) _____ . (p. 1026) The enzyme (11) _____ is required for this reaction. (p. 1026) This nucleic acid becomes integrated into the host's chromosome and is called a (12) _____ . (p. 1024) Once the new virions have been assembled they may be released by (13) _____ of the cell or by (14) _____ through the cell membrane. (p. 1026)

Recently a new kind of infectious agent called a (15) _____ has been found. It consists only of RNA with no protein coat. (p. 1027)

The kingdom Monera consists of the (16) _____ and _____ . (pp. 1030−31) The cells of all these groups lack a nuclear membrane and most other membranous organelles; they are called (17) _____ organisms. (p. 1030)

The bacterial cell is surrounded by a cell wall made of (18) _____ . (p. 1033)

Some bacteria have (19) _____ , which move the cell by their rotary motion. (p. 1035) Under certain conditions some bacteria form (20) _____ , which are resistant to destruction. (p. 1034) Bacteria have (21) _____ within the

cytoplasm to carry out protein synthesis. The three basic shapes of bacteria are
(22) _____,
_____ , and
_____ . (p. 1032) Bacteria reproduce by (23) _____ .
(p. 1036)

Most bacteria show the
(24) _____ mode of nutrition.
(p. 1036) Bacteria vary in their O_2 requirements; those that cannot survive in the presence of oxygen are called (25) _____ .
(p. 1037) Most of these organisms obtain their energy by the process of
(26) _____ . (p. 1037) Some bacteria contain chlorophyll but do not have
(27) _____ , which is the chief light-trapping pigment in higher plants. (p. 1039)

The Cyanobacteria resemble the higher plants in possessing the light-trapping pigment
(28) _____ . (p. 1039) This pigment is located in membranous sacs called
(29) _____ . (p. 1039) Special cells called (30) _____ fix atmospheric nitrogen. (p. 1040). The Cyano-bacteria divide by (31) _____ .
(p. 1036) Recently new procaryotic organisms have been found that have a pigment system like that of higher plants. These organisms, called the
(32) _____ , may be the ancestors of the chloroplasts in eucaryotic cells.
(p. 1040)

Many bacteria cause disease. The rules of procedure for proving that a particular organism causes a disease are called
(33) _____ . (p. 1041) Some bacteria harm their host by producing poisons called (34) _____ . (p. 1041)
Many diseases can be prevented by inoculating a patient with a (35) _____
containing an antigen that stimulates the patient to produce (36) _____ against the disease-causing organism, conferring an
(37) _____ immunity. (p. 1042)

Testing knowledge and understanding

Choose the one best answer.

38 Viruses are usually not classified as living organisms because they do not

 a contain DNA or RNA.
 b self-replicate.
 c possess chlorophyll.
 d possess nuclear membranes.
 e possess enzymes.   (p. 1024)

39 The coat of a virus is generally made of

 a nucleic acid.   *d* murein.
 b protein.   *e* glucose.
 c cellulose.   (p. 1023)

40 Which one of the following statements concerning viruses is *false*?

 a Viruses have no metabolic machinery of their own.
 b Viral genes are composed only of DNA.
 c Viruses cannot be cultured on artificial media.
 d Viruses will form crystals under certain conditions.   (p. 1023)

41 Viruses are

 a naked, circular pieces of RNA.
 b procaryotic cells without membranes.
 c infectious particles composed of nucleic acid cores and protein coats.
 d fragments of dead procaryotic or eucaryotic cells.
 e infectious proteins.   (p. 1023)

42 Which entity consists only of DNA?

 a a bacteriophage
 b a provirus
 c a human chromosome
 d a bacterial spore
 e a viral capsid   (p. 1024)

43 Which one of the following statements concerning viruses is *false*?

 a Viruses contain either DNA or RNA but not both.
 b Viruses can be crystallized.
 c Some viruses have a membranous envelope surrounding the protein coat.
 d Some viruses have multienzyme systems to generate ATP.
 e Viruses require living host cells for replication.   (p. 1024)

44 Which one of the following statements concerning viruses is *false*?

 a The nucleic acid in viruses may be DNA or RNA, either single-stranded or double-stranded.
 b Some viruses reproduce in the cytoplasm, others in the nucleus.
 c When a virus attacks a plant or animal cell, the entire virus particle usually enters the cell.
 d The nucleic acid in the virus codes for the proteins that make up the coat of the virus and for some enzymes.
 e Viruses are sensitive to most antibiotics, particularly penicillin.
            (pp. 1022, 1024–25)

45 Plant and animal viruses *differ* from bacteriophages in that

 a the whole virus of plant and animal viruses enters the host cell.
 b only bacteriophages can exist as proviruses.

c only plant and animal viral infections result in lysis of the host cell.

d only plant and animal viruses show great host cell specificity. (p. 1025)

46 Reverse transcriptase synthesizes DNA on an RNA template. This enzyme is characteristic of

a animal cells.
b bacterial viruses.
c fungal cells.
d certain DNA viruses.
e certain RNA viruses. (p. 1026)

47 In which *one* of the following ways do viruses resemble cellular organisms? Viruses

a divide by mitosis.
b can undergo mutation.
c exhibit aerobic respiration.
d have an extensive endoplasmic reticulum.
e have a membrane composed largely of phospholipids. (pp. 1023–29)

48 Prions are infectious agents made up of

a self-replicating proteins.
b double-stranded RNA with a protein coat.
c single-stranded DNA with a capsid.
d naked, circular bits of RNA.
e bacterial cells lacking a cell wall. (p. 1028)

49 Animal cells produce an antiviral protein known as

a interferon.   *c* lysozyme.
b viralase.   *d* penicillin.
(p. 1029)

50 Which one of the following bacteria belongs to the section Archaebacteria?

a Prochlorophyta
b Cyanobacteria
c Eubacteria
d green photosynthetic bacteria
e methanogens (p. 1030)

51 The cells of Monera differ from those of eucaryotic organisms in many ways. Which of the following is *not* one of those ways?

a Cells of Monera lack a nuclear membrane.
b When chlorophyll *a* is present in Monera cells, it is located in thylakoids, but these are not contained in chloroplasts.
c Cells of Monera lack mitochondria and an endoplasmic reticulum.
d Cells of Monera lack ribosomes.
e When the cells of Monera possess flagella, those flagella lack a 9 + 2 micro-tubular structure. (p. 1030)

52 A bacterial cell does *not* possess

a a cell wall.
b DNA.
c a plasma membrane.
d mitochondria.
e ribosomes. (p. 1030)

53 The cell walls of bacteria

a enable the bacteria to survive in a hypotonic medium.
b contain cellulose.
c are called capsules.
d are composed of chitin.
e are composed of protein. (p. 1033)

54 The antibiotic penicillin inhibits the ability of bacteria to

a synthesize proteins.
b replicate DNA.
c synthesize cell-wall material.
d perform respiration. (p. 1033)

55 Bacteria generally reproduce by

a conjugation.   *d* mitosis.
b transformation.   *e* meiosis.
c binary fission.   (p. 1034)

56 Which one of the following statements is *true*?

a All bacteria are heterotrophs.
b The flagella of bacteria move by rotary motion.
c The genetic material of bacteria is RNA.
d The cell walls of bacteria contain cellulose.
e All bacteria are aerobic. (p. 1035)

57 Bacteria that are poisoned by oxygen are termed

a aerobic.
b obligate anaerobes.
c facultative anaerobes.
d obligate parasites. (p. 1036)

58 Most bacteria are

a ingestive heterotrophs.
b absorptive heterotrophs.
c photosynthetic heterotrophs.
d photosynthetic autotrophs.
e chemosynthetic autotrophs. (p. 1036)

59 Which one of the following divisions of Monera possesses chlorophyll *a* and releases molecular oxygen as a by-product of photo-synthesis?

a Cyanobacteria
b halophilic Archaebacteria
c purple bacteria
d green bacteria (p. 1039)

60 Cyanobacteria are

    a heterotrophic procaryotes.
    b heterotrophic eucaryotes.
    c autotrophic procaryotes.
    d autotrophic eucaryotes.
    e chemosynthetic procaryotes.   (p. 1039)

61 Cyanobacteria differ from bacteria in

    a having a nuclear membrane.
    b possessing chlorophyll *a*.
    c having amino acids in their cell walls.
    d lacking lysosomes.
    e exhibiting no mitosis.   (p. 1039)

62 Cyanobacteria do *not* have

    a chlorophyll *a*.
    b chlorophyll *b*.
    c photosynthetic membranes.
    d autotrophic nutrition.
    e ribosomes.   (p. 1039)

63 Prochlorophyta have

    a chlorophyll *a* only.
    b chlorophyll *b* only.
    c both chlorophyll *a* and chlorophyll *b*.
    d chloroplasts.
    e a nuclear membrane.   (p. 1039)

64 Penicillin would be *least* effective against

    a scarlet fever.   *d* tuberculosis.
    b gonorrhea.   *e* strep throat.
    c AIDS.   (pp. 1029, 1041)

65 Which one of the following is never exhibited by any kind of bacteria?

    a conjugation   *d* spore formation
    b photosynthesis   *e* chemosynthesis
    c meiosis   (pp. 1032–39)

66 In which case is *passive* immunity acquired?

    a A child contracts measles.
    b An adult gets a common cold.
    c Antibodies are passed from mother to baby in colostrum.
    d An adult gets a case of gonorrhea.
    e A child receives a tetanus "booster" shot.   (p. 1042)

For further thought

1 Would you classify viruses as living or non-living? Justify your answer. (pp. 1022–32)

2 Antibiotics are effective against many bacterial diseases, but most viral infections do not respond to treatment with them. However, a few antibiotics are somewhat effective against certain viruses. Below are listed four antibiotics and their mode of action. For each, determine whether it would be effective against bacteria and/or viruses and explain your answer.

    a penicillin—affects procaryotic cell-wall synthesis
    b streptomycin—inhibits protein synthesis and causes misreadings of the mRNA code at the procaryotic ribosome
    c tetracycline—inhibits protein synthesis on the procaryotic ribosome
    d actinomycin—inhibits mRNA synthesis

3 When penicillin was first introduced to treat syphilis and gonorrhea, it was extraordinarily effective, but certain physicians predicted that eventually it would become useless. In time, the spirochete causing syphilis did become resistant to penicillin, and recently a strain of *Gonococcus* (the organism causing gonorrhea) began to show resistance. Why have these organisms become resistant? (see Chapter 33, p. 856)

4 In the late 1940s it was discovered that feeding antibiotics to farm animals not only controlled infection but also stimulated growth. Today, half the antibiotics produced in the United States are used for animal feeds. Some scientists view this practice with alarm; in their opinion the widespread use of antibiotics is dangerous, and we may eventually return to the pre-antibiotic era in the treatment of bacterial infections. Can you think of reasons for the scientists' concern?

5 Certain bacteria are essential components of the scheme of life. Discuss the role of bacteria in the cycling of inorganic nutrients. Could life survive if all bacteria were eliminated? (see Chapter 36, pp. 957–63)

ANSWERS

Testing recall

1 DNA, RNA
2 protein coat
3 bacteriophages
4 nucleic acid
5 ATP
6 nucleic acid and protein
7 lysis
8 entire virus
9 RNA replicase
10 DNA
11 reverse transcriptase
12 provirus
13 lysis
14 extrusion
15 viroid
16 Archaebacteria, Eubacteria
17 procaryotic
18 murein

19 flagella
20 endospores
21 ribosomes
22 coccus, spirillum, bacillus
23 binary fission
24 heterotrophic
25 obligate anaerobes
26 fermentation
27 chlorophyll *a*
28 chlorophyll *b*
29 thylakoids
30 heterocysts
31 binary fission
32 Prochlorophyta
33 Koch's postulates
34 toxins
35 vaccine
36 antibodies
37 active

Testing knowledge and understanding

38	*b*	46	*e*	53	*a*	60	*c*
39	*b*	47	*b*	54	*c*	61	*b*
40	*b*	48	*a*	55	*c*	62	*b*
41	*c*	49	*a*	56	*b*	63	*c*
42	*b*	50	*e*	57	*b*	64	*c*
43	*d*	51	*d*	58	*b*	65	*c*
44	*e*	52	*d*	59	*a*	66	*c*
45	*a*						

THE PROTISTAN KINGDOM

A GENERAL GUIDE TO THE READING

You will remember from Chapter 37 that the unicellular eucaryotic organisms are frequently difficult to classify as either plants or animals, and have consequently been placed in a kingdom of their own, the kingdom Protista. As you read about this kingdom in Chapter 39 in your text, you will want to focus on the following topics.

1   Animal-like protists. Be sure you understand why these organisms, discussed on pages 1046–53, are referred to as acellular. Also, for each of the phyla described, learn the form of locomotion (if any) characteristic of its members.

2   Funguslike protists. You need to understand which part of the life cycle is plantlike, and which part is animal-like (pp. 1053–56). Figures 39.14 (p. 1054) and 39.15 (p. 1055) will help you learn the differences between the cellular slime mold and the true slime mold.

3   Plantlike protists. In studying the material on these organisms (pp. 1057–60), be sure to note the important role of the diatoms and the dinoflagellates in aquatic food webs.

KEY CONCEPTS

1   The Protista include the eucaryotic organisms that are primarily unicellular or colonial. (p. 1044)

2   The Protista can be separated into three evolutionary lines: the animal-like, funguslike, and plantlike Protista. However, there are also many similarities among them. (pp. 1044–45)

3   Each protozoan should be regarded not as equivalent to a cell of a more complex animal but as a complete organism with the same properties and characteristics as cellular animals. (p. 1045)

4 The Mastigophora (Zooflagellata) appear to be the most primitive of all the Protozoa and may have given rise to the other protozoan groups, and possibly to the multicellular animals. (p. 1046)

5 The slime molds are funguslike protists that are animal-like in some stages of their life cycle and plantlike in others. (pp. 1053–56)

6 The plantlike protistans show a combination of plantlike and animal-like characteristics. (pp. 1057–60)

OBJECTIVES

After studying this chapter and reflecting on it, you should be able to carry out the following objectives.

1 For each of the three different lines of protists, explain how the members generally obtain the high-energy organic compounds necessary to carry out their life processes. (pp. 1044–45)

2 List the four protozoan phyla discussed in your text, give two distinguishing characteristics for each, and explain why many biologists prefer to consider the protozoans acellular rather than unicellular organisms. (pp. 1046–53)

3 Name the protozoan phylum that contains organisms believed to be ancestral to the others, and state the basis for this belief. (p. 1046)

4 Name the phylum that contains organisms believed to be the most evolutionarily advanced of the protozoans, and state the basis for this belief. (pp. 1051–53)

5 Using a diagram such as Figure 39.7 (p. 1049), describe the life cycle of the organism that causes malaria. (pp. 1048–49)

6 Describe the structure and reproductive pattern of the Protomycota. (p. 1053)

7 Contrast the life cycle of the true slime mold with that of the cellular slime mold, mentioning the similarities and differences. (pp. 1053–56)

8 Give ways in which the Euglenophyta resemble plants and ways in which they resemble animals. (p. 1057)

9 List the type of pigments, distinguishing characteristics, and economic importance of the Chrysophyta and Pyrrophyta. (pp. 1058–60)

KEY TERMS

The following terms are important in this chapter; you should become familiar with them.

organelle (p. 1045)
acellular (p. 1045)
malaria (p. 1048)
sporozoites (p. 1049)
merozoites (p. 1049)
trichocysts (p. 1052)
conjugation (p. 1052)
zoospores (p. 1053)
true slime mold (p. 1054)
plasmodium (p. 1054)
fruiting bodies (p. 1054)
sporangia (p. 1054)
cellular slime molds (p. 1054)
pseudoplasmodium (p. 1054)
euglenoid (p. 1057)
stigma (p. 1057)
pyrenoid (p. 1057)
diatom (p. 1058)
plankton (p. 1059)
phytoplankton (p. 1059)
zooplankton (p. 1059)
dinoflagellate (p. 1059)

SUMMARY

Recently the kingdom Protista has become restricted to those eucaryotic groups that are primarily unicellular or colonial. Though some of the protistans tend to be plantlike, others animal-like, and still others funguslike, they share many characteristics. There are conflicting interpretations of protist evolution. The protists may be regarded as consisting of three separate groups which represent the precursors of plants, animals, and fungi, or they can be placed in a separate evolutionary line that arose prior to the appearance of multicellular organisms. In this chapter we follow the precursor perspective.

Animal-like Protista: the Protozoa Although protozoans are usually said to be unicellular, many biologists consider them *acellular* since they are far more complex than other individual cells. They live in a variety of aquatic and moist habitats, and exhibit great diversity of form. Most are heterotrophic. Reproduction is usually asexual but may be sexual. Many forms encyst under unfavorable conditions. The Protozoa are often divided into five phyla: Mastigophora, Sarcodina, Sporozoa, Cnidospora, and Ciliata.

The *Mastigophora*, or zooflagellates, move by flagella and appear to be the most primitive of the

Protozoa. They may have given rise to other protozoan groups, and possibly to the multicellular animals. Most Mastigophora live as symbionts in the bodies of higher plants and animals. Some are parasites; for example, the genus *Trypanosoma* causes several diseases in human beings and domestic animals. One species causes African sleeping sickness, which makes large parts of Africa uninhabitable for humans.

The *Sarcodina* are the amoeboid Protozoa; they move by means of pseudopods. They are thought to be closely related to the zooflagellates because some zooflagellates undergo amoeboid phases, and a few sarcodines have flagellated stages. Several groups of sarcodines secrete hard calcareous or siliceous shells. The shells of two of these groups, the Foraminifera and Radiolaria, have formed much of the limestone and some of the siliceous rocks on the earth's surface.

The parasitic *Sporozoa* usually have a sporelike infective cyst stage in their complex life cycle. All are nonmotile. Malaria is caused by species of Plasmodium; the pathogen is transmitted from host to host by female *Anopheles* mosquitoes.

The *Ciliata* possess numerous cilia for movement and feeding. Their other organelles are greatly elaborated. The Ciliata differ from other protozoans in having a macronucleus and one or more micronuclei. The polyploid macronucleus controls normal cell metabolism; the micronuclei are concerned with reproduction and produce the macronucleus. Many ciliates reproduce occasionally by a sexual process called *conjugation*.

Funguslike Protista: Protomycota and Gymnomycota Two categories of fungal-type organisms can be recognized at the protistan level of complexity: the *Protomycota* (the true funguslike protists) and the *Gymnomycota* (the slime molds).

The Protomycota are saprophytes or parasites. Their haploid bodies may be a simple sac within the cell of a host, a sac with rootlike rhizoids, or a filamentous structure with a reproductive sac. During reproduction, flagellated cells are formed; they may function as gametes or as *zoospores*.

The slime molds are decidedly plantlike at some stages and animal-like at others. The life cycle of the *true slime molds* (division Myxomycota) proceeds from diploid multinucleate amoeboid plasmodium to stationary spore-producing plasmodium, to haploid spores, to flagellated gametes, to zygote, and back to the amoeboid plasmodium. The life cycle of the *cellular slime molds* (division Acrasiomycota) is quite different. The haploid spores form haploid free-living amoeboid cells. When food becomes scarce, some begin secreting cAMP, which causes the slime mold amoebae to aggregate and form a *pseudoplasmodium*, within which the individual cells remain distinct. Eventually a fruiting body develops and produces spores, beginning the cycle again. Under very wet conditions, however, a form of sexual recombination occurs between amoebae of different mating strains. The true slime molds and the cellular slime molds are probably not closely related.

The plantlike Protista Several groups of unicellular organisms show a combination of plantlike and animal-like characteristics. They are plantlike because many have chlorophyll and often a cell wall; they are animal-like in being highly motile.

The *Euglenophyta* are unicellular flagellates that lack cell walls. Like higher plants, their chlorophylls are *a* and *b*, but they store paramylum rather than starch. Many euglenoid species lack chlorophyll and are obligate heterotrophs; those that do possess chlorophyll are facultative heterotrophs. Euglenoids reproduce by longitudinal mitotic cell division; they have no sexual reproduction.

The yellow-green algae, the golden-brown algae, and the diatoms belong to the division *Chrysophyta*. These predominantly unicellular or colonial algae are shades of yellow or brown and possess chlorophylls *a* and *c* (no *b*). The walls of many are impregnated with silica or calcium. The undecomposed shells of diatoms form deposits of diatomaceous earth. The diatoms are an extremely important link in most freshwater and marine food chains; they are the most abundant component of the marine *plankton*. The phytoplankton are the principal photosynthetic producers in marine communities.

The second most important group of marine producers are the *Pyrrophyta*, or dinoflagellates. They are small, usually unicellular, organisms that typically possess two unequal flagella. The photosynthetic species possess chlorophylls *a* and *c*. Some species are luminescent, and some are poisonous. The so-called red tides that sometimes kill millions of fish are caused by species of red-pigmented dinoflagellates.

QUESTIONS

Testing recall

1   *The chart lists the various groups of Protista and some important characteristics. After studying this chapter you should be able to fill in the chart using the directions below. The correctly completed chart will be a good study aid.*

Column 1: Indicate whether the phylum or division contains organisms that are autotrophic, heterotrophic, or both.

Column 2: Indicate whether flagella are typically present or absent.

Column 3: List any mechanism of locomotion other than flagella.

Column 4: Indicate whether chlorophyll is present or absent. (If present give type(s).)

Column 5: List any other important or distinguishing characteristics.

Group	1 Autotrophic or heterotrophic	2 Flagella	3 Other means of movement	4 Chlorophyll pigments	5 Other distinguishing characteristics
Mastigophora					
Sarcodina					
Sporozoa					
Ciliata					
Protomycota					
Gymnomycota					
Euglenophyta					
Chrysophyta					
Pyrrophyta					

Testing knowledge and understanding

Choose the one best answer.

2 Which one of the following statements is *false* concerning the Protozoa?

   a Their food is usually digested in food vacuoles.

   b Movement is by pseudopods, or by the beating of cilia or flagella.

   c Reproduction is usually by binary fission.

   d Many can encyst under unfavorable conditions.

   e Most freshwater forms have contractile vacuoles. (p. 1036)

3 The most "primitive" protozoans are thought to be

   a flagellates.

   b amoeboid protozoans.

   c ciliates.

   d sporozoans. (p. 1046)

4 The organism that causes malaria is a(n)

   a ciliate.

   b flagellate.

   c dinoflagellate.

   d sporozoan.

   e amoeboid protozoan. (p. 1048)

5 Which one of the following groups has no flagellated cells (except in the gametes)?

   a Chrysophyta   *d* Sporozoa

   b Mastigophora   *e* Pyrrophyta

   c Euglenophyta   (p. 1048)

6 A piece of chalk is made of the shells of

   a ciliates.   *d* flagellates.

   b sarcodines.   *e* dinoflagellates.

   c diatoms.   (p. 1048)

7 When a female *Anopheles* mosquito bites and transmits the *Plasmodium* into a human being

   a the sporozoites are discharged into the body.

   b the merozoites are released from the mosquito's salivary glands.

   c the gametes fuse and form an amoeboid zygote.

   d the injected sporozoites produce spores.

   e the amoeboid *Plasmodium* moves to the liver. (pp. 1048–49)

8 The protozoans showing the most complex and well-developed organelles belong to the phylum

   a Mastigophora.   *c* Ciliata.

   b Sarcodina.   *d* Sporozoa.

                             (p. 1051)

9 Suppose you are given an unknown organism to identify. You find that it is unicellular and lacks a cell wall and locomotor organelles. It is an internal parasite. You conclude that it is most likely to be a

   a ciliate.

   b sporozoan.

   c slime mold.

   d flagellate.

   e amoeboid protozoan. (p. 1048)

10 A true slime mold differs from a cellular slime mold in that the true slime mold

   a has a large diploid, coenocytic plasmodium.

   b has a fruiting body in its life cycle.

   c produces haploid spores.

   d secretes pulses of cAMP for aggregation.

   e reproduces only asexually. (pp. 1044–45)

11 The movement of slime molds most closely resembles that of

   a ciliates.

   b amoeboid protozoans.

   c golden-brown algae.

   d flagellates.

   e euglenoids. (pp. 1047, 1054)

12 All the animal-like and fungal Protista

   a are parasitic.

   b are heterotrophic.

   c have cell walls.

   d move by pseudopods.

   e reproduce by spores. (pp. 1045, 1053)

13 Which one of the following groups includes both autotrophic and heterotrophic forms?

   a euglenoids

   b flagellates

   c sporozoans

   d amoeboid protozoans

   e slime molds (pp. 1057–58)

14 Which one of the following characteristics is unique to the euglenoids?

   a flagella

   b absence of a cell wall

   c pyrenoids

   d chlorophylls *a* and *b*

   e photosynthetic and nonphotosynthetic forms (p. 1057)

15 Which one of the following statements is *false* concerning the dinoflagellates?

   a Some can produce light.

   b Some cause the so-called red tides.

   c They have chlorophyll and are photosynthetic.

   d They are an important component of the phytoplankton.

e Their boxlike shells contain silica.
(pp. 1059–60)

16 Red tides, which sometimes kill fish, result from blooms of

a euglenoids.   *d* diatoms.
b dinoflagellates.   *e* ciliates.
c Cyanobacteria.   (p. 1060)

17 A unicellular organism has a siliceous (glass-like) shell and is autotrophic. It is most likely

a a dinoflagellate.   *d* *Amoeba.*
b *Euglena.*   *e* a ciliate.
c a diatom.   (p. 1058)

For further thought

1 There is increasing evidence that persistent pesticides such as DDT may affect the activity of the marine phytoplankton. What ecological problems might result?

2 The World Health Organization lists malaria as the infectious disease that kills more individuals than any other. Malarial control programs center around the control of *Anopheles* mosquitoes. What stages of the life cycle of the *Plasmodium* take place in the mosquito? If all mosquito populations were exterminated, would malaria be eliminated?

3 In what ways do the plantlike Protista resemble higher plants? How are they different?

4 In Africa, humans and domestic cattle are parasitized by a species of *Trypanosoma* that causes the fatal African sleeping sickness in humans and nagana in cattle. Large areas of Africa are uninhabitable for humans and cattle because of these diseases. Although the native wild animals are infected with the trypanosomes, they show no symptoms of disease. The trypanosome is spread from host to host by the bite of the tsetse fly. At present, intensive research is going on with the intention of eventually eliminating the tsetse fly. What might be the ecological effect of this act?

ANSWERS

Testing recall

1 Information for the chart can be found as follows: Mastigophora, pp. 1046–47; Sarcodina, pp. 1047–48; Sporozoa, pp. 1048–50; Ciliata, pp. 1051–53; Protomycota, p. 1053; Gymnomycota, pp. 1053–56; Euglenophyta, pp. 1057–58; Chrysophyta, pp. 1058–59; Pyrrophyta, pp. 1059–60.

Testing knowledge and understanding

2	*c*	6	*b*	10	*a*	14	*c*
3	*a*	7	*a*	11	*b*	15	*e*
4	*d*	8	*c*	12	*b*	16	*b*
5	*d*	9	*b*	13	*a*	17	*c*

THE PLANT KINGDOM

A GENERAL GUIDE TO THE READING

This chapter describes the diverse organisms within the plant kingdom and the evolutionary relationships among them. Many students find learning this material difficult, possibly because it includes a good many unfamiliar terms. The terms are easier to master if you learn them one at a time, as you encounter them in your reading. Another stumbling block can be plant life cycles, which may at first seem confusing. Some hints that may help you learn the life cycles are presented in the discussion of this topic just below.

As you read Chapter 40 in your text, you will want to concentrate on the following topics.

1 Differences between the Thallophyta and the Embryophyta. You need to learn the three fundamental differences between the two groups (p. 1061).

2 The volvocine series. This line of green algae, discussed on pages 1064–66, exemplifies a number of evolutionary trends and provides a clue as to how multicellularity may have

arisen in green plants. Some new terms are presented—heterogamy, anisogamy, and oogamy; you will want to learn these, along with the term isogamy, which was presented in the preceding section (p. 1064).

3 Plant life cycles. Plant life cycles were introduced in Chapter 23; you may wish to review briefly this discussion (p. 642). Next, learn the life cycle shown in both Figure 23.32, B (p. 642) and Figure 40.14 (p. 1069). Most plants have a life cycle similar to this, showing alternation of generations: a diploid sporophyte (spore-producing) stage alternates with a haploid gametophyte (gamete-producing) stage. The life cycles of the different plant groups are simply variations on this basic cycle. One variation is shown in Figures 40.2 (p. 1063) and 40.10 (p. 1068). Here there is no sporophyte stage, the zygote being the only diploid stage; this is believed to be the ancestral or primitive condition. Another variation is shown in Figure 40.19 (p. 1071). Here there is no gametophyte stage; this is considered to be the advanced condition. All the other life

cycles as shown in Figures 40.22 (p. 1075), 40.38 (p. 1083), 40.45 (p. 1088), and 40.51 (p. 1093), exhibit the more typical alternation of generations. You will notice a major evolutionary trend in the life cycles of plant groups —a progressive decrease in the size and importance of the gametophyte and a corresponding increase in the size and importance of the sporophyte.

4 The brown algae. The brown algae (pp. 1068–72) are complex plants that have convergently evolved many similarities to vascular plants. Learn these similarities and also the differences between the brown algae and vascular plants.

5 The red algae. In reading about the red algae (p. 1072), pay special attention to the role of their accessory pigments.

6 The movement onto land. Seven basic problems faced by plants on land are listed (p. 1073); you need to understand these problems, since much of the evolution of the embryophyte plants can best be understood in terms of adaptations that help solve these problems.

7 Liverworts, hornworts, and mosses. The bryophytes (p. 1075) are the most primitive of the land plants. You will want to know why they have never become totally free of their dependence upon an aquatic environment.

8 Characteristics of the tracheophytes. Learn the four basic characteristics of the tracheophytes, or vascular plants. (p. 1078).

9 Psilopsida, club mosses, and horsetails. These groups, described on pages 1078–81, are the earliest tracheophytes; from them we can infer the evolutionary history of the higher plants. Notice in particular the evolutionary advances exhibited by each group.

10 Ferns. The characteristics of ferns (pp. 1081–84), and their life cycle (Fig. 40.38, p. 1083), are important.

11 Evolutionary advances in the seed plants. Several important advances are outlined briefly on page 1084. They are explained in more detail in the subsequent specific descriptions of the gymnosperms and the angiosperms.

12 Gymnosperms. You need to learn the life cycle of the pine (pp. 1086–89 and Fig. 40.45, p. 1088) and to understand the advances in the life cycle of a pine as compared with that of a typical fern.

13 Angiosperms. Pay special attention to the flowering plants (pp. 1090–94), since they are the most abundant plants and hence the dominant group in the plant kingdom. You need to learn the parts of the flower and the details of the reproductive process, and to understand how the life cycle of the angiosperm differs from that of the gymnosperm. Be sure you know the differences between monocots and dicots (p. 1093).

14 Summary of evolutionary trends. This important section (p. 1094) provides an excellent summary of much that was discussed in the chapter. Figure 40.53 (p. 1094) is a helpful pictorial representation of the evolutionary trends among plant groups.

KEY CONCEPTS

1 The divisions of the plant kingdom have traditionally been separated into two groups, the Thallophyta and the Embryophyta. The thallophytes are more primitive; they show little tissue differentiation, generally have unicellular reproductive structures without a wall of sterile jacket cells, and have external development of the embryo. (p. 1061)

2 The green algae (Chlorophyta) are regarded as the group from which the land plants arose. (pp. 1062–68)

3 In the life cycle of the most primitive plants, the haploid stages are dominant; this apparently was the ancestral condition. (p. 1068)

4 The evolution of most plant groups shows a tendency toward reduction of the gametophyte (multicellular haploid stage) and increasing importance of the sporophyte (multicellular diploid stage). (pp. 1068–94)

5 The evolutionary move from an aquatic existence to a terrestrial one was not simple, for the terrestrial environment is in many ways hostile to life. Plants evolved a number of characteristics that enabled them to survive on land. (p. 1073)

6 The bryophytes represent the most conspicuous exception to the evolutionary trend toward reduction of the gametophyte and increasing importance of the sporophyte. (pp. 1074–77)

7 The evolution of the embryophytes, and especially the tracheophytes, is best understood in terms of adaptations for life in a terrestrial environment. (pp. 1077–92)

8 The angiosperms are the most successful land plants because they have evolved the most efficient adaptations for living and reproducing on land. (pp. 1090–94)

OBJECTIVES

After studying this chapter and reflecting on it, you should be able to carry out the following objectives.

1 List the divisions belonging to the Thallophyta and those belonging to the Embryophyta. Then compare the two groups with respect to degree of tissue differentiation, type of reproductive structure (unicellular or multicellular, and lacking or possessing a protective wall), and place where the early development stages occur (inside or outside of the female reproductive structure). (pp. 1061, 1073–74)

2 List two characteristics of the Chlorophyta (green algae); then list three evolutionary trends exemplified by this group. (pp. 1062–64)

3 Using a diagram such as Figure 40.2 (p. 1063), describe the life cycle of *Chlamydomonas*, a representative unicellular green alga. Indicate whether the dominant stage is haploid or diploid. Explain what the term isogamy means, and why *Chlamydomonas* is regarded as isogamous. (pp. 1063–64)

4 Trace the evolutionary changes in the volvocine series from *Chlamydomonas* through *Gonium, Pandorina,* and *Pleodorina,* to *Volvox.* In doing so, specify the evolutionary advance(s) at each stage and use correctly the following terms: heterogamy, anisogamy, and oogamy. (pp. 1063–65)

5 Using a diagram such as Figure 40.2,B (p. 1063), describe the life cycle typical of very primitive plants, and compare it to the life cycle of a more advanced alga, as shown in a diagram such as Figure 40.14 (p. 1069). Explain what is meant by alternation of generations, and define the terms sporophyte, gametophyte, and sporangium. (pp. 1062–68)

6 Contrast the Phaeophyta (brown algae) and the Rhodophyta (red algae) with respect to where they grow and to their size, type of cell wall, storage product, and types of chlorophyll and other pigments. (pp. 1068–72)

7 Using a diagram such as Figure 40.19 (p. 1071), describe the life cycle of *Fucus,* and indicate how it differs from the life cycle of the green algae. List three ways in which the brown algae resemble higher plants. (pp. 1070–72)

8 Discuss at least five problems plants face in a terrestrial environment, and indicate how the embryophytes have "solved" these problems.

Define the terms antheridium and archegonium. (pp. 1073–74)

9 List the major characteristics of the bryophytes; then, using a diagram such as Figure 40.22 (p. 1075), describe their life cycle and name the dominant stage. On a drawing of a bryophyte such as Figure 40.23 (p. 1076), point out the sporophyte stage and the gametophyte stage. (pp. 1074–77)

10 List four distinctive characteristics of the tracheophytes, and trace the evolutionary advances of the tracheophytes from the psilopsids through the club mosses, horsetails, and ferns, to the seed plants. In doing so, indicate the order in which these groups appear in the fossil record. (pp. 1077–78)

11 Define and use correctly each of the following terms: sporophyll, homosporous, heterosporous, megaspore, microspore, antheridium, archegonium. (pp. 1071, 1080)

12 Using a diagram such as Figure 40.38 (p. 1083), describe the life cycle of a typical fern and name the dominant stage; then indicate whether the ferns are heterosporous or homosporous. Explain why the ferns are in some respects no better adapted than the bryophytes for life on land. (pp. 1081–84)

13 Describe three adaptations that probably contributed to the great evolutionary success of the seed plants. (pp. 1084, 1092)

14 Using a diagram such as Figure 40.45 (p. 1088), describe the life cycle of a pine tree and discuss the ways in which this life cycle is more advanced than that of a typical fern. Then indicate whether the pine is homosporous or heterosporous. (pp. 1087–89)

15 Using a diagram such as Figure 40.46 (p. 1090), identify the parts of a flower and describe the process of reproduction (Figures 40.47, p. 1090, and 40.49, p. 1091, may be helpful). Summarize the main ways in which the life cycle of an angiosperm differs from that of the pine, a representative gymnosperm. (pp. 1090–92)

16 List at least five differences between monocotyledons and dicotyledons. (p. 1093)

17 Explain how the flowering plants are adapted to meet the several problems of a terrestrial environment referred to in objective 8. (pp. 1090–93)

18 Using Figure 40.53 (p. 1094) and the material on page 1094, summarize the evolutionary trends among plant groups. (pp. 1092–94)

19 For each of the following pairs, indicate which of the specified conditions is primitive, and which is advanced:

   isogamy—heterogamy
   heterospory—homospory
   diploid stage dominant—haploid stage dominant
   no protection for gametes—protection for gametes
   multicellularity—unicellularity
   water needed for fertilization—no water needed for fertilization
   tissue differentiation into roots, stems, leaves—little tissue differentiation
   gametophyte independent—gametophyte dependent on sporophyte
   reproductive structure multicellular—reproductive structure unicellular
                    (pp. 1080, 1092, 1094)

KEY TERMS

The following terms are important in this chapter; you should become familiar with them.

Thallophyta (p. 1061)
Embryophyta (p. 1061)
thallus (p. 1061)
Chlorophyta (p. 1062)
zoospore (p. 1063)
isogamy (p. 1064)
heterogamy (p. 1065)
anisogamy (p. 1065)
oogamy (p. 1065)
holdfast (p. 1067)
sporangium (p. 1067)
alternation of generations (p. 1068)
gametophyte (p. 1068)
sporophyte (p. 1068)
Phaeophyta (p. 1068)
fucoxanthin (p. 1070)
gametangia (p. 1070)
antheridium ⎫
oogonium ⎭ types of gametangia (p. 1071)
stipe (p. 1072)
blade (p. 1072)
Rhodophyta (p. 1072)
archegonium (p. 1073)
Bryophyta (p. 1074)
Tracheophyta (p. 1077)
Psilopsida (p. 1078)
Lycopsida (p. 1079)
sporophyll (p. 1080)
megaspore (p. 1080)
microspore (p. 1080)
homosporous (p. 1080)
heterosporous (p. 1080)

Sphenopsida (p. 1081)
Pteropsida (p. 1081)
Spermopsida (p. 1084)
angiosperm (p. 1084)
gymnosperm (p. 1085)
conifer (p. 1085)
cone (p. 1086)
pollen grain (p. 1087)
micropyle (p. 1087)
ovule (p. 1087)
pollen tube (p. 1088)
seed (p. 1089)
sepal ⎫
petal ⎪
corolla ⎪
stamen ⎪
filament ⎪
anther ⎬ parts of a flower (pp. 1090–91)
ovary ⎪
pistil ⎪
style ⎪
stigma ⎪
carpel ⎪
ovule ⎭
fusion nucleus (p. 1092)
endosperm (p. 1092)
fruit (p. 1092)
dicotyledon (p. 1093)
monocotyledon (p. 1093)

SUMMARY

The divisions of the plant kingdom have traditionally been separated into two groups: *Thallophyta* and *Embryophyta*. The thallophytes show little tissue differentiation, their reproductive structures are frequently unicellular and always lack a jacket of sterile cells, and embryonic development takes place outside the female reproductive structure. By contrast, the embryophytes show more tissue differentiation, their multicellular reproductive organs are surrounded by sterile jacket cells, and their early embryonic development occurs within the female reproductive organ.

In this classification system, the algal groups with many multicellular members are included in the Thallophyta; they belong in the divisions Chlorophyta, Phaeophyta, and Rhodophyta. The Embryophyta include two divisions, the Bryophyta and Tracheophyta.

Thallophyta The *Chlorophyta* (green algae) are probably the group from which the land plants arose. The green algae possess chlorophylls *a* and *b* and carotenoids. Many divergent evolutionary tendencies can be perceived within the Chlorophyta, including: the evolution of motile colonies, a change to nonmotile unicells and colonies, the evolution of coenocytic (multinucleated) organisms, and the evolution of multicellular filaments and three-dimensional leaflike thalluses (plant bodies with little tissue differentiation).

The unicellular green algae such as *Chlamydomonas* reproduce asexually by *zoospores* and sometimes show a simple form of sexual reproduction. There are no separate male and female individuals, and all the gametes are alike (*isogamy*). Isogamy is probably the primitive condition in plants. The zygote is the only diploid stage in the life cycle; dominance of the haploid stage is characteristic of most primitive plants.

The *volvocine series* is a series of genera showing gradual progression from the unicellular condition to an elaborate colonial organization. A number of evolutionary changes are seen in this series: a change from unicellular to colonial life; increased coordination of activity and increased interdependence among the cells; increased division of labor; and a gradual change from isogamy to *anisogamy* (gametes of two different sizes), to *oogamy* (motile male gamete, nonmotile female gamete).

The life cycles of many green algae include a multicellular stage. Often this stage is a filamentous thallus, which may be branching or nonbranching, depending on the species. Sexual reproduction is usually isogamous. The multicellular stage is haploid; the zygote is the only diploid stage.

Ulva is a multicellular green alga with an expanded leaflike thallus two cells thick. Its life cycle includes both multicellular haploid (*gametophyte*) and multicellular diploid (*sporophyte*) stages. In the life cycle, the haploid zoospores (stage 1) divide mitotically to produce haploid multicellular thalluses (stage 2). These may reproduce asexually by zoospores or sexually by gametes (stage 3). Fusion of two gametes produces diploid zygotes (stage 4), which divide mitotically to produce diploid multicellular thalluses (stage 5). Certain reproductive cells of the diploid thallus divide by meiosis to form haploid zoospores, which begin a new cycle. This type of life cycle exhibits *alternation of generations* in that a haploid multicellular phase alternates with a diploid multicellular phase.

Multicellularity in plants arose first in the gametophyte and many green algae have no sporophyte stage, but *Ulva* shows a more advanced life cycle in that both sporophyte and gametophyte stages are present and equally important.

The *Phaeophyta* (brown algae) are multicellular and almost exclusively marine. The thallus may be a filament or a large, complex, three-dimensional structure. The Phaeophyta possess chlorophyll *a* and *c* and the brownish pigment *fucoxanthin*. Reproduction may be asexual or sexual. The life cycle usually exhibits alternation of generations. In many forms the sporophyte and gametophyte are essentially equal, but in other forms the sporophyte is the dominant stage. In a few forms the gametophyte stage is completely absent.

Although the Phaeophyta are thallophytes, they do show some tissue differentiation. They are complex plants that have convergently evolved many similarities to the vascular plants.

The *Rhodophyta* (red algae) are mostly marine seaweeds. Most are multicellular. The Rhodophyta possess chlorophylls *a* and *d*, phycoerythrins, and phycocyanins. The accessory pigments are important in absorbing light and transferring the energy to chlorophyll *a* for photosynthesis. The life cycles of the red algae are complex; alternation of generations is usual. Flagellated cells never occur.

The movement onto land The Embryophyta have evolved numerous adaptations for life on land. Life probably arose in the water, and the evolutionary move to land was not a simple one. A terrestrial environment poses many problems for plants, including obtaining enough water, transporting water and dissolved materials from one part of the plant to another, preventing excessive evaporation, maintaining a moist surface for gas exchange, supporting the plant against gravity, carrying out reproduction, and withstanding extreme environmental fluctuations.

Much of the evolution of the embryophyte plants can best be understood in terms of adaptations to help solve these problems. The sterile jacket cells around the multicellular sex organs help protect the enclosed gametes from desiccation. Such male and female sex organs are known as *antheridia* and *archegonia*. All embryophytes are oogamous, and fertilization and early embryonic development take place within the moist environment of the archegonia. The aerial surfaces of the plant are often covered by a waxy cuticle, which prevents excessive water loss.

These plants are biochemically similar to the green algae; they possess chlorophylls *a* and *b*, and the reserve material is starch.

Bryophyta The *Bryophyta* (liverworts, hornworts, and mosses) are small plants that grow only in moist places. Moisture must be available for the flagellated sperm to swim to the egg. Most bryophytes lack vascular tissues for transport and support. The bryophytes show alternation of generations; the haploid stage is dominant. The sporophyte is attached to the gametophyte; it consists of a foot embedded in the gametophyte, a stalk, and a capsule (sporangium). Meiosis occurs within the sporangium, producing haploid spores that fall to the ground, germinate, and eventually develop into mature gametophytes.

Tracheophyta All members of the division *Tracheophyta* possess four important attributes lacking in the algae: a protective layer of sterile jacket cells around the reproductive organs; multicellular embryos retained within the archegonia; cuticles on aerial parts; and xylem. All four are fundamental adaptations for a terrestrial existence.

The *Psilopsida* are the most primitive of the vascular plants. They lack leaves and have no true roots, although they do have underground stems with rhizoids. Photosynthesis occurs in the green, branching stems, some of which bear sporangia at their tips. The life cycle shows an alternation of generations.

The *Lycopsida* (club mosses) show many advances over the Psilopsida. They have true roots, stems, and leaves. Some of the leaves have become specialized for reproduction and bear sporangia on their surfaces; these leaves are called *sporophylls*. In many Lycopsida the sporophylls are compacted together to form a club-shaped structure (strobilus). Some lycopsids produce only one kind of spore (i.e. they are *homosporous*); these develop into a gametophyte that bears both antheridia and archegonia. Others produce two types of spores, large *megaspores* and smaller *microspores*; these plants are *heterosporous*. The megaspores develop into female gametophytes, the microspores into male gametophytes.

Like the lycopsids, the *Sphenopsida* (horsetails) have true roots, stems, and leaves. Some ancient sphenopsids were large trees; much of today's coal was formed from these plants.

The *Pteropsida* (ferns) probably evolved from the Psilopsida. The ferns are fairly advanced plants with a well developed vascular system. The large, leafy fern plant is the diploid sporophyte phase. Most ferns are homosporous; the spores are produced in sporangia on the fertile leaves (sporophylls). Upon germination the spores develop into small, independent, nonvascular gametophytes bearing antheridia and archegonia. The flagellated sperm must swim to the archegonium. The gametophytes therefore require a moist habitat; thus ferns are restricted to a moist environment.

The dominant and best adapted land plants are those belonging to the *Spermopsida* (seed plants). The gametophyte stage of these plants has been much reduced, and the sperm are not free-swimming, flagellated cells. In addition, the embryo is protected from desiccation by a seed coat. Heterospory is characteristic of all seed plants. There are six classes of Spermopsida: five of these are often referred to as the gymnosperms, and the sixth class is the Angiospermae.

The *gymnosperms* appear to be older and less specialized than the angiosperms. The best-known group of gymnosperms is the conifers (class Coniferae), or evergreens. The leaves of most of these plants are small needles or scales.

The pine provides a good example of the seed method of reproduction. The pine tree is the diploid sporophyte stage. It bears two kinds of *cones*: large female cones, which produce spores that develop into female gametophytes; and small male cones, which produce spores that develop into *pollen grains* (the male gametophyte). The pollen grains are released into the air; some may land on the scales of a female cone. The pollen grain develops a *pollen tube*, which grows through the tissues of the female sporangium into one of the archegonia. There it discharges its two sperm nuclei, one of which fertilizes the egg. The zygote develops into an embryo, and a tough seed coat surrounds it and its stored food material. Finally, the *seed* is shed from the cone.

The Angiospermae are the dominant land plants. Their reproductive structures are the flowers. In a flower the outer leaflike *sepals* protect the inner floral parts; the *petals* often attract animal pollinators; the *stamens* (filament and anther) are the male reproductive organs; and the central *pistil* is the female reproductive organ. Each pistil consists of a *stigma, style,* and *ovary*. Within the ovary are one or more sporangia called *ovules*. Meiosis occurs in each ovule, producing one functional megaspore, which divides to produce (in most species) a haploid, seven-celled, eight-nucleate, female gametophyte.

Each anther has sporangia in which meiosis occurs, producing haploid microspores. These develop into thick-walled, two-nucleate pollen grains—the male gametophytes. A pollen grain germinates when it lands on the pistil; a pollen tube grows through the pistil, enters the ovary, and discharges two sperm nuclei into the female gametophyte. "Double fertilization" occurs: one sperm nucleus fertilizes the egg, the other combines with the two polar nuclei to form a triploid nucleus, which will give rise to the *endosperm*.

After fertilization the ovule matures into a seed, which consists of a seed coat, stored food (endosperm), and embryo. The seeds are enclosed in fruits that develop from the ovaries and associated structures.

Angiosperms differ from gymnosperms in possessing xylem vessels and in many reproductive characteristics: angiosperms have greatly reduced gametophytes; their pollen tubes have farther to grow; they have *"double fertilization"*; their endosperm is triploid; and the seeds are enclosed within fruits.

For these reasons the angiosperms are the most successful and dominant land plants. They are divided into two subclasses: the Monocotyledonae (embryo has one cotyledon) and the Dicotyledonae (embryo has two cotyledons). There are many basic differences between the two groups.

QUESTIONS

Testing recall

Match each statement below with the division to which it most clearly applies. Answers may be used once, more than once, or not at all.

a   Chlorophyta (green algae)
b   Phaeophyta (brown algae)
c   Rhodophyta (red algae)

1   The division contains many unicellular and multicellular forms. (p. 1062)

2   The division members possess the pigment fucoxanthin. (p. 1070)

3   The division members possess the pigments chlorophyll *a* and chlorophyll *b*. (p. 1062)

4   Some members of the division show a primitive type of tissue differentiation, having phloemlike tissue. (p. 1072)

5   Many forms show alternation of generations with a dominant sporophyte. (p. 1071)

6   Accessory pigments enable plants in this division to survive in deep water. (p. 1072)

7   Land plants probably evolved from this division. (p. 1062)

8   Members of this division typically possess the red pigment phycoerythrin in addition to chlorophylls *a* and *d*. (p. 1072)

9   The members of this division lack flagellated cells in life cycle. (p. 1073)

10   The members of this division store starch as reserve material. (p. 1062)

Mark each characteristic below T *if it is found in the Thallophyta, and* E *if it is found in the Embryophyta. (Some characteristics may be found in both groups.)*

11   tissue differentiation (p. 1061)

12   isogamy (p. 1064)

13   multicellular sex organs (pp. 1061, 1070)

14   sterile jacket cells around reproductive organs (p. 1061)

15   dominant gametophyte (pp. 1068, 1075)

16   zygote development outside female reproductive organs (p. 1061)

17   cuticle on aerial parts (p. 1074)

18   embryo development within archegonium (pp. 1061, 1076)

19   The table below lists some of the problems faced by land plants. Complete the chart by filling in the features the land plants evolved to solve these problems. The completed chart will be a useful study aid.

Problem	Bryophyta	Psilopsida	Lycopsida	Sphenopsida	Pteropsida	Spermopsida
Obtaining enough water						
Transport of materials						
Supporting plant against the pull of gravity						
Preventing excessive water loss						
Carrying out reproduction with little water						
Protection for the embryo						

Testing knowledge and understanding

Choose the one best answer. (Questions 20–24 include groups covered in other chapters.)

20   The evolutionary trend toward multicellularity is best demonstrated by comparing various members of the

a   Phaeophyta (brown algae).
b   Bryophyta.
c   Cyanobacteria.
d   Chlorophyta (green algae).
e   Rhodophyta (red algae).   (p. 1066)

21   In which one of the following divisions would you *least* expect to find a multicellular sporophyte stage in the life cycle?

a   Euglenophyta   d   Rhodophyta
b   Chlorophyta   e   Bryophyta
c   Phaeophyta   (pp. 1062, 1057)

22   In the ocean, which algae would normally be found in the deepest water?

a   Chlorophyta (green)
b   Chrysophyta
c   Phaeophyta (brown)
d   Rhodophyta (red)   (p. 1072)

23   The algal division that has evolved the most complex multicellular thallus is

a   Chlorophyta (green).
b   Chrysophyta.
c   Phaeophyta (brown).
d   Rhodophyta (red).   (p. 1072)

24   "Higher" plants have probably evolved from which algal line?

a   Chlorophyta (green)
b   Rhodophyta (red)
c   Phaeophyta (brown)
d   Chrysophyta
e   Pyrrophyta (dinoflagellates)   (p. 1075)

25 Which one of the following best describes the type of sexual reproduction in which the gametes are all motile but are of two sizes, one type being smaller than the other?

a homospory   d oogamy
b heterospory   e isogamy
c anisogamy   (p. 1065)

26 A certain plant is photosynthetic, does not use true starch as its principal carbohydrate storage product, has a rudimentary phloem-like tissue but no xylem, and possesses multicellular sex organs not enclosed by sterile jacket cells. To which group does the plant probably belong?

a Bryophyta
b Tracheophyta
c Rhodophyta (red algae)
d Chlorophyta (green algae)
e Phaeophyta (brown algae) (pp. 1070–72)

27 Green algal, red algal, and brown algal species are classified as belonging to the kingdom Plantae. They are not included in the kingdom Protista because

a most of these species are multicellular.
b all of these species are multicellular.
c the chlorophyll pigments of these species are unlike those of the Protista.
d all of these algae are marine organisms, whereas all of the Protista are freshwater organisms.
e all algae in the Protista are microscopic in size, whereas all of these species are macroscopic in size.   (p. 1062)

Questions 28–30 refer to the following diagram of a moss plant.

28 Which of the labeled structures (*a, b, c, d, e*) is (are) composed of diploid cells? (p. 1075)

29 In which part(s) of the plant does photosynthesis primarily occur? (p. 1075)

30 In which part of the plant are the spores produced? (p. 1076)

31 The movement of plants from water to land and their subsequent increase to large size was made possible by the appearance of

a collenchyma.   d epidermis.
b parenchyma.   e cortex.
c xylem.   (p. 1078)

32 As plants occupied drier habitats they could no longer rely on the transport of sperm in water. The gymnosperms overcame this problem by evolving

a pollen.
b cones.
c motile sperm.
d eggs and sperm in the same plant.
e mature sperm produced only in the rainy season.   (p. 1087)

33 Which one of the following is *not* an embryophyte?

a moss
b fern
c horsetail
d giant kelp (a "seaweed")
e more than one of the above
                        (pp. 1070, 1078–81)

34 Algae often have characteristic accessory pigments. The accessory pigment in the Rhodophyta (red algae) is adaptive because

a   it serves as protective coloration.
b   it absorbs light that penetrates deep into water.
c   these algae have no chlorophyll.
d   it makes the algae inedible.   (p. 1072)

35 In plants, the gametophyte generation is produced by

a   fertilization of an egg cell by a sperm cell.
b   the formation of four gametes by meiosis.
c   mitosis of sporophyte cells.
d   haploid spores that divide by mitosis.
e   fusion of male and female sporophyte cells.   (p. 1068)

36 In plants, the sporophyte generation is produced by

a   fusion of two diploid cells.
b   growth of a sperm cell.
c   mitosis of haploid cells.
d   mitosis of diploid cells.
e   meiosis.   (p. 1068)

37 Which one of the following does *not* function specifically as an adaptation to the land environment?

a   multicellular sex organ with jacket cells
b   structures to attach the plant to the substrate
c   cuticle
d   embryonic development within the archegonium
e   vascular tissue   (p. 1072)

38 You are given an unknown plant to study in the laboratory. You find that it has chlorophyll, but no xylem. Its multicellular sex organs are enclosed in a layer of jacket cells. Its gametophyte stage is free-living. The plant probably belongs in the

a   Chlorophyta (green algae).
b   Phaeophyta (brown algae).
c   Bryophyta.
d   Rhodophyta (red algae).
e   Tracheophyta.   (pp. 1074−76)

39 The earliest tracheophytes, which probably gave rise to the other vascular plants, belong to the subdivision

a   Spermopsida (seed plants).
b   Lycopsida (club mosses).
c   Psilopsida (psilopsids).
d   Pteropsida (ferns).   (p. 1078)

40 The sporophylls of ferns produce spores by

a   mitosis.
b   meiosis.   (p. 1082)

41 A plant has vascular tissue, scalelike leaves arranged in whorls, a jointed stem, and a reproductive cone but no seeds. The plant probably should be classified as a

a   moss.   d   horsetail.
b   club moss.   e   conifer.
c   fern.   (p. 1081)

42 The seedless vascular plants are regarded as less highly adapted to life on land than the seed plants. Which one of the following traits of seedless plants illustrates this condition?

a   distinctive diploid and haploid stages
b   presence of vascular tissue
c   dominant sporophyte generation
d   sperm cells with flagella
e   sporophyte attached to gametophyte
                                           (p. 1084)

43 Along with the seed, the seed plants have evolved several additional adaptations to the land environment. Which one of the following is *not* such an adaptation?

a   Flagellated gametes are not required for seed formation.
b   The female gametophyte is protected from desiccation by the surrounding tissues of the sporophyte.
c   The seed and/or associated structures serve as a means of dispersal.
d   The method of seed formation introduces a new type of genetic recombination.
e   The seed has its own food supply for the enclosed embryo.   (pp. 1089−93)

44 Which one of the following statements is *true*?

a   Gymnosperms have a free-living gametophyte stage in their life cycle.
b   Most gymnosperms have flagellated sperm cells.
c   The endosperm of a spruce seed is haploid.
d   The spruce seed is protected by a fruit.
                                      (pp. 1085−89)

45 Place the following groups of plants in order beginning with those that first appeared on the earth and progressing toward those that appeared most recently in time.

a   gymnosperms, angiosperms, ferns, moss, algae
b   algae, moss, ferns, gymnosperms, angiosperms
c   algae, ferns, moss, gymnosperms, angiosperms
d   algae, moss, ferns, angiosperms, gymnosperms
e   moss, algae, angiosperms, gymnosperms, ferns   (pp. 1062−90)

46 In angiosperms, the sporophylls are the

a stamens and pistils.
b sepals and petals.
c sepals and stamens.
d anthers and petals.
e sepals and pistils. (p. 1090)

47 Which one of the following traits is characteristic of *both* the gymnosperms and the angiosperms?

a double fertilization to form embryo and endosperm
b haploid endosperm
c fruit derived from ovaries
d sporophylls modified into stamens and pistils
e heterospory (pp. 1090–93)

48 The female gametophyte of a flowering plant

a contains a single haploid nucleus.
b gives rise to the ovary.
c fuses with a pollen grain during pollination.
d is produced in the ovule.
e is fertilized by the filament. (p. 1091)

49 In angiosperms, the gametophytes are produced within the

a stamens and pistils.
b sepals and petals.
c sepals and stamens.
d anthers and petals.
e sepals and pistils. (p. 1091)

50 A flower has six petals, three sepals, nine stamens, three ovules, and one ovary. How many seeds will this flower produce?

a one   d nine
b three   e 27
c six   (p. 1091)

51 The plant in question is a

a monocot.
b dicot.   (p. 1093)

52 The endosperm of a flowering plant is triploid because

a it results from the fusion of a normal gamete and a diploid gamete.
b two male nuclei fuse with one female nucleus.
c two polar nuclei fuse with a sperm nucleus.
d none of the above (p. 1092)

53 Which one of the following is a correct sequence of processes that take place when a flowering plant reproduces?

a meiosis—fertilization—ovulation—germination

b fertilization—meiosis—nuclear fusion—formation of endosperm
c meiosis—pollination—nuclear fusion—formation of embryo and endosperm
d growth of pollen tube—pollination—germination—fertilization
e meiosis—mitosis—nuclear fusion—pollination (p. 1091)

54 A fruit is a mature

a embryo.   d stem.
b seed.   e root.
c ovary.   (p. 1092)

55 In tobacco plants, the diploid number is 48. How many chromosomes do tobacco endosperm cells have?

a 12   d 72
b 24   e 96
c 48   (p. 1092)

56 Which one of the following is *not* a characteristic of monocots?

a absence of secondary growth
b scattered vascular bundles
c flower parts in fours or fives
d one cotyledon
e parallel leaf veins (p. 1093)

Match each function below with the associated part of the flower. Answers may be used once, more than once, or not at all.

a anther   d sepal
b ovary   e stigma
c ovule

57 contains the female gametophyte (p. 1090)

58 produces the male gametophyte (p. 1090)

59 matures to form the fruit (p. 1092)

60 site on which pollen is deposited by pollinator (p. 1090)

61 forms the seed (p. 1092)

For further thought

1 Draw an evolutionary tree for the kingdom Plantae.

2 The Phaeophyta are complex plants that have convergently evolved many similarities to the vascular plants. Many live in the intertidal zone, where they are subject to wave action and the danger of desiccation when the tide is out. What adaptations have they evolved to survive in such a habitat?

3 As tracheophytes evolved, the sporophyte generation became more dominant, and the

gametophyte smaller and more dependent on the sporophyte. What is the adaptive advantage of the sporophyte stage over the gametophyte?

4 What is the relationship between the gametophyte and sporophyte of a liverwort? a fern? a pine tree? a flowering plant?

5 Discuss some of the adaptations by means of which cross-pollination is encouraged in the flowering plants.

6 Mendel, in his experiments with the garden pea, found that peas are normally self-fertilized since maturation of the floral organs occurs before the flower has opened. What procedure would you follow to bring about cross-fertilization in such a plant?

7 The movement to land from an aquatic environment presented a number of problems for plant reproduction. Describe these problems and explain how the angiosperms have adapted to survive and reproduce on land. In your answer, diagram the life cycle of the angiosperms and describe in some detail the process of sexual reproduction.

11 E—and a little in the kelps
12 T
13 E—and in some brown algae
14 E
15 T, E
16 T
17 E
18 E
19 Chart: Information can be found on the following pages: Bryophyte, pp. 1074–77; Psilopsida, p. 1079; Lycopsida, pp. 1079–80; Sphenopsida, p. 1081; Pteropsida, pp. 1081–84; Spermopsida, pp. 1084–94. See also Tables 40.1 (p. 1074) and 40.2 (p. 1084).

Testing knowledge and understanding

20	d	31	c	42	d	52	c
21	a	32	a	43	d	53	c
22	d	33	d	44	c	54	c
23	c	34	b	45	b	55	d
24	a	35	d	46	a	56	c
25	c	36	d	47	e	57	c
26	e	37	b	48	d	58	a
27	a	38	c	49	a	59	b
28	d, e	39	c	50	b	60	e
29	c	40	b	51	a	61	c
30	e	41	d				

ANSWERS

Testing recall

1	a	4	b	7	a	9	c
2	b	5	b	8	c	10	a
3	a	6	c				

Chapter 41

THE FUNGAL KINGDOM

A GENERAL GUIDE TO THE READING

This chapter provides a brief overview of the fungal kingdom. Fungi used to be included in the plant kingdom because they are predominantly sedentary and because their cells have walls. However, they differ from plants in so many ways that they are now placed in a kingdom of their own. Your understanding of fungi will be enhanced if you concentrate on the following topics.

1   General characteristics of fungi. The main characteristics of fungi are described on page 1095.

2   Forms of sexual reproduction. Since the characteristics of sexual reproduction are important in distinguishing among the fungi, you will want to learn these characteristics for each division. Figures 41.3 (p. 1097), 41.5 (p. 1099), and 41.10 (p. 1102) are the key to understanding the various forms of reproduction in fungi.

KEY CONCEPTS

1   The fungi are a diverse group of sedentary, absorptive heterotrophic organisms; they have great economic importance, and play a vital role as decomposers. (pp. 1095–96)

2   The cells of fungi are eucaryotic. They have walls, but are usually organized into thread-like hyphae in which the cellular partitions may be absent or incomplete. (p. 1095)

3   The characteristics of sexual reproduction are significant in distinguishing among the several divisions of fungi. (p. 1096)

OBJECTIVES

After studying this chapter and reflecting on it, you should be able to carry out the following objectives.

1   Give the main characteristics of the fungi. (p. 1095)

2 Compare the four phyla with respect to: occurrence, presence or absence of cellular partitions in their hyphae, mode of reproduction, and economic importance. (pp. 1096–1102)

3 Give four ways in which the Oomycota differ from the other fungi. (p. 1096)

4 Using a diagram such as Figure 41.3 (p. 1097), describe the life cycle of *Rhizopus*. Explain why fungi like *Rhizopus* are referred to as conjugation fungi. (pp. 1096–97)

5 Using a diagram such as Figure 41.5 (p. 1099), describe the process of sexual reproduction in a cup fungus. Then, using a diagram such as Figure 41.6 (p. 1099), describe how a sac fungus reproduces asexually. (pp. 1098–99)

6 Using a diagram such as Figure 41.10 (p. 1102), describe the process of sexual reproduction in a club fungus. (pp. 1101–2)

KEY TERMS

The following terms are important in this chapter; you should become familiar with them.

hypha (p. 1095)
mycelium (p. 1095)
chitin (p. 1095)
Oomycota (p. 1096)
Zygomycota (p. 1097)
rhizoid (p. 1097)
Ascomycota (p. 1098)
conidium (p. 1098)
budding (p. 1098)
lichen (p. 1101)
Basidiomycota (p. 1101)
basidium (p. 1102)

SUMMARY

The fungi are saprophytic or parasitic organisms that are primarily multicellular or multinucleate. The partitions between the cells are generally either absent or partial, so the cytoplasm is continuous. The cell wall usually contains *chitin*. The cells are usually organized into branched filaments called *hyphae*, which form a mass called a *mycelium*. Reproduction may be sexual or asexual, but the haploid stages are usually dominant. The fungi are important decomposers and many are of great economic importance. Four phyla of true fungi are recognized: Oomycota, Zygomycota, Ascomycota, and Basidiomycota. Characteristics related to sexual reproduction are important in defining the groups.

The *Oomycota* (water molds) are generally aquatic organisms that differ markedly from the other phyla in that they have flagellated zoospores, are oogamous and diploid, and lack chitin in their cell walls. The hyphae lack cellular partitions (septa).

Members of the *Zygomycota* (conjugation fungi) are widespread as saprophytes in soil and dung. Their hyphae characteristically lack cross-walls, so the multinuclear cytoplasm is continuous (coenocytic). Reproduction may be asexual (by spores) or sexual. Sexual reproduction occurs by conjugation between two cells from hyphae of two different mycelia. These gamete cells fuse to form a zygote, which develops a thick wall and becomes dominant. At germination, the nucleus undergoes meiosis and a hypha develops. It releases asexual spores that grow into new mycelia. *Rhizopus* (black bread mold) is an example of this phylum.

The *Ascomycota* (sac fungi) are a diverse group, varying from unicellular yeasts through powdery mildews to cup fungi. The hyphae are septate, but the septa usually are incomplete and have large holes. All produce a reproductive structure called an *ascus* during their sexual cycle. An ascus is a sac within which haploid spores, usually eight, are produced. These fungi also reproduce asexually by means of special spores called *conidia*. The fungal member of a *lichen* is generally an Ascomycota.

The *Basidiomycota* (club fungi) include the puffballs, mushrooms, toadstools, and bracket fungi. The large fruiting bodies of these fungi are composed of compacted, septate hyphae. During sexual reproduction, hyphae from two different mycelia with uninucleate cells unite and give rise to hyphae with binucleate cells. Certain terminal cells become zygotes when their two nuclei fuse. The zygote then becomes a *basidium*, a club-shaped reproductive structure. Meiosis occurs within the basidium, and four haploid spores are produced. Each spore may give rise to a new mycelium.

QUESTIONS

Testing recall

1 *Complete the following chart. It will be a useful study aid.*

	Hyphae (septate or non-septate)	Flagellated Cells (yes or no)	Dominant Stage (haploid or diploid)	Sexual Reproductive Structure	Representative Members	Economic Importance
Oomycota						
Zygomycota						
Ascomycota						
Basidiomycota						

Testing knowledge and understanding

Choose the one best answer.

2 In most of the fungi, the dominant stage is

  a haploid.
  b diploid.   (p. 1096)

3 Which one of the following pairs is *incorrectly* matched?

  a Oomycota–diploid
  b Zygomycota–*Rhizopus*
  c Zygomycota–septate hyphae
  d Ascomycota–lichens
  e Ascomycota–conidia   (p. 1096)

4 Which one of the following is *not* a characteristic of fungi?

  a absorptive heterotrophic forms
  b body made up of threadlike hyphae
  c reproduction by spores
  d both saprophytic and autotrophic forms
  e cell wall containing chitin   (p. 1095)

5 Fungi are

  a heterotrophic.
  b autotrophic.
  c photosynthetic.
  d more than one of the above.   (p. 1095)

6 Zygotes of fungi are

  a haploid.
  b diploid.   (p. 1102)

7 In the fungi, the spores are

  a haploid.
  b diploid.   (p. 1097)

8 During their sexual cycle, the Zygomycota (conjugation fungi) produce a(n)

  a conidium.
  b ascus.
  c basidium.
  d flagellated gamete.
  e thick-walled zygote.   (p. 1097)

9 The fungi that produce flagellated zoospores and have oogamous reproduction belong to the

   a Ascomycota (sac fungi).
   b Oomycota (water molds).
   c Zygomycota (conjugation fungi).
   d Basidiomycota (club fungi).   (p. 1096)

10 Yeast is classified as one of the

   a Protista.
   b Ascomycota (sac fungi).
   c Basidiomycota (club fungi).
   d Monera.
   e Zygomycota (conjugation fungi).
                               (p. 1098)

11 In the stalk of a mushroom each of the cells

   a is haploid.
   b is diploid.
   c has two separate haploid nuclei.
   d has two separate diploid nuclei.
   e has four separate haploid nuclei.
                                 (p. 1102)

12 Which one of the following is *not* found in the Basidiomycota (club fungi)?

   a a mycelium
   b a basidium
   c conidia
   d binucleate hyphae
   e hyphae with cross walls   (pp. 1101–2)

13 When we eat a mushroom we are eating

   a a mycelium.   *d* hyphae.
   b the fruiting body. *e* all of the above
   c basidia.   (pp. 1101–2)

14 A student was given an unknown organism to identify. Though it looked like a solid mass of tissue, the student found it was made up of a mass of filaments, the cells of which lacked chlorophyll. Cross walls were present in the filaments, however, and two nuclei were found in many of the cells. The reproductive structures were club shaped, and four spores were found on the tip of these structures. The student decided that this plant probably belonged in the

   a bacteria.
   b Ascomycota (sac fungi).
   c Oomycota (water molds).
   d Basidiomycota (club fungi).
   e Zygomycota (conjugation fungi).
                           (pp. 1101–2)

ANSWERS

Testing recall

1 Information of the chart can be found as follows: Oomycota, p. 1096; Zygomycota, pp. 1096–97; Ascomycota, pp. 1098–1101; Basidiomycota, pp. 1101–2.

Testing knowledge and understanding

2	*a*	6	*b*	9	*b*	12	*c*
3	*c*	7	*a*	10	*b*	13	*e*
4	*d*	8	*e*	11	*c*	14	*d*
5	*a*						

Chapter 42

THE ANIMAL KINGDOM: INVERTEBRATES

A GENERAL GUIDE TO THE READING

This chapter and the next provide an overview of the diversity of animals and the evolutionary relationships among them. Much of the material should be familiar to you; the different organ systems and how they work were discussed in detail in Part II, "The Biology of Organisms," and your text includes cross references to this material. The emphasis in this chapter is on evolutionary relationships, so you will want to concentrate on them as you read. One interpretation of these relationships is shown in the phylogenetic tree in Figure 42.30 (p. 1118), and Table 42.1 (p. 1146) provides an enormously helpful summary and comparison of the major characteristics of the animal phyla. In addition, the following topics in Chapter 42 deserve special attention.

1   Sponges. The sponges (pp. 1103–4) are believed to have evolved from the Protista independently of other multicellular animals, as Figure 42.30 (p. 1118) makes clear.

2   Radial symmetry vs. bilateral symmetry. Since you will be studying both the radiate phyla and the Bilateria, you need to understand the differences between radial and bilateral symmetry. The definition of radial symmetry in Chapter 6 (p. 161) may be helpful.

3   The origin of multicellularity. Two hypotheses for the origin of multicellularity are described (pp. 1108–12). You will note that the sequence of organisms proposed by the planuloid hypothesis bears some resemblance to the volvocine series in the evolution of multicellularity in plants, as discussed in Chapter 40 (pp. 1064–66).

4   Presence of a coelom. A coelom is a cavity between the digestive tract and the body wall. The absence or presence of a coelom, and the type of coelom, when present, are important in animal classification. Four different types of body are discussed: the acoelomate (p. 1112), the pseudocoelomate (p. 1119), and

those having one or the other of the true coeloms—the coelom that arises as a split in an initially solid mass of mesoderm and the coelom that arises as cavities in mesodermal pouches (pp. 1117–19). Figure 42.32 (p. 1119) is a useful summary of body types.

5 Divergence of the Protostomia and Deuterostomia. The two main branches of the animal kingdom are the Protostomia and the Deuterostomia. Because the differences between the protostomes and the deuterostomes are fundamental to our classification scheme, you will want to make sure you understand the characteristics separating them (pp. 1117–19).

6 Molluscs. Learn the fundamental molluscan body plan, and then see how it is modified in the various classes (pp. 1123–28).

7 Segmented worms. Look closely at the polychaetes, which are considered to show the more primitive features of this phylum, discussed on pages 1128–29. The earthworms and their relatives probably evolved from the polychaetes.

8 Arthropods. In studying the arthropods (pp. 1130–39), pay particular attention to the insects, since they are an enormously successful group of organisms.

9 The relationships between echinoderms, hemichordates, and chordates. You need to learn the evolutionary relationships linking these three diverse groups of animals (pp. 1145–46).

Metazoa (multicellular animals that develop from embryos). (pp. 1108–112)

5 The Platyhelminthes (flatworms) and the Nemertina (proboscis worms) are thought to be the most primitive bilaterally symmetrical animals; in each, the body is composed of three well-developed germ layers, and is acoelomate. (pp. 1112–16)

6 A major split probably occurred in the animal kingdom soon after the emergence of a bilateral organism. One evolutionary line led to the phyla in which the blastopore becomes the mouth; these are called Protostomia. The other line led to the phyla in which the blastopore becomes the anus and a new mouth is formed; these are called the Deuterostomia. (pp. 1116–19)

7 In several protostome phyla, the body cavity is not completely enclosed by mesoderm; it is a pseudocoelom. All the other protostome phyla have true coeloms (at least at some stage of development); in most groups these arise as a split in the initially solid mass of mesoderm. (pp. 1119–22)

8 The Arthropoda are generally regarded as the most highly evolved representatives of the protostome line. (pp. 1130–39)

9 The Deuterostomia form the second main branch of the animal kingdom. (p. 1139)

10 The phylum Echinodermata, the most primitive of the major deuterostome phyla, is linked in important ways to the Hemichordata and the Chordata. (pp. 1139–46)

KEY CONCEPTS

1 Characteristics such as level of organization, type of symmetry, presence or absence of segmentation, nature of embryonic development, and form of larva, if any, have been used to establish hypothetical phylogenetic relationships among the various animal groups. (pp. 1103–46)

2 The Porifera (sponges) differ greatly from other multicellular animals, and are believed to have evolved on a line of their own. (pp. 1103–4)

3 The two radiate phyla—the Cnidaria and the Ctenophora—comprise radially symmetrical animals with bodies that are relatively simple in structure. (pp. 1104–8)

4 Several different hypotheses have been proposed for the origin of multicellularity in the

OBJECTIVES

After studying this chapter and reflecting on it, you should be able to carry out the following objectives.

1 Describe three characteristics of the Porifera (sponges), and explain why they are believed to have evolved independently of other multicellular organisms and hence to be not closely related to the other animal groups. Indicate a possible evolutionary relationship between the sponges and the animal-like protists. (pp. 1103–4)

2 Describe the distinguishing characteristics of the Coelenterata (also called the Cnidaria); then name the three coelenterate classes, and list the identifying features of each. (pp. 1105–8)

3 Using a diagram such as Figure 42.6 (p. 1106), describe the life cycle of a typical hydrozoan such as *Obelia*; then explain how the life cycles of the jellyfish (a scyphozoan) and the sea anemone (an anthozoan) would differ from that of *Obelia*.

4 Using diagrams such as Figures 42.13 (p. 1109), 42.14 (p. 1109), 42.15 (p. 1110), and 42.19 (p. 1111), compare the gastraeae, planuloid, ciliate, and plakula hypotheses of the origin of the Eumetazoa. Give the advantages and disadvantages of each hypothesis. (pp. 1109–12)

5 Discuss the advantages of bilateral symmetry over radial symmetry. (p. 1112. See also p. 460)

6 Describe the major characteristics of the Platyhelminthes; then name the three classes in this phylum, and list the distinguishing features of each. Discuss the tapeworm's special adaptations for parasitism. (pp. 1112–16)

7 Specify two important evolutionary advances found in the nemertines.

8 Complete the following chart, comparing the Protostomia and the Deuterostomia. (pp. 1116–19)

	Protostomia	Deuterostomia
Fate of the blastopore		
Type of early cleavage (determinate or indeterminate, radial or spiral)		
Origin of the mesoderm		
Formation of the coelom (as a split in a solid mass of mesoderm or as cavities in mesodermal pouches)		
Major phyla		

9 Describe three distinguishing characteristics of the aschelminthes group, and using diagrams such as Figures 42.32 (p. 1119) and 42.35 (p. 1121) explain why the body cavity is called a pseudocoelom rather than a true coelom. Then list the identifying features of the nematodes and discuss the economic importance of this phylum. (pp. 1120–22)

10 Describe and give the function of the lophophore, and name three phyla possessing this structure. (pp. 1122–23)

11 Describe the body plan common to all Mollusca, and indicate the modifications it exhibits in the Gastropoda (snails), Polyplacophora (chitins), Monoplacophora, Scaphopoda (tusk shells), Bivalvia (bivalves), and Cephalopoda (squids and octopuses). (pp. 1123–28)

12 Describe the distinguishing characteristics of the Annelida, and compare the polychaetes, the earthworms, and the leeches with respect to the presence or absence of a distinct head, the degree of segmentation, the presence or absence of parapodia, and the number of setae, or bristles. (pp. 1128–30)

13 Describe the ways in which the Onychophora resemble the annelids and ways in which they resemble the arthropods. (p. 1130)

14 Describe the distinguishing characteristics of the arthropods, and indicate the major modifications of the ancestral body plan that have arisen among the various arthropod subphyla. Distinguish between chelicerae and mandibles. (pp. 1130–32)

15 Describe the distinguishing characteristics of the Crustacea, Chilopoda, Diplopoda, and Insecta, and list representative examples of each. Briefly describe the structure of the generalized insect body. (pp. 1135−38)

16 Name the three major phyla of the Deuterostomia, and list representative examples of each. (p. 1139)

17 Describe the distinguishing characteristics of the echinoderms, and list representative examples of each class. (pp. 1139−41)

18 Discuss the evolutionary relationships linking the echinoderms, hemichordates, and chordates. (pp. 1145−46)

KEY TERMS

The following terms are important in this chapter; you should become familiar with them.

Porifera (p. 1103)
collar cell (p. 1104)
spicule (p. 1104)
Coelenterata (cnidaria) (p. 1105)
mesoglea (p. 1105)
nematocyst (p. 1105)
polyp stage (p. 1106)
medusa stage (p. 1106)
planula (p. 1107)
Hydrozoa (p. 1105)
Scyphozoa (p. 1107)
Anthozoa (p. 1108)
Ctenophora (p. 1108)
Gastraea hypothesis (p. 1109)
planuloid hypothesis (p. 1109)
planuloid larva (p. 1109)
plakula hypothesis (p. 1110)
ciliate hypothesis (p. 1112)
acoelomate (p. 1112)
bilateral symmetry (p. 1112)
Platyhelminthes (p. 1112)
Turbellaria (p. 1113)
acoeloid (p. 1113)
Trematoda (p. 1113)
schistosomiasis (p. 1114)
Cestoda (p. 1114)
scolex (p. 1115)
proglottid (p. 1115)
Nemertina (p. 1116)
complete digestive tract (p. 1116)
Gnathostomulida (p. 1116)
Protostomia (p. 1117)
Deuterostomia (p. 1117)
true coelom (p. 1119)
pseudocoelom (p. 1119)
Aschelminthian phyla (p. 1120)

Rotifera (p. 1120)
Nemata (p. 1120)
trichinosis (p. 1122)
lophophore (p. 1122)
foot (p. 1123)
visceral mass (p. 1123)
mantle (p. 1123)
mantle cavity (p. 1123)
radula (p. 1124)
Polyplacophora (p. 1124)
Monoplacophora (p. 1125)
Gastropoda (p. 1125)
Scaphopoda (p. 1126)
Bivalvia (p. 1127)
Cephalopoda (p. 1127)
Polychaeta (p. 1128)
parapodium (p. 1128)
trochophore (p. 1129)
Oligochaeta (p. 1129)
Hirudinoidea (p. 1129)
Onychophora (p. 1130)
Arthropoda (p. 1130)
hemocoel (p. 1131)
Trilobita (p. 1132)
Chelicerata (p. 1133)
chelicera (p. 1133)
pedipalp (p. 1133)
antenna (p. 1134)
Mandibulata or Uniramia (p. 1134)
mandible (p. 1134)
maxilla (p. 1135)
Crustacea (p. 1135)
Chilopoda (p. 1136)
Diplopoda (p. 1137)
Insecta (p. 1137)
labium (p. 1137)
labrum (p. 1137)
Echinodermata (p. 1139)
water vascular system (p. 1140)
tube feet (p. 1141)
Stelleroidea or Asteroidea (p. 1141)
disc (p. 1141)
arms (p. 1141)
Echinoidea (p. 1143)
Holothuroidea (p. 1143)
Crinoidea (p. 1144)
Hemichordata (p. 1144)
pharyngeal slits (p. 1144)
dipleurula larva (p. 1145)

SUMMARY

The animal kingdom includes many diverse groups of organisms.

The *Porifera* (sponges) differ greatly from other multicellular animals and are believed to have evolved on a line of their own. They are multicellular animals at a primitive, pre-tissue level of organization. Sponges are filter feeders; their bodies are perforated sacs through which water flows. The flagellated collar cells lining the interior are remarkably similar to protozoan collared flagellates.

The radiate phyla These phyla—Cnidoria (Coelenterata) and Ctenophora—comprise radially symmetrical animals whose bodies have definite tissue layers but no distinct internal organs. They have a gastrovascular cavity and lack a coelom. Their bodies consist of two well-developed tissue layers, an outer epidermis (ectoderm) and an inner gastrodermis (endoderm), with a third layer of mesoglea (variable in its development) between them.

The *Cnidaria* or *Coelenterata* include the hydra, jellyfishes, sea anemones, and corals. There is some division of labor among their tissues, but it is never as complete as in most bilateral multicellular animals, and most functions performed by mesodermal tissues in other animals are performed by ectodermal or endodermal cells in coelenterates. There are three classes: Hydrozoa, Scyphozoa, and Anthozoa.

Many hydrozoans have a sedentary *polyp* stage that alternates with a free-swimming jellyfishlike *medusa* stage. Certain cells of the medusa produce gametes. The zygote develops into an elongate ciliated *planula* larva, which eventually settles and gives rise to the polyp. In the scyphozoans (jellyfishes), the medusa stage is dominant; in the anthozoans (sea anemones and corals), it is absent altogether. The Anthozoa are the most advanced of the Coelenterata.

The *Ctenophora* differ from the coelenterates in having mesodermal muscles and in possessing eight rows of ciliary plates; they lack the coelenterate's nematocysts and polymorphic life cycle.

Origin of the Metazoa Many biologists hypothesize that the *Metazoa* evolved from a hollow-sphere colonial flagellate (*blastaea*) with an intermediate *planuloid* stage. Others have suggested that the Metazoa evolved from a two-layered, flattened, creeping *plakula*. Its ventral surface may have been specialized for nutrition; a temporary digestive system could have been formed by elevating part of its body. Support for this hypothesis has come from the rediscovery of the organism *Trichoplax*, which bears a marked resemblance to the hypothetical plakula. Still other biologists have suggested that the multicellular animals arose from multinucleate ciliates in which cellular membranes formed around each nucleus.

Acoelomate Bilateria The *Platyhelminthes* and *Nemertina* are regarded as the most primitive bilaterally symmetrical animals. Their bodies are composed of three well-developed tissue layers; there is no coelom.

The *Platyhelminthes* (flatworms) are dorso-ventrally flattened, elongate animals with (usually) a gastrovascular cavity. The flatworms have advanced to the organ level of construction. The more extensive development of mesoderm probably made this advance possible. The phylum is divided into three classes: Turbellaria, Trematoda, and Cestoda.

The *Turbellaria*, of which planarians are examples, are free living. Since one order, the Acoela, is very primitive, some biologists hypothesize that an *acoeloid* organism could have arisen from a planuloid ancestor, or from a multinucleate ciliate, and that the more complex flatworms and other metazoan phyla then evolved from such an acoeloid organism.

The *Trematoda* (flukes) and *Cestoda* (tapeworms) are entirely parasitic. They have evolved many adaptations for parasitism: a resistant cuticle, loss or reduction of structures, extremely well-developed reproductive systems, and elaborate life cycles.

The members of the phylum *Nemertina* probably evolved from the turbellarian flatworms. However, they have a *complete digestive system* and a simple blood circulatory system.

Divergence of the Protostomia and Deuterostomia The fate of the embryonic blastopore differs in various animal groups. In many animals the blastopore becomes the mouth, and a new anus develops; in others the situation is reversed. This fundamental difference in development suggests that a major evolutionary split occurred soon after the origin of a bilateral ancestor. One evolutionary line led to all the phyla in which the blastopore becomes the mouth, the *Protostomia* (first mouth). The other line led to the phyla in which the blastopore forms the anus and a new mouth is formed, the *Deuterostomia* (second mouth). The protostomes and deuterostomes also differ in the determinateness and pattern of the initial cleavages, the mode of origin of the mesoderm and coelom, and the type of larva.

The pseudocoelomate Protostomia A true coelom is a cavity enclosed entirely by mesoderm and located between the digestive tract and body wall. In several protostome phyla the body cavity is partly bounded by ectoderm and endoderm; it is called a *pseudocoelom*.

The largest and most important pseudocoelomate phyla are those of the aschelminthes group. These are small, wormlike animals without a definite head that have a complete digestive tract. The members of two phyla are very abundant: the Rotifera (wheel animals) and the Nemata (roundworms). Many nematodes are parasitic in plants and animals. *Trichinella*, *Ascaris*, hookworms, pinworms, and filaria worms parasitize humans.

Coelomate Protostomia Sometime during their development, all the other protostome phyla possess true coeloms, which usually develop as a split in the mesoderm. All have a complete digestive tract, and most have well-developed circulatory, excretory, and nervous systems.

Three small phyla, *Phoronida, Ectoprocta,* and *Brachiopoda*, resemble one another in having a feeding device called a *lophopore*—a fold that encircles the mouth and bears numerous ciliated tentacles.

The phylum *Mollusca* is the second largest in the animal kingdom; it includes the snails and slugs, clams and oysters, and squids and octopuses. The molluscs are soft-bodied animals composed of a ventral muscular *foot*, a *visceral mass*, and a *mantle*, which covers the visceral mass and usually contains glands that secrete a shell. The mantle often overhangs the visceral mass, thus enclosing a *mantle cavity*, which frequently contains gills.

Molluscs have an open circulatory system. Most marine molluscs pass through one or more free-swimming larval stages, but the freshwater and land snails complete the corresponding developmental stages in the egg. The molluscs are usually divided into eight classes, of which we considered five: Polyplacophora (chitins), Monoplacophora, Gastropoda (snails), Scaphopoda (tusk shells), Bivalvia (bivalves), and Cephalopoda (squids, octopuses). The cephalopods are specialized for rapid movement and predation; they have convergently evolved many similarities to vertebrates.

The phylum *Annelida* (segmented worms) is usually divided into three classes: Polychaeta, Oligochaeta, and Hirudinoidea. The polychaetes are marine animals with a well-defined head bearing eyes and antennae. Each body segment bears a pair of *parapodia*, which function in movement and gas exchange. The Oligochaeta include earthworms and freshwater species; they lack a well-developed head and parapodia, and have few setae. The Hirudinea are the leeches. They are the most specialized of the annelids and show little internal segmentation.

The members of the phylum *Onychophora* are of special interest because they combine annelid and arthropod characters and are regarded as an early evolutionary offshoot from the line leading to the arthropods from an ancient annelidlike ancestor.

The phylum *Arthropoda* is the largest of the phyla. Arthropods are characterized by a jointed chitinous exoskeleton and jointed legs. They have elaborate musculature, a well-developed nervous system and sense organs, and an open circulatory system.

Arthropods are believed to have evolved from a polychaete annelid or from the ancestor of the polychaetes. The arthropod body may be an elaboration and specialization of the annelid's segmented body. Evidence indicates that the first arthropods had long wormlike bodies composed of nearly identical segments, each bearing a pair of legs. Among evolutionary trends seen in the arthropods are reduction in the number of segments; groupings of segments into body regions; increasing cephalization; and specialization of some legs for functions other than movement, with loss of legs from other segments.

The Arthropoda can be divided into three subphyla: Trilobita, Chelicerata, and Uniramia (or Mandibulata). The *Trilobita* are now extinct; they resembled the hypothetical arthropod ancestor. Fossils show that every segment bore a pair of nearly identical legs. In both the Chelicerata and Uniramia the tendency towards specialization of some appendages and loss of others is apparent.

The *Chelicerata* include the horseshoe crabs and arachnids (spiders, ticks, mites, scorpions). The body is usually divided into a cephalothorax and an abdomen. The first pair of postoral legs is modified as mouthparts called *chelicerae*. There are five other pairs of appendages on the cephalothorax; in some the first pair is modified as feeding devices called *pedipalps*.

The members of the *Uniramia* have *mandibles* as their first pair of mouthparts, and two additional pairs of mouthparts called *maxillae*. The Uniramia include the classes Crustacea (lobsters, crabs, shrimps, water fleas, sow bugs), Chilopoda (centipedes), Diplopoda (millipedes), and Insecta.

The insects are an enormous group of diverse organisms; there are more species of insects than of all other animal groups combined. The insect body is divided into a head, thorax, and abdomen. The head bears many sensory receptors, usually including compound eyes; one pair of antennae; and three mouthparts (mandibles, maxillae, and *labium*) derived from ancestral legs.

The upper lip, or *labrum*, may also be derived from ancestral legs. The thorax is composed of three segments, each bearing a pair of legs. In many insects both the second and third segments bear a pair of wings. The abdomen is composed of a variable number (12 or fewer) of segments; the terminal parts may be modified for reproduction.

The Deuterostomia The phyla *Echinodermata, Hemichordata,* and *Chordata* are in the Deuterostomia. Members of the phylum *Echinodermata* are exclusively marine, mostly bottom-dwelling animals. The adults are radially symmetrical, but the larvae are bilateral, and it is generally held that the echinoderm evolved from bilateral ancestors. All members have an internal skeleton made of bony plates, a well-developed coelom, and a unique *water-vascular system*. All show penta-radiate symmetry. The echinoderms are divided into four classes: Stelleroidea, Echinoidea, Holothuroidea, and Crinoidea.

The *Stelleroidea* or *Asteroidea* (which include sea stars and brittle stars) have a body that consists of a central disc with five arms. The brittle stars have a smaller central disc and five long thin arms. The *Echinoidea* (sea urchins and sand dollars) have no arms, but do have five bands of tube feet, movable spines, and a complex chewing apparatus. *Holothuroidea* (sea cucumbers) have a much reduced endoskeleton and a leathery body. Gas exchange is by a respiratory tree. *Crinoidea* (sea lilies) are often stalked and sessile.

Members of the phylum *Hemichordata* (acorn worms) show affinities with both the echinoderms and chordates. They resemble the chordates in their pharyngeal gill slits and dorsal nerve cord. However, their ciliated larva (a *dipleurula*) is almost identical to that of the echinoderms. The hemichordates' ties to both these groups have helped clarify the relationships between these two major phyla. The chordates and echinoderms are believed to have evolved from a common ancestor at some remote time.

The chart on page 1146 of your text provides a useful summary of some important characteristics of the animal phyla.

QUESTIONS

Testing recall

The following is a chart of the animal kingdom, set up as a phylogenetic tree. (It shows only one of many possibilities, since much is uncertain about the evolution of animal groups.) A phylogenetic tree can be useful in learning the various groups, particularly if you label the sites of various evolutionary advances. The dashed lines on the tree represent certain evolutionary advances. For example, above line a, *all organisms are multicellular. Place each of the characteristics listed below on the appropriate line. Answers may be used once, more than once, or not at all.*

1   complete digestive tract (p. 1116)

2   body cavity that is either a pseudocoelom or a true coelom (p. 1119)

3   true coelom (p. 1117)

4   jointed appendages and exoskeleton (p. 1130)

5   segmentation in adult (p. 1128)

6   three well-developed germ layers (p. 1112)

7   fundamental bilateral symmetry (p. 1112)

8   Identify the Deuterostomia and the Protostomia. (pp. 1116–19)

Match each organism below with the section in which it belongs. Answers may be used once, more than once, or not at all.

a   Radiata   c   Protostomia
b   Deuterostomia

9   jellyfish (p. 1107)

10   sea star (p. 1141)

11   earthworm (p. 1129)

12   snail (p. 1125)

13   hydra (p. 1105)

14   nematode worm (p. 1121)

15   cockroach (p. 1137)

16   squid (p. 1127)

17   acorn worm (p. 1145)

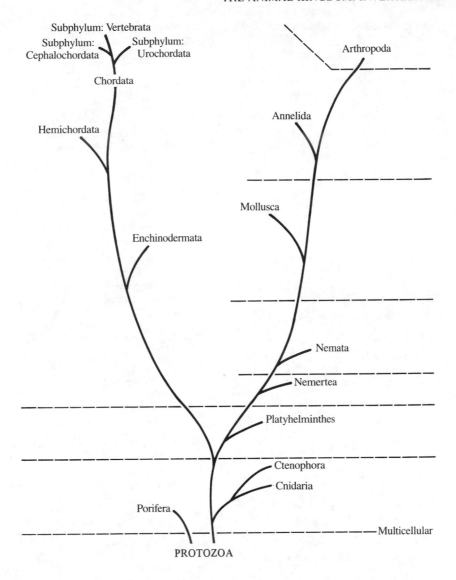

Subphylum: Vertebrata
Subphylum: Cephalochordata
Subphylum: Urochordata
Chordata
Hemichordata
Enchinodermata
Arthropoda
Annelida
Mollusca
Nemata
Nemertea
Platyhelminthes
Ctenophora
Cnidaria
Porifera
Multicellular
PROTOZOA

Match each of the important animal characteristics below with the phylum or phyla whose members possess that characteristic. Answers may be used once, more than once, or not at all.

a Annelida
b Arthropoda
c Aschelminthes
d Chordata
e Coelenterata (Cnidoria)
f Echinodermata
g Hemichordata
h Mollusca
i Platyhelminthes
j Porifera

18 radial symmetry (p. 1105)

19 water-vascular system (p. 1140)

20 jointed legs and exoskeleton (p. 1130)

21 segments with parapodia (p. 1128)

22 proboscis, collar, and pharyngeal slits (pp. 1144−45)

23 mantle, foot, visceral mass (p. 1123)

24 pseudocoelom (p. 1119)

25 dipleurula larva (p. 1145)

26 bilateral symmetry, gastrovascular cavity (p. 1112)

27 saclike body lined with collar cells (p. 1103)

Testing knowledge and understanding

The following key lists five animals, each representing a different phylum. The order in which they are listed is random and does not reflect the evolutionary placement of the phyla. Recalling the evolutionary relationships postulated in your text, match each characteristic below with the animal representing the most "primitive" phylum in which the characteristic appears. Answers may be used once, more than once, or not at all.

a   sponge
b   nematode worm
c   hydra
d   clam
e   flatworm

28   true coelom (p. 1119)

29   circulatory system (p. 1124)

30   complete digestive tract (p. 1120)

31   well-defined tissue organization (p. 1112)

32   bilateral symmetry (p. 1112)

33   organ level of development (p. 1113)

34   nervous system (p. 1105)

35   gastrovascular activity (p. 1105)

Choose the one best answer.

36   The evidence linking the sponges to the flagellated protozoans is

a   the sponge's collar cells.
b   similar photosynthetic pigments.
c   a common reserve food, starch.
d   the volvoxlike sponge blastula.
e   There is no such evidence.   (p. 1104)

37   A major difference between a sponge and a hydra is that

a   the sponge, but not the hydra, has a primitive nerve net.
b   the hydra, but not the sponge, has three cell layers.
c   water enters and exits through separate openings in the sponge, but through a single opening (mouth) in the hydra.
d   the hydra, but not the sponge, is bounded on the outside by a layer of epidermal cells.   (pp. 1104–7)

38   Adult sea anemones have which coelenterate body form?

a   polyp
b   medusa
c   planula   (p. 1107)

39   Which one of the following is *not* a characteristic of the coelenterates?

a   flame-cell excretory system
b   nerve net
c   digestive tract with one opening
d   nematocysts
e   tissue level of development   (p. 1105)

40   Which one of the following animals does *not* have a well-developed mesoderm?

a   housefly
b   jellyfish
c   snail
d   planarian
e   earthworm   (p. 1105)

41   A major evolutionary advance seen in the Platyhelminthes (flatworms) but not in the Coelenterata is

a   a gastrovascular cavity.
b   a coelom.
c   bilateral symmetry.
d   endoderm.
e   ectoderm   (p. 1112)

42   Which one of the following has a pseudocoelom?

a   roundworm (nematode)
b   earthworm
c   clam
d   cockroach
e   planarian   (p. 1119)

43   Which one of the following has a planula larva?

a   echinoderms
b   mollusca
c   arthropods
d   coelenterates
e   hemichordates   (p. 1107)

44   Trichinosis, a disease human beings can get from eating insufficiently cooked pork, is caused by a

a   virus.
b   fungus.
c   fluke.
d   tapeworm.
e   nematode worm.   (p. 1122)

45   Which one of the following is *not* characteristic of molluscs?

a   open circulatory system
b   foot with visceral mass
c   shell secreted by the mantle
d   anus developed from the blastopore   (pp. 1117, 1123)

46   *All* molluscs and annelids have

a   gills for gas exchange.
b   an open circulatory system.
c   segmentation.
d   a true coelom.
e   trochophore larvae.   (p. 1122)

47 All annelids have

  a segmentation.
  b parapodia.
  c setae.
  d suckers.
  e trochophore larvae.   (p. 1128)

48 Annelids differ from flatworms in that

  a all annelids live in the soil, whereas all
     flatworms live in freshwater.
  b annelids have a digestive tract, whereas
     flatworms do not.
  c annelids have a cavity between the
     digestive tract and the body wall, whereas
     flatworms do not.
  d annelids have a nervous system, whereas
     flatworms do not.
                (pp. 1112, 1113, 1128–30)

49 Annelids differ from nematode worms in that

  a annelids have a complete digestive tract,
     whereas nematodes do not.
  b annelids have circular muscles, whereas
     nematodes have only longitudinal
     muscles.
  c annelids are exclusively free-living,
     whereas nematodes are exclusively
     parasitic.
  d annelids have a pseudocoelom, whereas
     nematodes are acoelomate.
  e annelids have a mouth derived from the
     blastopore, whereas nematodes have an
     anus derived from the blastopore.
                (pp. 1120–21, 1128–30)

50 Which one of the following animal phyla is
    least closely related to all the others?

  a Coelenterata   *d* Arthropoda
  b Porifera (sponges) *e* Mollusca
  c Annelida   (p. 1105)

51 Which one of the following statements is true
    of both annelids and arthropods?

  a Both exhibit indeterminate cleavage in
     early development.
  b The blastopore becomes the anus in both.
  c Both exhibit spiral cleavage.
  d The mesoderm arises as outpocketings of
     the endoderm in both.   (pp. 1117, 1119)

52 Which one of the following *arthropod*
    characteristics was probably the most impor-
    tant adaptation for the land environment?

  a coelom
  b segmentation
  c complete digestive tract
  d exoskeleton
  e circulatory system   (pp. 1130–32)

53 A spider is an arachnid, not an insect.
    Arachnids differ from insects in that

  a arachnids have eight walking legs,
     whereas insects have six.
  b arachnids have jointed exoskeletons and
     legs, whereas insects do not.
  c arachnids have an open circulatory
     system, whereas insects do not.
  d insects always have one pair of flying
     wings, whereas arachnids have two pairs.
  e arachnids, but not insects, have a mantle
     that covers the foot.   (pp. 1133–35)

54 Which one of the following is *not* typical of
    insects?

  a body divided into a head, thorax, and
     abdomen
  b compound eyes
  c pincerlike chelicerae
  d one pair of antennae
  e three pairs of legs   (p. 1137)

55 In humans, the blastopore becomes the anus,
    and cleavage is radial. An invertebrate
    phylum sharing these characteristics is the

  a Porifera (sponges).
  b Arthropoda.
  c Hemichordata.
  d Mollusca.
  e none of the above   (p. 1145)

56 In which pair do the organisms named have
    very similar larval stages?

  a starfish–vertebrate
  b starfish–annelid
  c hemichordate–arthropod
  d mollusc–starfish
  e starfish–hemichordate   (p. 1145)

For further thought

1 Explain and discuss the pattern of increasing
   complexity within the animal kingdom (at the
   phylum level).

2 Contrast the annelids and arthropods with
   respect to the degree of segmentation,
   development of the nervous system, adaptations for digestion, and the gas-exchange and
   circulatory systems.

3 Compare the molluscs and annelids with
   respect to the basic body plan, the gas-
   exchange systems, circulatory systems, and
   excretory systems.

ANSWERS

1	c	4	g	6	b
2	d	5	f	7	b
3	e				

8 The Deuterostomia are shown on the branch at left (Echinodermata–Chordata); the Protostomia are shown on the branch at right (Platyhelminthes–Arthropoda).

9	a	14	c	19	f	24	c
10	b	15	c	20	b	25	f, g
11	c	16	b	21	a	26	i
12	c	17	c	22	g	27	j
13	a	18	e, f	23	h		

28	d	36	a	43	d	50	b
29	d	37	c	44	e	51	c
30	b	38	a	45	e	52	d
31	c	39	a	46	d	53	a
32	e	40	b	47	a	54	c
33	e	41	c	48	c	55	c
34	c	42	a	49	b	56	e
35	c						

Chapter 43

THE ANIMAL KINGDOM: CHORDATES

A GENERAL GUIDE TO THE READING

This last chapter is devoted to the phylum Chordata—the phylum to which we belong. The emphasis is on the evolutionary relationships of the various groups. Much of the material will be familiar to you since we have discussed many different aspects of vertebrate biology in earlier chapters. Cross references are supplied to earlier chapters so you can refresh your memory. You will want to concentrate on the following topics.

1 Chordate characteristics. Begin by memorizing the three basic chordate characteristics on page 1147.

2 Tunicates. In reading about the tunicates (pp. 1147–48), notice that though the adult tunicate does not resemble a chordate in any way, the larva (Figure 43.2, p. 1149) bears some resemblance to what we might expect a primitive chordate to look like.

3 Vertebrate evolution. The vertebrate classes are discussed on pages 1150–63. You will

want to focus on the evolutionary relationships among them, shown graphically in Figures 43.4 (p. 1150) and 43.6 (p. 1151), and on the key advances in each vertebrate class. Try to understand which of the advances were adaptations to overcome the difficulties of life on land.

4 Evolution of the primates. Especially important in the discussion of this topic (pp. 1163–68) are the basic primate characteristics listed on page 1163. You will want to learn them.

5 Evolution of human beings. Learn the anatomical changes that occurred in the evolution of human beings from an apelike ancestor (pp. 1168–73). The relationships among the fossil hominids are summarized in Figure 43.38 (p. 1173).

6 The interaction of cultural and biological evolution. You will want to understand how human beings are able to influence not only the evolution of other species but also our own biological evolution. (pp. 1173–75)

KEY CONCEPTS

1 All chordates possess, at some time in their life cycle, a notochord, pharyngeal gill pouches, and a dorsal hollow nerve cord. (p. 1147)

2 The members of the subphylum Vertebrata may be regarded as the most highly evolved representatives of the deuterostome line. (pp. 1150–63)

3 The acquisition of hinged jaws was one of the most important events in the history of vertebrates since it made possible the development of a variety of feeding methods and life styles. (p. 1152)

4 Any fish that had appendages better suited for land locomotion than those of their fellows would have been able to exploit the ecological opportunities open to them on land more fully; through selection pressure exerted over millions of years, the fins of these first vertebrates to walk on land would slowly have evolved into legs. (p. 1154)

5 The evolution of the amniotic egg, which provides a fluid-filled chamber in which the embryo may develop even when the egg itself is in a dry place, was an important evolutionary advance in the conquest of land. (p. 1156)

6 At the end of the Cretaceous era, an asteroid collision, comet impact, or cycles of intense volcanic activity may have produced enough dust and smoke to alter the earth's climate and decrease photosynthesis, resulting in mass extinctions of the dinosaurs, swimming and flying reptiles, and many invertebrates. (p. 1160)

7 Fossil evidence indicates that the members of the mammalian order Primates arose from a tree-dwelling stock of small, shrewlike insectivores. (p. 1163)

8 Many of the traits most important to us as human beings first evolved because our distant ancestors lived in trees. (pp. 1163–69)

OBJECTIVES

After studying this chapter and reflecting on it, you should be able to carry out the following objectives.

1 Give three distinguishing features of the Chordata, list the three subphyla, and give an example of each. (p. 1147)

2 Describe the larval and adult forms of the tunicates, and suggest an evolutionary relationship between the tunicates and the vertebrate line. (p. 1148)

3 List the seven classes of modern vertebrates, and indicate the evolutionary relationships among them. State the order in which they appear in the fossil record. (pp. 1150–63)

4 Give a possible explanation for the mass extinctions of the dinosaurs and other animals at the end of the Cretaceous.

5 Describe the key innovations of each vertebrate class, and specify which of these were adaptations to overcome the difficulties of life on land. (pp. 1150–63)

6 List eight characteristics shared by all primates. (p. 1163)

7 Discuss current ideas of the evolutionary history of man. In doing so, describe some of the anatomical changes that occurred in the course of evolution from ape ancestor to modern human, and discuss the roles that bipedalism, tool use, and increased brain size may have played in human evolution. State the order in which early hominids appear in the fossil record. (pp. 1163–73)

KEY TERMS

The following terms are important in this chapter; you should become familiar with them.

notochord (p. 1147)
pharyngeal slits (p. 1147)
Tunicata (p. 1147)
Cephalochordata (p. 1148)
lancelet (p. 1149)
segmentation (p. 1150)
Agnatha (p. 1150)
Placodermi (p. 1152)
hinged jaws (p. 1152)
Chondrichthyes (p. 1152)
cartilaginous fish (p. 1152)
Osteichthyes (p. 1152)
bony fish (p. 1152)
fleshy-finned fishes (p. 1153)
Amphibia (p. 1154)
Reptilia (p. 1155)
amniotic egg (p. 1156)
Aves (p. 1160)
Mammalia (p. 1161)
stem reptile (p. 1161)
monotreme (p. 1162)
marsupial (p. 1162)
placental (p. 1162)

SUMMARY

The phylum *Chordata* contains both invertebrate and vertebrate members. It is divided into three subphyla: Tunicata, Cephalochordata, and Chordata. At some time in their life cycle, all chordates have a *notochord*, pharyngeal slits, and a dorsal hollow nerve cord.

The adults of the *Tunicata* (tunicates) are sessile filter feeders. The larvae, however, are free-swimming, bilaterally symmetrical, tadpolelike organisms with pharyngeal slits, notochords, and dorsal hollow nerve cords.

The members of the *Cephalochordata* (lancelets) are small marine organisms with a permanent dorsal hollow nerve cord and notochord. They are filter feeders and show segmentation.

The *Vertebrata* are characterized by an endoskeleton that includes a segmented backbone composed of a series of vertebrae that develop around the embryonic notochord.

The first vertebrates belonged to the class *Agnatha*; they were fishlike animals encased in an armor of bony plates. They lacked jaws and paired fins. The lamprey and hagfishes are the only jawless fishes found today, and are very unlike their ancient ancestors.

Three classes of fish arose from the Agnatha: the Placodermi, Chondrichthyes, and Osteichthyes. All evolved from a common ancestor with hinged jaws and paired fins. The evolution of hinged jaws was an important advance; it allowed alternative feeding methods and life styles to develop. The jaw was developed from a set of gill support bars. The Placodermi were an important group of armored fishes; all are now extinct.

The modern *Chondrichthyes* (sharks, skates, rays) have a cartilaginous skeleton, whereas the *Osteichthyes* have a bony skeleton. The primitive bony fishes had lungs in addition to gills. Early in their evolution the class split into two divergent groups. One group underwent great radiation, giving rise to most of the bony fishes alive today; the other group, now mostly extinct, gave rise to the present-day lungfishes and lobe-fin fishes. The lobe-fins had lungs and leglike fleshy paired fins that may have enabled them to crawl. They appear to be ancestral to the amphibians.

The first members of the class *Amphibia* were quite fishlike, but they became a large and diverse group as they exploited the ecological opportunities on land. Most of the amphibians became extinct by the end of the Triassic; the few that survived were the ancestors of the salamanders, apodes, and frogs and toads. Although the amphibians are terrestrial they must return to water to reproduce since they utilize external reproduction, lay fishlike eggs, and have an aquatic larva.

Evolution of the primates Fossil evidence indicates that the members of the mammalian order *Primata* arose from an arboreal stock of small shrewlike insectivores. Important characteristics of the primates include grasping hands (with nails), increased mobility of the shoulder joint (braced by the clavicle), rotational movement of the elbow, good binocular vision, and expansion of the brain. These traits are associated with an arboreal way of life.

The living primates are classified in two suborders: Prosimii and Anthropoidea. The prosimians (lemurs, tarsiers, and others) are more primitive; they are small arboreal animals with opposable first digits. Two lines of Anthropoidea evolved from the Prosimii; one line led to the New World monkeys, and the other, which split, to the Old World monkeys and the apes.

The living great apes (Pongidae) fall into four groups: gibbons, orangutans, gorillas, and chimpanzees. All are large animals with no tail, a large skull and brain, and long arms. All have a tendency to walk semi-erect. The earliest members of the family Hominidae probably arose from the same pongid stock that produced the gorillas and chimpanzees. Fossils of *Dryopithecus* suggest that the common ancestor of apes and humans had a low rounded cranium, moderate superorbital ridges, and moderate forward projection of face and jaw. In the course of human evolution the jaw became shorter, the skull more balanced on top of the vertebral column, the brain case larger, the nose more prominent, the arms shorter, the feet flatter, and the big toe ceased being opposable. The various fossil humans are intermediate in these characteristics.

Evolution of human beings The upright posture and bipedal locomotion of human beings may have evolved as adaptations for groundfeeding on plant food or for freeing the hands to carry objects or manipulate tools. These adaptations would also have made it easier to maintain surveillance.

The first truly hominid fossils belong to the genus *Australopithecus*. There were at least three species of australopithecines and a third form, more humanlike, designated *Homo habilis*. *Homo habilis* was later replaced by *Homo erectus*. *Homo sapiens* probably evolved from this form about 250,000 years ago.

Cultural and biological evolution are interwoven, although cultural evolution can proceed more rapidly than biological evolution. Human beings profoundly influence the evolution of other species and now have the ability to influence and control their own evolution.

The *Reptilia* are the first truly terrestrial group. They use internal fertilization, lay amniotic shelled eggs, have no larval stage, and have scaly, relatively impermeable skin. In addition, their legs are larger and stronger than those of amphibians, their lungs are better developed, and their heart is almost four-chambered. All the living reptiles (turtles, snakes, lizards, crocodiles, alligators, and the tuatara) except the crocodiles are directly descended from the stem reptiles. One line that evolved from the stem reptiles (thecodonts) gave rise to the crocodiles, flying reptiles, and the dinosaurs. All the dinosaurs became extinct except for one group of specialized descendants: the birds. Another line of reptiles (therapsids) gave rise to the mammals.

At the end of the Cretaceous, there was a mass extinction of the dinosaurs and swimming and flying reptiles, as well as many invertebrates. Many scientists believe that an asteroid collision, comet impact, or intense volcanic activity produced enough dust and smoke to alter the earth's climate and decrease photosynthesis.

The *Aves* (birds) have evolved many adaptations for their active way of life. Birds are warm-blooded, their heart is four-chambered, they have feathers for insulation, their bones are light, and they have an extensive system of air sacs.

The *Mammalia* also are warm-blooded and have a four-chambered heart. They have a diaphragm, hair for insulation, a one-bone lower jaw, differentiated teeth, and an enlarged neocortex. Embryonic development occurs within the female; the young are born alive and are nourished by milk from the mammary glands. Early in their evolution the mammals split into two groups: the monotremes (egg-laying mammals) and the group that includes the marsupials and placentals. Marsupial embryos remain in the uterus for a short time and then undergo further development attached to a nipple in the mother's abdominal pouch. Placental embryos complete their development in the uterus.

QUESTIONS

Testing recall

Match each item with the one best answer.
Each answer may be used more than once.

a	Agnatha	*g*	Hemichordata
b	Amphibia	*h*	Mammalia
c	Aves	*i*	Osteichthyes
d	Cephalochordata	*j*	Reptilia
e	Chondrichthyes	*k*	Tunicata
f	Echinodermata		

1 feathers a characteristic (p. 1161)

2 jawless fish with cartilaginous skeletons and scaleless skin (p. 1151)

3 saclike marine forms whose larval stage has well-developed notochord and pharyngeal slits (p. 1148)

4 body hair a characteristic (p. 1162)

5 primitive animals in which the adult form possesses pharyngeal slits but no notochord (p. 1144)

6 first amniotic vertebrates (p. 1156)

7 aquatic animals with bony skeletons and respiration by gills (pp. 1152–53)

8 slender elongate animals whose body form closely approaches that of early fishlike vertebrates; have fins, distinct tail, show segmentation, and are filter feeders. (p. 1149)

9 endothermic animals with a muscular diaphragm (p. 1161)

10 marine animals with hinged jaws and cartilaginous skeletons believed to have evolved from ancient jawed fishes (p. 1152)

Testing knowledge and understanding

Choose the one best answer.

11 All of the following statements are true of both chordates and echinoderms *except:*

a Both exhibit indeterminate cleavage in early development.

b The embryonic blastopore becomes the anus in both.
c Both have true coeloms.
d Both exhibit radial cleavage.
e Both develop pharyngeal slits and a dorsal hollow nerve cord.
(pp. 1117–19, 1147)

12 According to the classification in your text, the lowest level of classification shared by a frog and a goldfish is the

a kingdom.   *d* order.
b class.   *e* subphylum.
c phylum.   (pp. 1152–55)

13 In both arthropods and vertebrates

a the blastopore becomes the mouth.
b cleavage is indeterminate.
c a ciliated larval form is present.
d the mesoderm originates as outpocketings of the endoderm.
e a true coelom is present.   (pp. 1117–19)

14 Both the living jawless fishes and the cartilaginous fishes lack true bone. The ancestors of both these groups

a had cartilaginous skeletons rather than bony skeletons.
b had bony skeletons.
c had no skeleton of any kind.
d lived on land.
e were capable of flight.   (pp. 1151–52)

15 Segmentation is a characteristic found in all of the following organisms *except*

a fish.   *d* frog.
b lancelet.   *e* snake.
c tunicate.   (p. 1150)

16 Arthropods and vertebrates may both be regarded, in a sense, as the most highly evolved representatives of their evolutionary lines. They have convergently evolved many striking similarities, presumably in response to similar selection pressures. Which one of the following is *not* a true statement about a similarity between the two groups of animals?

a Most members of both groups possess jointed appendages.
b At least some members of both groups have evolved wings.
c At least some members of both groups have a structurally distinct and highly specialized head region.
d In both groups the circulatory system has become highly specialized for oxygen transport, as is necessary in such active animals.
e Hinged jaws occur in both groups, though their evolutionary derivations are quite different.   (pp. 1130–32, 1150–61)

17 Two fossil vertebrates, each representing a different class, are found in the undisturbed rock layers of a cliff. One is an early amphibian. The other fossil, found in an older rock layer below the amphibian, is most likely to be

a a dinosaur.
b a snake.
c an insectivorous mammal.
d a fish.
e a bird.   (p. 1155)

18 The first vertebrate land colonizers were

a amphibians.   *d* clams.
b reptiles.   *e* birds.
c mammals.   (p. 1155)

19 At some stage of development, all animals in the phylum Chordata

a have a notochord.
b have a backbone.
c have pharyngeal slits.
d all of the above
e *a* and *c*   (p. 1147)

20 Which one of the following is the correct sequence of evolution in the subphylum Vertebrata?

a cartilaginous fishes–jawless fishes–bony fishes–amphibians–reptiles
b jawless fishes–cartilaginous fishes–bony fishes–birds–reptiles
c jawless fishes–bony fishes–amphibians–reptiles–mammals
d jawless fishes–cartilaginous fishes–bony fishes–reptiles–amphibians–mammals
(pp. 1150–61)

21 The jawless fishes

a are now extinct.
b were the direct ancestors of the lobe-finned fish and the lungfish.
c gave rise to the placoderms.
d were very large fish with paired fins and scales, much like our modern fish.
(p. 1151)

22 The fleshy-finned fishes appear to have been the direct ancestors of present-day

a freshwater bony fish.
b stem reptiles.
c amphibians.
d cartilaginous fishes.
e placoderms.   (p. 1153)

23 All members of three vertebrate classes *must* have internal fertilization. These three classes are

a Agnatha, Aves, Mammalia.
b Amphibia, Reptilia, Mammalia.

c Reptilia, Aves, Mammalia.
d Osteichthyes, Amphibia, Reptilia.
e Amphibia, Reptilia, Aves. (pp. 1154–61)

24 In which one of the following animals is the embryo *not* protected by amniotic fluid?

a robin
b alligator
c bullfrog
d human being
e lion
(p. 1156)

25 A study of fossils suggests that mammals evolved directly from

a therapsid reptiles.
b dinosaurs.
c amphibians.
d lobe-finned fishes.
e modern reptiles. (p. 1161)

26 A study of fossils suggests that the most immediate ancestors of birds were

a amphibians.
b placoderms.
c bony fishes.
d mammals.
e reptiles.
(p. 1160)

27 Which one of the following human characteristics evolved primarily as an adaptation for an arboreal (tree-dwelling) life?

a hair
b freely rotating shoulder joint
c bipedalism
d bowl-shaped pelvis
e head balanced over the vertebral column
(p. 1163)

28 The evolution of which one of the following human characteristics probably *cannot* be explained in part, either directly or indirectly, by the fact that the ancestors of human beings were adapted for an arboreal existence?

a binocular (3-D) vision
b freely rotating shoulder joint
c opposable thumb
d great reduction of body hair
e enhanced eye–hand coordination
(p. 1163)

29 According to the best available evidence, which one of the following was the most recent ancestor of *Homo sapiens*?

a *Homo habilis*
b *Homo erectus*
c *Dryopithecus*
d *Australopithecus africanus*
e *Australopithecus afarensis* (pp. 1170–72)

30 Which of the following is *not* a characteristic associated with primates in general?

a location of the foramen magnum, the hole in the skull through which the spine passes, in the bottom of the skull

b mobile digits (fingers and toes)
c eyes placed toward the front of the head
d fingernails rather than claws
e two mammary glands (p. 1163)

31 The animal most closely related to *Homo sapiens* is the

a gibbon.
b chimpanzee.
c gorilla.
d orangutan.
e monkey. (p. 1168)

32 What is the scientific name given to Neanderthal man?

a *Australopithecus africanus*
b *Australopithecus robustus*
c *Homo erectus*
d *Homo habilis*
e *Homo sapiens* (p. 1172)

Match each description below to the item to which it applies. Answers may be used once, more than once, or not at all.

a *Australopithecus*
b Cro-Magnon man
c *Dryopithecus*
d *Homo erectus*
e tree-dwelling insectivore

33 earliest representative of our own subspecies (p. 1172)

34 *first* true hominid to walk upright (p. 1170)

35 common ancestor of apes and humans (p. 1168)

36 common ancestor of all of the primates (p. 1163)

For further thought

1 Discuss the problems faced by animals in terrestrial life, and explain important adaptations in the insects and terrestrial vertebrates that enable them to "solve" these problems.

2 Arthropods and vertebrates may both be regarded, in a sense, as the most highly evolved representatives of their respective evolutionary lines. They have independently evolved many similar "solutions" to the problems of survival and reproduction. Discuss ways in which the groups resemble each other and ways in which they differ.

3 Movement onto land presented problems for both plants and animals related to reproduction. a) Describe the processes of reproduction in the primitive plants and in lower animals. b) Compare and contrast the manner in which the terrestrial higher plants and animals "solved" the problem of no

longer being able to fertilize eggs in water and the mechanisms that evolved to supply nutrition to the growing embryo and protect it against dessication.

4 Draw a phylogenetic tree of the seven vertebrate classes. Explain in some detail the evolutionary relationships among the ostracoderms, lobe-finned fishes, archosaurs, and therapsids.

5 The move to land and the evolution of lungs and air breathing necessitated changes in the circulatory system. Describe in some detail the changes in how the fish circulatory system was altered in the amphibians and the mammals.

6 Describe how an aquatic vertebrate such as a perch carries on gas exchange and discuss the anatomical changes in the gas-exchange systems of lungfish, amphibians, reptiles, and mammals that occurred as a result of the move to a terrestrial existence.

ANSWERS

Testing recall

1	c	4	h	7	i	9	h
2	a	5	g	8	d	10	e
3	k	6	j				

Testing knowledge and understanding

11	e	18	a	25	a	31	b
12	e	19	d	26	e	32	e
13	e	20	c	27	b	33	b
14	b	21	c	28	d	34	a
15	c	22	c	29	b	35	c
16	d	23	c	30	a	36	e
17	d	24	c